Optimization Techniques and Applications with Examples

# Optimization Techniques and Applications with Examples

*Xin-She Yang*
*Middlesex University London*

WILEY

This edition first published 2018


*Registered Office*
John Wiley & Sons, Inc., 111 River Street, Hoboken, NJ 07030, USA

*Editorial Office*
111 River Street, Hoboken, NJ 07030, USA

For details of our global editorial offices, customer services, and more information about Wiley products visit us at www.wiley.com.

Wiley also publishes its books in a variety of electronic formats and by print-on-demand. Some content that appears in standard print versions of this book may not be available in other formats.



*Library of Congress Cataloging-in-Publication Data*

Names: Yang, Xin-She, author.
Title: Optimization techniques and applications with examples / Xin-She Yang.
Description: Hoboken, New Jersey : John Wiley & Sons, 2018. | Includes bibliographical references and index. |
Identifiers: LCCN 2018019716 (print) | LCCN 2018033280 (ebook) | ISBN 9781119490609 (Adobe PDF) | ISBN 9781119490623 (ePub) | ISBN 9781119490548 (hardcover)
Subjects: LCSH: Mathematical optimization.
Classification: LCC QA402.5 (ebook) | LCC QA402.5 .Y366 2018 (print) | DDC 519.6–dc23
LC record available at https://lccn.loc.gov/2018019716

Cover image: © MirageC/Getty Images
Cover design: Wiley

Set in 10/12pt Warnock by SPi Global, Pondicherry, India

Printed in the United States of America

V10003543_081118

# Contents

## List of Figures

# List of Tables

# Preface

Optimization and operations research are part of many university courses because they are important to many disciplines and applications such as engineering design, business planning, computer science, data mining, machine learning, artificial intelligence, and industries.

There are many books on the market, but most books are very specialized in the sense that they target at a specialized audience in a given subject such as applied mathematics, business management, computer science, or engineering. Most of these books, though with much details and often more than 500 pages, have focused on the traditional optimization techniques, commonly used in operations research. In some sense, some courses in the traditional operations research areas have focused too much on many traditional topics, and these topics have many excellent textbooks. However, optimization techniques have evolved over the last few decades, and some new techniques start to show their effectiveness and have become an integrated part of new mainstream methods. Though important traditional topics will be still introduced in this book, the contents on some techniques may be brief instead of overlapping with other books, we will provide relevant references to appropriate literature when necessary. Therefore, this book tries to address modern topics, as used mostly by researchers working in these newly emerging areas, spanning optimization, data science, machine intelligence, engineering, and computer sciences, so that the topics covered in this book are most relevant to the current areas of research and syllabus topics in optimization, data mining, machine learning, and operations research.

This book aims at an introductory level, introducing all the fundamentals of all the commonly used techniques so that students can see the broadness and diversity of the methods and algorithms and sample the flavor of each method (both traditional and new). The book covers almost all major optimization techniques with many worked examples. Topics include mathematical foundations, optimization formulation, optimality conditions, algorithmic complexity, linear programming, convex optimization, integer programming, artificial neural network, clustering and classifications, regression and least squares, constraint-handling, queueing theory, support vector machine and multiobjective optimization, evolutionary computation, nature-inspired algorithms, and others. Each topic is explained in detail with step-by-step examples to show how each method works, and exercises are also provided to test the acquired knowledge and potentially apply to real problem solving.

With an informal approach and more than 100 worked examples and exercises, this introductory book is especially suitable for both undergraduates and graduates to rapidly acquire the basic knowledge in optimization, operational research, machine learning, and data mining. This book can also serve as a reference for researchers and professionals in engineering, computer sciences, operations research, data mining, and management science.

*Cambridge and London* Xin-She Yang
March 2018

# Acknowledgements

I would like to thank my students who took my relevant courses and provided some feedback on the contents concerning optimization algorithms and examples. I also thank the participants for their comments at international conferences where I gave tutorials on nature-inspired algorithms. I would also like to thank the editors, Kathleen Pagliaro, Vishnu Narayanan, Mindy Okura-Marszycki, and staff at Wiley for their help and professionalism. Last but not least, I thank my family for their support.

Xin-She Yang

## Acronyms

| | |
|---|---|
| AIC | Akaike information criterion |
| ANN | Artificial neural network |
| APSO | Accelerated particle swarm optimization |
| BA | Bat algorithm |
| BFGS | Broyden–Fletcher–Goldfarb–Shanno |
| BFR | Bradley–Fayyad–Reina |
| BIC | Bayesian information criterion |
| BPNN | Back propagation neural network |
| CNN | Convolutionary neural network |
| CPF | Cumulative probability function |
| CS | Cuckoo search |
| CURE | Clustering using representative |
| DE | Differential evolution |
| DL | Deep learning |
| ES | Evolution strategy |
| FA | Firefly algorithm |
| FPA | Flower pollination algorithm |
| GA | Genetic algorithm |
| IP | Integer programming |
| KKT | Karush–Kuhn–Tucker |
| LP | Linear programming |
| MIP | Mixed integer programming |
| ML | Machine learning |
| NSGA | Non-dominated sorting genetic algorithm |
| NP | Non-deterministic polynomial-time |
| PCA | Principal component analysis |
| PDF | Probability density function |
| PSO | Particle swarm optimization |
| QP | Quadratic programming |
| RBM | Restricted Boltzmann machine |
| SA | Simulated annealing |

| | |
|---|---|
| SGA | Stochastic gradient ascent |
| SGD | Stochastic gradient descent |
| SQP | Sequential quadratic programming |
| SVM | Support vector machine |
| TSP | Traveling salesman problem |

# Introduction

Optimization is everywhere, though it can mean different things from different perspectives. From basic calculus, optimization can be simply used to find the maximum or minimum of a function such as $f(x) = x^4 + 2x^2 + 1$ in the real domain $x \in \mathbb{R}$. In this case, we can either use a gradient-based method or simply spot the solution due to the fact that $x^4$ and $x^2$ are always nonnegative, the minimum values of $x^4$ and $x^2$ are zero, thus $f(x) = x^4 + 2x^2 + 1$ has a minimum $f_{\min} = 1$ at $x_* = 0$. This can be confirmed easily by taking the first derivative of $f(x)$ with respect to $x$, we have

$$f'(x) = 4x^3 + 4x = 4x(x^2 + 1) = 0, \tag{I.1}$$

which has only one solution $x = 0$ because $x$ is a real number. The condition $f'(x) = 0$ seems to be sufficient to determine the optimal solution in this case. In fact, this function is convex with only one minimum in the whole real domain.

However, things become more complicated when $f(x)$ is highly nonlinear with multiple optima. For example, if we try to find the maximum value of $f(x) = \text{sinc}(x) = \sin(x)/x$ in the real domain, we can naively use

$$f'(x) = \left[\frac{\sin(x)}{x}\right]' = \frac{x\cos(x) - \sin(x)}{x^2} = 0, \tag{I.2}$$

which has an infinite number of solutions for $x \neq 0$. There is no simple formula for these solutions, thus a numerical method has to be used to calculate these solutions. In addition, even with all the efforts to find these solutions, care has to be taken because the actual global maximum $f_{\max} = 1$ occurs at $x_* = 0$. However, this solution can only be found by taking the limit $x \to 0$, and it is not part of the solutions from the above condition of $f'(x) = 0$. This highlights the potential difficulty for nonlinear problems with multiple optima or multi-modality.

Furthermore, not all functions are smooth. For example, if we try to use $f'(x) = 0$ to find the minimum of

$$f(x) = |x| e^{-\sin(x^2)}, \tag{I.3}$$

we will realize that $f(x)$ is not differentiable at $x = 0$, though the global minimum $f_{\min} = 0$ occurs at $x_* = 0$. In this case, optimization techniques that require the calculation of derivatives will not work.

Problems become more challenging in higher-dimensional spaces. For example, the nonlinear function

$$f(\boldsymbol{x}) = \left\{ \left[ \sum_{i=1}^{n} \sin^2(x_i) \right] - \exp\left( -\sum_{i=1}^{n} x_i^2 \right) \right\} \cdot \exp\left( -\sum_{i=1}^{n} \sin^2 \sqrt{|x_i|} \right), \tag{I.4}$$

where $-10 \leq x_i \leq 10$ for $i = 1, 2, \ldots, n$, has a minimum $f_{\min} = -1$ at $\boldsymbol{x}_* = (0, 0, \ldots, 0)$, but this function is not differentiable at $\boldsymbol{x}_*$.

Therefore, optimization techniques have to be diverse to use gradient information when appropriate, and not to use it when it is not defined or not easily calculated. Though the above nonlinear optimization problems can be challenging to solve, constraints on the search domain and certain independent variables can make the search domain much more complicated, which can consequently make the problem even harder to solve. In addition, sometime, we have to optimize several objective functions instead of just one function, which will in turn make a problem more challenging to solve.

In general, it is possible to write most optimization problems in the general form mathematically

$$\underset{\boldsymbol{x} \in \mathbb{R}^n}{\text{minimize}} \quad f_i(\boldsymbol{x}) \qquad (i = 1, 2, \ldots, M), \tag{I.5}$$

$$\text{subject to} \quad \phi_j(\boldsymbol{x}) = 0 \qquad (j = 1, 2, \ldots, J), \tag{I.6}$$

$$\psi_k(\boldsymbol{x}) \leq 0 \qquad (k = 1, 2, \ldots, K), \tag{I.7}$$

where $f_i(\boldsymbol{x})$, $\phi_j(\boldsymbol{x})$, and $\psi_k(\boldsymbol{x})$ are functions of the design vector

$$\boldsymbol{x} = (x_1, x_2, \ldots, x_n)^{\mathrm{T}}. \tag{I.8}$$

Here, the components $x_i$ of $\boldsymbol{x}$ are called design or decision variables, and they can be real continuous, discrete, or the mixture of these two. The functions $f_i(\boldsymbol{x})$ where $i = 1, 2, \ldots, M$ are called the objective functions, and in the case of $M = 1$, there is only a single objective, and the problem becomes single-objective optimization. The objective function is sometimes called the cost function, loss function, or energy function in the literature. The space spanned by the decision variables is called the design space or search space $\mathbb{R}^n$, while the space formed by the objective function values is called the response space or objective landscape. The equalities for $\phi_j$ and inequalities for $\psi_k$ are called constraints. It is worth pointing out that we can also write the inequalities in the other way $\geq 0$, and we can also formulate the objectives as maximization.

If a solution vector $\boldsymbol{x}$ satisfies all the constraints $\phi_j(\boldsymbol{x})$ and $\psi_k(\boldsymbol{x})$, it is called a feasible solution; otherwise, it is infeasible. All the feasible points or vectors form the feasible regions in the search space.

In a rare but extreme case where there is no objective at all, there are only constraints. Such a problem is called a feasibility problem and any feasible solution is an optimal solution to such problems.

If all the problem functions are linear, the problem becomes a linear programming (LP) problem. Here, the term "programming" means planning, it has nothing to do with computer programming in this context. Thus, mathematical optimization is also called mathematical programming. In addition, if the variables in LP only take integer values, the LP becomes an integer LP or simple integer programming (IP). If the variables can only take two values (0 or 1), such an LP is called binary IP.

If there are no constraints (i.e. $J = 0$ and $K = 0$), the problem becomes an unconstrained optimization problem. If $J \geq 1$ but $K = 0$, it becomes equality-constrained optimization. Conversely, if $J = 0$ but $K \geq 1$, it becomes the inequality-constrained problem.

The classification of optimization problems can also be done by looking at the modality of the objective landscape, which can be divided into multimodal problems and unimodal problems including convex optimization. In addition, classification can also be about the determinacy of the problem. If there is NO randomness in the formulation, the problem is called deterministic and in fact all the above problems are essentially deterministic. However, if there is uncertainty in the variables or function forms, then optimization involves probability distribution and expectation, such problems are often called stochastic optimization or robust optimization.

Classification of optimization problems and the terminologies used in the optimization literature can be diverse and confusing. We summarize most of these terms in Figure I.1.

Whether an optimization problem is considered easy or hard, it can depend on many factors and the actual perspective of mathematical formulations. In fact, three factors that make a problem more challenging are: nonlinearity of the objective function, the high dimensionality of the problem, and the complex shape of the search domain.

- Nonliearity and multimodality: The high nonlinearity of a function to be optimized usually means multimodality with multiple local modes with multiple (even infinite) optima. In most cases, algorithms to solve such problems are more likely to get trapped in local modes.
- High dimensionality: High dimensionality is associated with the so-called curse of dimensionality where the distance becomes large, while the volume of any algorithm that can actually search becomes insignificant compared to the vast feasible space.

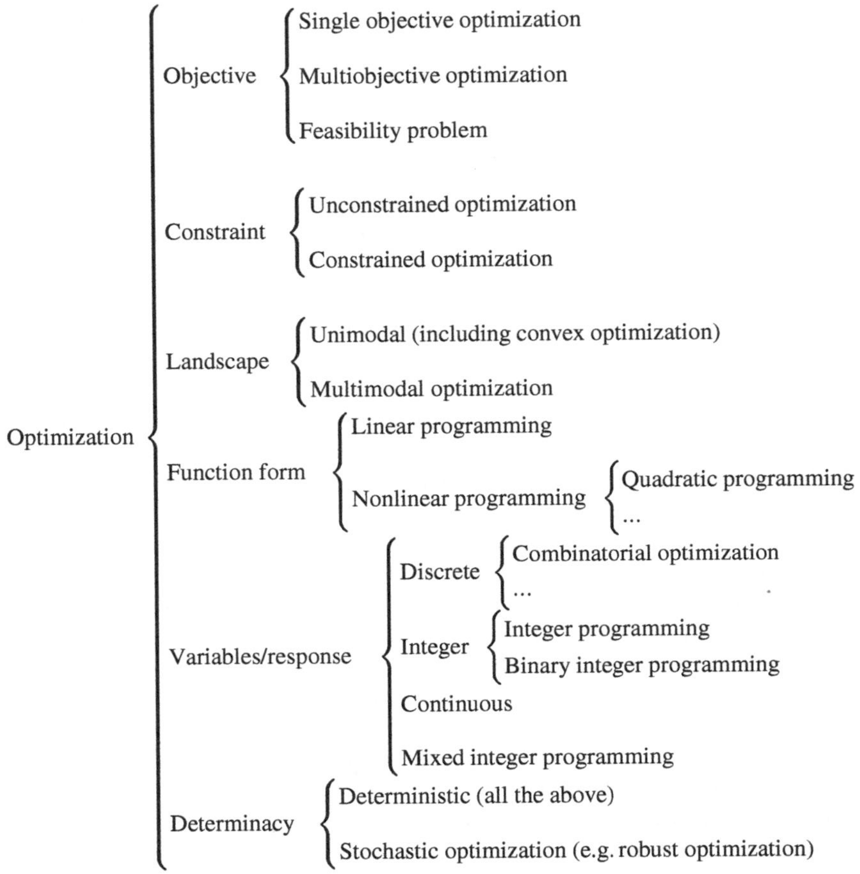


**Figure I.1** Classification of optimization problems.

- Constraints: The multiple complex constraints can make the feasible region complicated with irregular geometry or shapes. In some cases, feasible regions can be split into multiple disconnected regions with isolated islands, which makes it harder for algorithms to search all the feasible regions (thus potentially missing the true optimality).

Other factors such as the evaluation time of an objective are also important. In many applications such as protein folding, bio-informatics, aero-space engineering, and deep machine learning (ML), the evaluation of a single objective can take a long time (from a few hours to days or even weeks), therefore the computational costs can be very high. Thus, the choice of efficient algorithms becomes paramount.

Algorithms for solving optimization problems tend to be iterative, and thus multiple evaluations of objectives are needed, typically hundreds or thousands (or even millions) of evaluations. Different algorithms may have different efficiency and different requirements. For example, Newton's method, which is gradient-based, is very efficient for solving smooth objective functions, but they can get stuck in local modes if the objective is highly multimodal. If the objective is not smooth or has a kink, then the Nelder–Mead simplex method can be used because it is a gradient-free method, and can work well even for problems with discontinuities, but it can become slow and get stuck in a local mode. Algorithms for solving nonlinear optimization are diverse, including the trust-region method, interior-point method, and others, but they are mostly local search methods. In a special case when the objective becomes convex, they can be very efficient. Quadratic programming (QP) and sequential quadratic programming use such convexity properties to their advantage.

The simplex method for solving LP can be efficient in practice, but it requires all the problem functions to be linear. But, if an LP problem has integer variables, the simplex method will not work directly, it has to be combined with branch and bound to solve IP problems.

As traditional methods are usually local search algorithms, one of the current trends is to use heuristic and metaheuristic algorithms. Here *meta-* means "beyond" or "higher level," and they generally perform better than simple heuristics. In addition, all metaheuristic algorithms use certain tradeoff of randomization and local search. It is worth pointing out that there are no agreed definitions of heuristics and metaheuristics in the literature, some use "heuristics" and "metaheuristics" interchangeably. However, recent trends tend to name all stochastic algorithms with randomization and local search as metaheuristic. Here, we will also use this convention. Randomization provides a good way to move away from local search to the search on the global scale. Therefore, almost all metaheuristic algorithms intend to be suitable for global optimization, though global optimality may be still challenging to achieve for most problems in practice.

Most metaheuristic algorithms are nature-inspired as they have been developed based on some abstraction of nature. Nature has evolved over millions of years and has found perfect solutions to almost all the problems she met. We can thus learn the success of problem-solving from nature and develop nature-inspired heuristic and/or metaheuristic algorithms. More specifically, some nature-inspired algorithms are inspired by Darwin's evolutionary theory. Consequently, they are said to be biology-inspired or simply bio-inspired.

Two major components of any metaheuristic algorithms are: selection of the best solutions and randomization. The selection of the best ensures that the solutions will converge to the optimality, while the randomization avoids the solutions being trapped at local optima and, at the same time, increases

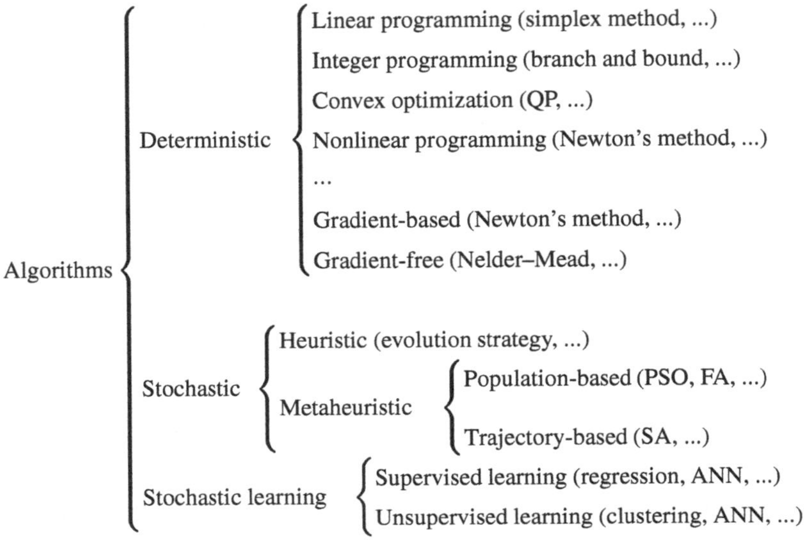


**Figure I.2** Classification of optimization algorithms.

the diversity of the solutions. The good combination of these two components will usually ensure that the global optimality is achievable.

Metaheuristic algorithms can be classified into many ways. One way is to classify them as: population-based and trajectory-based. For example, genetic algorithms are population-based as they use a set of strings, so are the particle swarm optimization (PSO) and the firefly algorithm (FA) which use multiple agents or particles. PSO and FA are also referred to as agent-based algorithms.

On the other hand, simulated annealing (SA) uses a single agent or solution which moves through the design space or search space in a piecewise style. A better move or solution is always accepted, while a not-so-good move can be accepted with certain probability. The steps or moves trace a trajectory in the search space, with a nonzero probability that this trajectory can reach the global optimum.

The types of algorithms that solve relevant optimization problems are summarized in Figure I.2 where we also include some machine learning algorithms such as artificial neural network (ANN), regression, and support vector machine (SVM).

We will introduce all the above optimization problems and related algorithms in this book. Therefore, the first two chapters review the fundamental mathematics relevant to optimization, complexity and algorithms. Then, Chapters 3–5 in Part II cover the formulation of optimization problems and traditional techniques. Chapters 6–10 in Part III focus on the applied optimization, including LP, IP, regression, and machine learning algorithms such as ANN, SVM as well as data mining techniques, and queueing theory and management. Part IV has two chapters, introducing the multiobjective optimization and handling of constraints, and Part V introduces the latest techniques of evolutionary algorithms and nature-inspired optimization algorithms. Finally, the appendices briefly discuss the commonly used optimization software packages. In addition, each chapter has some exercises with answers. Detailed Bibliography or Further Reading also provided at the end of each chapter.

## Further Reading

Boyd, S.P. and Vandenberghe, L. (2004). *Convex Optimization*. Cambridge, UK: Cambridge University Press.

Fletcher, R. (2000). *Practical Methods of Optimization*, 2e. New York: Wiley.

Gill, P.E., Murray, W., and Wright, M.H. (1982). *Practical Optimization*. Bingley: Emerald Publishing.

Nocedal, J. and Wright, S.J. (2006). *Numerical Optimization*, 2e. New York: Springer.

Yang, X.S. (2010). *Engineering Optimization: An Introduction with Metaheuristic Applications*. Hoboken, NJ: Wiley.

Yang, X.S. (2014). *Nature-Inspired Optimization Algorithms*. London: Elsevier.

# Part I

# Fundamentals

# 1

# Mathematical Foundations

The basic requirements of this book are the fundamental knowledge of functions, basic calculus, and vector algebra. However, we will review here the most relevant fundamentals of functions, vectors, differentiation, and integration. Then, we will introduce some useful concepts such as eigenvalues, complexity, convexity, probability distributions, and optimality conditions.

## 1.1 Functions and Continuity

### 1.1.1 Functions

Loosely speaking, a function is a quantity (say $y$) which varies with another independent quantity or variable $x$ in a deterministic way. For example, a simple quadratic function is

$$y = x^2. \tag{1.1}$$

For any given value of $x$, there is a unique corresponding value of $y$. By varying $x$ smoothly, we can vary $y$ in such a manner that the point $(x, y)$ will trace out a curve on the $x$–$y$ plane (see Figure 1.1). Thus, $x$ is called the independent variable, and $y$ is called the dependent variable or function. Sometimes, in order to emphasize the relationship as a function, we use $f(x)$ to express a generic function, showing that it is a function of $x$. This can also be written as $y = f(x)$.

The domain of a function is the set of numbers $x$ for which the function $f(x)$ is valid (that is, $f(x)$ gives a valid value for a corresponding value of $x$). If a function is defined over a range $a \leq x \leq b$, we say its domain is $[a, b]$ that is called a closed interval. If both $a$ and $b$ are not included, we have $a < x < b$, which is denoted by $(a, b)$, and we call this interval an open interval. If $b$ is included, while $a$ is not, we have $a < x \leq b$, and we often write this half-open and half-closed interval as $(a, b]$. Thus, the domain of function $f(x) = x^2$ is the whole set of all real numbers $\mathbb{R}$, so we have

$$f(x) = x^2 \qquad (-\infty < x < +\infty). \tag{1.2}$$

*Optimization Techniques and Applications with Examples*, First Edition. Xin-She Yang.

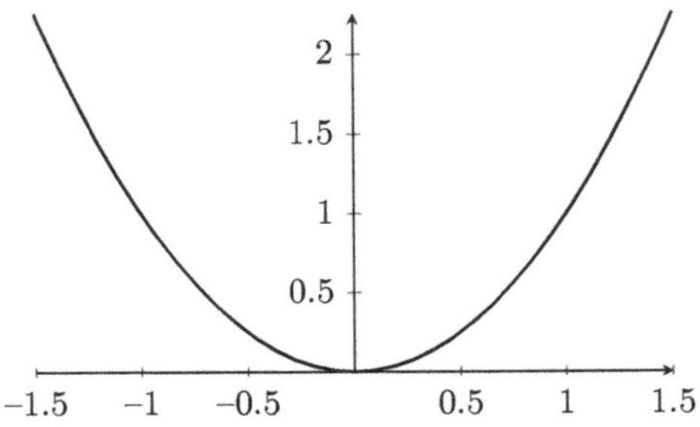


**Figure 1.1** A simple quadratic function $y = x^2$.

Here, the notation $\infty$ means infinity. In this case, the domain of $f = x^2$ is all the real numbers, which can be simply written as $\mathbb{R}$, that is, $x \in \mathbb{R}$.

All the values that a function can take for a given domain form the range of the function. Thus, the range of $y = f(x) = x^2$ is $0 \leq y < +\infty$ or $[0, +\infty)$. Here, the closed bracket "[" means that the value (here 0) is included as part of the range, while the open round bracket ")" means that the value (here $+\infty$) is not part of the range.

### 1.1.2 Continuity

A function is called continuous if an infinitely small change $\delta$ of the independent variable $x$ always lead to an infinitely small change in $f(x + \delta) - f(x)$. Alternatively, we can loosely view that the graph representing the function forms a single piece, unbroken curve. More formally, we can say that for any small change $\delta > 0$ of the independent variable in the domain, there is an $\epsilon > 0$ such that

$$|f(x + \delta) - f(x)| < \epsilon. \tag{1.3}$$

This is the continuity condition. Obviously, functions such as $x$, $|x|$, and $x^2$ are all continuous.

If a function does not satisfy the continuity condition, then the function is called discontinuous. For example, the Heaviside step function

$$H(x) = \begin{cases} 1 & \text{if} \quad x \geq 0, \\ 0 & \text{if} \quad 0 < 0, \end{cases} \tag{1.4}$$

is discontinuous at $x = 0$ as shown in Figure 1.2 where the solid dot means that $x = 0$ is included in the right branch $x \geq 0$, and the hollow dot means that it is not included. In this case, this function is called a right-continuous function.

### 1.1.3 Upper and Lower Bounds

For a given non-empty set $S \in \mathbb{R}$ of real numbers, we now introduce some important concepts such as the supremum and infimum. A number $U$ is called

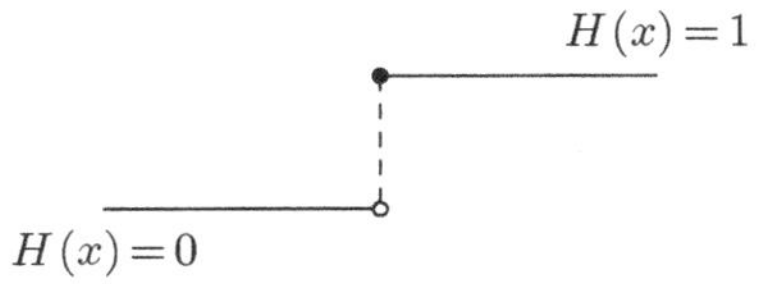


**Figure 1.2** Discontinuity of the Heaviside step function at $x = 0$.

an upper bound for $S$ if $x \leq U$ for all $x \in S$. An upper bound $\beta$ is said to be the least (or smallest) upper bound for $S$, or the supremum, if $\beta \leq U$ for any upper bound $U$. This is often written as

$$\beta \equiv \sup_{x \in S} x \equiv \sup S \equiv \sup(S). \tag{1.5}$$

All such notations are widely used in the literature of mathematical analysis. Here, "$\equiv$" denotes both equality and definition, which means "is identical to."

On the other hand, a number $L$ is called a lower bound for $S$ if $x \geq L$ for all $x$ in $S$ (that is, for all $x$, denoted by $\forall x \in S$). A lower bound $\alpha$ is referred to as the greatest (or largest) lower bound if $\alpha \geq L$ for any lower bound $L$, which is written as

$$\alpha \equiv \sup_{x \in S} x \equiv \inf S \equiv \inf(S). \tag{1.6}$$

In general, both the supremum $\beta$ and the infimum $\alpha$, if they exist, may or may not belong to $S$.

**Example 1.1** For example, any numbers greater than 5, say, 7.2 and 500 are an upper bound for the interval $-2 \leq x \leq 5$ or $[-2, 5]$. However, its smallest upper bound (or sup) is 5. Similarly, numbers such as $-10$ and $-10^5$ are lower bound of the interval, but $-2$ is the greatest lower bound (or inf). In addition, the interval $S = [15, \infty)$ has an infimum of 15 but it has no upper bound. That is to say, its supremum does not exist, or $\sup S \to \infty$.

There is an important completeness axiom which says that if a non-empty set $S \in \mathbb{R}$ of real numbers is bounded above, then it has a supremum. Similarly, if a non-empty set of real numbers is bounded below, then it has an infimum.

Furthermore, the maximum for $S$ is the largest value of all elements $s \in S$, and often written as $\max(S)$ or $\max S$, while the minimum, $\min(S)$ or $\min S$, is the smallest value among all $s \in S$. For the same interval $[-2, 5]$, the maximum of this interval is 5 which is equal to its supremum, while its minimum 5 is also equal to its infimum. Though the supremum and infimum are not necessarily part of the set $S$, however, the maximum and minimum (if they exist) always belong to the set.

However, the concepts of supremum (or infimum) and maximum (or minimum) are not the same, and maximum/minimum may not always exist.

**Example 1.2** For example, the interval $S = [-2, 7)$ or $-2 \leq x < 7$ has the supremum of $\sup S = 7$, but $S$ has no maximum.

Similarly, the interval $(-10, 15]$ does not have a minimum, though its infimum is $-10$. Furthermore, the open interval $(-2, 7)$ has no maximum or minimum; however, its supremum is 7, and infimum is $-2$.

It is worth pointing out that the problems we will discuss in this book will always have at least a maximum or a minimum.

## 1.2 Review of Calculus

### 1.2.1 Differentiation

The gradient or first derivative of a function $f(x)$ at point $x$ is defined by

$$f'(x) \equiv \frac{df(x)}{dx} = \lim_{\delta \to 0} \frac{f(x+\delta) - f(x)}{\delta}. \tag{1.7}$$

From the definition and basic function operations, it is straightforward to show that

$$(x^n)' = nx^{n-1} \quad (n = 1, 2, 3, \ldots). \tag{1.8}$$

In addition, for any two functions $f(x)$ and $g(x)$, we have

$$[af(x) + bg(x)]' = af'(x) + bg'(x), \tag{1.9}$$

where $a, b$ are two real constants. Therefore, it is easy to show that

$$(x^3 - x + k)' = (x^3)' - x' + k' = 3x^2 - 1 + 0 = 3x^2 - 1, \tag{1.10}$$

where $k$ is a constant. This means that a family of functions shifted by a different constant $k$ will have the same gradient at the same point $x$, as shown in Figure 1.3.

Some useful differentiation rules are the product rule

$$[f(x)g(x)]' = f'(x)g(x) + f(x)g'(x) \tag{1.11}$$

and the chair rule

$$\frac{df[g(x)]}{dx} = \frac{df(g)}{dg} \cdot \frac{dg(x)}{dx}, \tag{1.12}$$

where $f(g(x))$ is a composite function, which means that $f$ is a function of $g$, and $g$ is a function of $x$.

**Example 1.3** For example, from $[\sin(x)]' = \cos(x)$ and $(x^3)' = 3x^2$, we have

$$\frac{d\sin(x^3)}{dx} = \cos(x^3) \cdot (3x^2) = 3x^2 \sin(x^3).$$

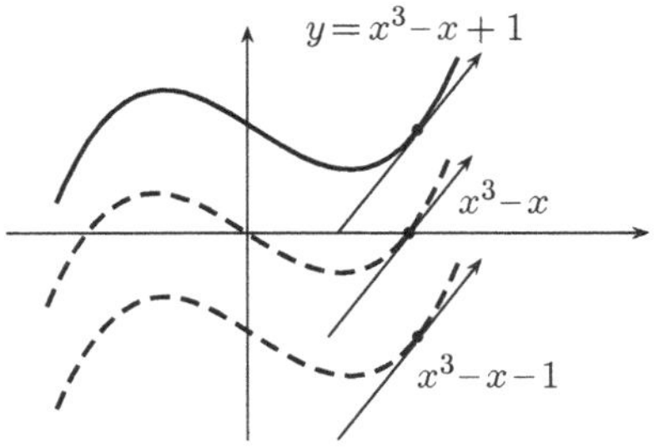


**Figure 1.3** The gradients of a family of curves $y = x^3 - x + k$ (where $k = 0, \pm 1$) at any point $x$ are the same $3x^2 - 1$.

Similarly, we have

$$\frac{\mathrm{d}\sin^n(x)}{\mathrm{d}x} = n\sin^{n-1}(x)\cdot\cos(x).$$

From the product rule (1.11), if we replace $g(x)$ by $1/g(x)$, we have

$$\frac{\mathrm{d}[g(x)^{-1}]}{\mathrm{d}x} = -1g^{-1-1}\cdot\frac{\mathrm{d}g(x)}{\mathrm{d}x} = -\frac{1}{[g(x)]^2}\frac{\mathrm{d}g(x)}{\mathrm{d}x} \tag{1.13}$$

and

$$\frac{\mathrm{d}[f(x)/g(x)]}{\mathrm{d}x} = \frac{\mathrm{d}[f(x)g(x)^{-1}]}{\mathrm{d}x}$$
$$= f'(x)g(x)^{-1} + f(x)\left\{\frac{-1}{[g(x)]^2}\frac{\mathrm{d}g(x)}{\mathrm{d}x}\right\} = \frac{g(x)f'(x) - f(x)g'(x)}{[g(x)]^2}, \tag{1.14}$$

which is the well-known quotient rule. For example, we have

$$\frac{\mathrm{d}\tan(x)}{\mathrm{d}x} = \frac{\mathrm{d}[\sin(x)/\cos(x)]}{\mathrm{d}x} = \frac{\cos(x)\sin'(x) - \sin(x)\cos'(x)}{\cos^2(x)}$$
$$= \frac{\cos^2(x) + \sin^2(x)}{\cos^2(x)} = \frac{1}{\cos^2(x)}. \tag{1.15}$$

It is worth pointing out that a continuous function may not have well-defined derivatives. For example, the absolute or modulus function

$$f(x) = |x| = \begin{cases} x & \text{if } x \geq 0, \\ -x & \text{if } x < 0, \end{cases} \tag{1.16}$$

does not have a well-defined gradient at $x$ because the gradient of $|x|$ is $+1$ if $x$ approaches 0 from $x > 0$ (using notation $0^+$). However, if $x$ approaches 0 from $x < 0$ (using notation $0^-$), the gradient of $|x|$ is $-1$. That is

$$\left.\frac{\mathrm{d}|x|}{\mathrm{d}x}\right|_{x\to 0^+} = +1, \quad \left.\frac{\mathrm{d}|x|}{\mathrm{d}x}\right|_{x\to 0^-} = -1. \tag{1.17}$$

In this case, we say that $|x|$ is not differentiable at $x = 0$. However, for $x \neq 0$, we can write

$$\frac{d|x|}{dx} = \frac{x}{|x|} \quad (x \neq 0). \tag{1.18}$$

**Example 1.4** A nature extension of this is that for a function $f(x)$, we have

$$\frac{d|f(x)|}{dx} = \frac{f(x)}{|f(x)|}\frac{df(x)}{dx}, \qquad \text{if} f(x) \neq 0. \tag{1.19}$$

As an example, for $f(x) = |x^3|$, we have

$$\frac{d|x^3|}{dx} = \frac{x^3}{|x^3|}\frac{dx^3}{dx} = \frac{x^3}{|x^3|}(3x^2) = \frac{3x^5}{|x^3|}$$
$$= \frac{3x^5}{|x^2| \cdot |x|} = \frac{3x^5}{x^2|x|} = \frac{3x^3}{|x|}, \qquad \text{if} \quad x \neq 0.$$

Higher derivatives of a univariate real function can be defined as

$$f''(x) \equiv \frac{d^2f(x)}{dx^2} \equiv \frac{df'(x)}{dx}, \; f'''(x) = [f''(x)]', \; \ldots, \; f^{(n)}(x) = \frac{d^nf(x)}{dx^n}, \tag{1.20}$$

for all positive integers $(n = 1, 2, \ldots)$.

**Example 1.5** The first, second, and third derivatives of $f(x) = xe^{-x}$ are

$$f'(x) = x'e^{-x} + x(e^{-x})' = e^{-x} + x(-1e^{-x}) = e^{-x} - xe^{-x},$$

$$f''(x) = [e^{-x} - xe^{-x}]' = xe^{-x} - 2e^{-x},$$

and

$$f'''(x) = [xe^{-x} - 2e^{-x}]' = -xe^{-x} + 3e^{-x}.$$

For a continuous function $f(x)$, if its first derivatives are well defined at every point in the whole domain, the function is called differentiable or a continuously differential function. A continuously differentiable function is said to be class $C^1$ if its first derivative exists and is continuous. Similarly, a function is said to be class $C^2$ if its both first and second derivatives exist, and are continuous. This can be extended to class $C^k$ in a similar manner. If the derivatives of all orders (all positive integers) everywhere in its domain, the function is called smooth.

It is straightforward to check that $f(x) = x$, $\sin(x)$, $\exp(x)$, and $xe^{-x}$ are all smooth functions, but $|x|$ and $|x|e^{-|x|}$ are not smooth. Some of these functions are shown in Figure 1.4.

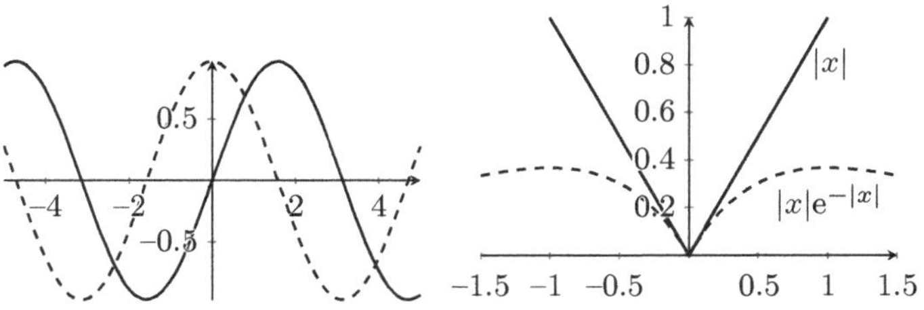


**Figure 1.4** Smooth functions (left) and non-smooth (but continuous) functions (right).

### 1.2.2 Taylor Expansions

In numerical methods and some mathematical analysis, series expansions make some calculations easier. For example, we can write the exponential function $e^x$ as a series about $x_0 = 0$ as

$$e^x = \alpha_0 + \alpha_1 x + \alpha_2 x^2 + \alpha_3 x^3 + \cdots + \alpha_n x^n. \tag{1.21}$$

Now let us try to determine these coefficients. At $x = 0$, we have

$$e^0 = 1 = \alpha_0 + \alpha_1 \times 0 + \alpha_2 \times 0^2 + \cdots + \alpha_n \times 0^n = \alpha_0, \tag{1.22}$$

which gives $\alpha_0 = 1$. In order to reduce the power or order of the expansion so that we can determine the next coefficient, we first differentiate both sides of Eq. (1.21) once; we have

$$e^x = \alpha_1 + 2\alpha_2 x + 3\alpha_3 x^2 + \cdots + n\alpha_n x^{n-1}. \tag{1.23}$$

By setting again $x = 0$, we have

$$e^0 = 1 = \alpha_1 + 2\alpha_2 \times 0 + \cdots + n\alpha_n \times 0^{n-1} = \alpha_1, \tag{1.24}$$

which gives $\alpha_1 = 1$. Similarly, differentiating it again, we have

$$e^x = (2 \times 1) \times \alpha_2 + 3 \times 2\alpha_3 x + \cdots + n(n-1)\alpha_n x^{n-2}. \tag{1.25}$$

At $x = 0$, we get

$$e^0 = (2 \times 1) \times \alpha_2 + 3 \times 2\alpha_3 \times 0 + \cdots + n(n-1)\alpha_n \times 0^{n-2} = 2\alpha_2, \tag{1.26}$$

or $\alpha_2 = 1/(2 \times 1) = 1/2!$. Here, $2! = 2 \times 1$ is the factorial of 2. In general, the factorial $n!$ is defined as $n! = n \times (n-1) \times (n-2) \times \cdots \times 2 \times 1$.

Following the same procedure and differentiating it $n$ times, we have

$$e^x = n!\alpha_n, \tag{1.27}$$

and $x = 0$ leads to $\alpha_n = 1/n!$. Therefore, the final series expansion can be written as

$$e^x = 1 + x + \frac{1}{2!}x^2 + \frac{1}{3!}x^3 + \cdots + \frac{1}{n!}x^n + \cdots, \tag{1.28}$$

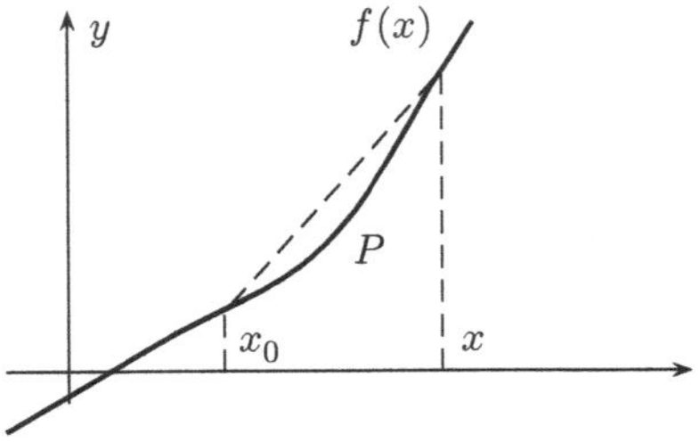


**Figure 1.5** Expansion and approximations for $f(x) = f(x_0 + h)$, where $h = x - x_0$.

which is an infinite series. Obviously, we can follow a similar process to expand other functions. We have seen here the importance of differentiation and derivatives.

If we know the value of $f(x)$ at $x_0$, we can use some approximations in a small interval $h = x - x_0$ (see Figure 1.5). Following the same idea as Eq. (1.21), we can first write the approximation in the following general form:

$$f(x) = a_0 + a_1(x - x_0) + a_2(x - x_0)^2 + \cdots + a_n(x - x_0)^n, \tag{1.29}$$

and then try to figure out the unknown coefficients $a_i (i = 0, 1, 2, \ldots)$. For the above approximation to be valid at $x = x_0$, we have

$$f(x_0) = a_0 + 0 \quad \text{(all the other terms are zeros)}, \tag{1.30}$$

so that $a_0 = f(x_0)$.

Now let us first take the first derivative of Eq. (1.29),

$$f'(x) = 0 + a_1 + 2a_2(x - x_0) + \cdots + na_n(x - x_0)^{n-1}. \tag{1.31}$$

By setting $x = x_0$, we have

$$f'(x_0) = 0 + a_1 + 0 + \cdots + na_n \times 0, \tag{1.32}$$

which gives

$$a_1 = f'(x_0). \tag{1.33}$$

Similarly, we differentiate Eq. (1.29) twice with respect to $x$ and we have

$$f''(x) = 0 + 0 + a_2 \times 2 \times 1 + \cdots + n(n-1)a_n(x - x_0)^2. \tag{1.34}$$

Setting $x = x_0$, we have

$$f''(x_0) = 2!a_2, \quad \text{or} \quad a_2 = \frac{f''(x_0)}{2!}. \tag{1.35}$$

Following the same procedure, we have

$$a_3 = \frac{f'''(x_0)}{3!}, \quad a_4 = \frac{f''''(x_0)}{4!}, \quad \ldots, \quad a_n = \frac{f^{(n)}(x_0)}{n!}. \tag{1.36}$$

Thus, we finally obtain

$$f(x) = f(x_0) + f'(x_0)(x - x_0) + \frac{f''(x_0)}{2!}(x - x_0)^2 + \frac{f'''(a)}{3!}(x - x_0)^3 + \cdots + \frac{f^{(n)}(x_0)}{n!}(x - x_0)^n, \tag{1.37}$$

which is the well-known Taylor series.

In a special case when $x_0 = 0$ and $h = x - x_0 = x$, the above Taylor series becomes zero centered, and such expansions are traditionally called Maclaurin series

$$f(x) = f(0) + f'(0)x + \frac{f''(0)}{2!}x^2 + \frac{f'''(0)}{3!}x^3 + \cdots + \frac{f^{(n)}}{n!}x^n + \cdots, \tag{1.38}$$

named after mathematician Colin Maclaurin.

In theory, we can use as many terms as possible, but in practice, the series converges very quickly and only a few terms are sufficient. It is straightforward to verify that the exponential series for $e^x$ is identical to the results given earlier. Now let us look at other examples.

**Example 1.6** Let us expand $f(x) = \sin x$ about $x_0 = 0$. We know that

$$f'(x) = \cos x, \quad f''(x) = -\sin x, \quad f'''(x) = -\cos x, \quad \ldots,$$

or $f'(0) = 1$, $f''(0) = 0$, $f'''(0) = -1$, $f''''(0) = 0, \ldots,$ which means that

$$\sin x = \sin 0 + xf'(0) + \frac{f''(0)}{2!}x^2 + \frac{f'''(0)}{3!}x^3 + \cdots$$
$$= x - \frac{x^3}{3!} + \frac{x^5}{5!} + \cdots,$$

where the angle $x$ is in radians.

For example, we know that $\sin 30° = \sin(\pi/6) = 1/2$. We now use the expansion to estimate it for $x = \pi/3 = 0.523\,598$,

$$\sin\frac{\pi}{6} \approx \frac{\pi}{6} - \frac{(\pi/6)^3}{3!} + \frac{(\pi/6)^5}{5!}$$
$$\approx 0.523\,599 - 0.023\,92 + 0.000\,032\,8 \approx 0.500\,002\,132\,6,$$

which is very close to the true value $1/2$.

If we continue the process to infinity, we then reach the infinite power series and the error $f^{(n)}(0)x^n/n!$ becomes negligibly small if the series converges. For example, some common series are

$$\frac{1}{1-x} = 1 + x + x^2 + x^3 + \cdots + x^n + \cdots, \quad x \in (-1, 1), \tag{1.39}$$

$$\sin x = x - \frac{x^3}{3!} + \frac{x^5}{5!} - \cdots, \quad \cos x = 1 - \frac{x^2}{2!} + \frac{x^4}{4!} - \cdots, \quad x \in \mathbb{R}, \tag{1.40}$$

$$\tan(x) = x + \frac{x^3}{3} + \frac{2x^5}{15} + \frac{17x^7}{315} + \cdots, \quad x \in \left(-\frac{\pi}{2}, \frac{\pi}{2}\right), \tag{1.41}$$

and

$$\ln(1+x) = x - \frac{x^2}{2} + \frac{x^3}{3} - \frac{x^4}{4} + \frac{x^5}{5} - \cdots, \quad x \in (-1, 1]. \tag{1.42}$$

As an exercise, we leave the reader to prove the above series.

### 1.2.3 Partial Derivatives

For multivariate functions, we can define the partial derivatives with respect to an independent variable by assuming that other independent variables are constants. For example, for a function $f(x, y)$ with two independent variables, we can define

$$\frac{\partial f(x,y)}{\partial x} \equiv \frac{\partial f}{\partial x} \equiv \left.\frac{\partial f}{\partial x}\right|_y = \lim_{\delta \to 0, y=\text{constant}} \frac{f(x+\delta, y) - f(x,y)}{\delta}, \tag{1.43}$$

and

$$\frac{\partial f(x,y)}{\partial y} \equiv \frac{\partial f}{\partial y} \equiv \left.\frac{\partial f}{\partial y}\right|_x = \lim_{\delta \to 0, x=\text{constant}} \frac{f(x, y+\delta) - f(x,y)}{\delta}. \tag{1.44}$$

Similar to the ordinary derivatives, partial derivative operations are also linear.

**Example 1.7** For $f(x, y) = x^2 + y^3 + 3xy^2$, its partial derivatives are

$$\frac{\partial f}{\partial x} = \frac{\partial x^2}{\partial x} + \frac{\partial y^3}{\partial x} + \frac{\partial(3xy^2)}{\partial x} = 2x + 0 + 3y^2 = 2x + 3y^2,$$

where we have treated $y$ as a constant and also used the fact that $dy/dx = 0$ because $x$ and $y$ are both independent variables. Similarly, we have

$$\frac{\partial f}{\partial y} = \frac{\partial x^2}{\partial y} + \frac{\partial y^3}{\partial y} + \frac{\partial(3xy^2)}{\partial y} = 0 + 3y^2 + 3x(2y) = 3y^2 + 6xy.$$

We can define higher-order derivatives as

$$\frac{\partial^2 f}{\partial x^2} = \frac{\partial}{\partial x}\left(\frac{\partial f}{\partial x}\right), \quad \frac{\partial^2 f}{\partial y^2} = \frac{\partial}{\partial y}\left(\frac{\partial f}{\partial y}\right), \tag{1.45}$$

and

$$\frac{\partial^2 f}{\partial x \partial y} = \frac{\partial^2 f}{\partial y \partial x} = \frac{\partial}{\partial x}\left(\frac{\partial f}{\partial y}\right) = \frac{\partial}{\partial y}\left(\frac{\partial f}{\partial x}\right). \tag{1.46}$$

Let us revisit the previous example.

**Example 1.8** For $f(x, y) = x^2 + y^3 + 3xy^2$, we have

$$\frac{\partial^2 f}{\partial x^2} = \frac{\partial(2x + 3y^2)}{\partial x} = 2 + 0 = 2, \quad \frac{\partial^2 f}{\partial y^2} = \frac{\partial(3y^2 + 6xy)}{\partial y} = 6y + 6x.$$

In addition, we have

$$\frac{\partial^2 f}{\partial x \partial y} = \frac{\partial(3y^2 + 6xy)}{\partial x} = 0 + 6y = 6y,$$

or

$$\frac{\partial^2 f}{\partial y \partial x} = \frac{\partial(2x + 3y^2)}{\partial y} = 0 + 3(2y) = 6y,$$

which shows that

$$\frac{\partial^2 f}{\partial x \partial y} = \frac{\partial^2 f}{\partial y \partial x}.$$

Other higher-order partial derivatives can be defined in a similar manner.

### 1.2.4 Lipschitz Continuity

A very important concept related to optimization and convergence analysis is the Lipschitz continuity of a function $f(x)$,

$$|f(x_1) - f(x_2)| \leq L|x_1 - x_2|, \tag{1.47}$$

for any $x_1$ and $x_2$ in the domain of $f(x)$. Here, $L \geq 0$ is called the Lipschitz constant or the modulus of uniform continuity, which is independent of $x_1$ and $x_2$. This is equivalent to that case that the absolute derivative is finite, that is

$$\frac{|f(x_1) - f(x_2)|}{|x_1 - x_2|} \leq L < \infty, \tag{1.48}$$

which limits the rate of change of function. This means the small change in the independent variable (input) can lead to the arbitrarily small change in the function (output). However, when the Lipschitz constant is sufficiently large, a small change in $x$ could lead to a much larger change in $f(x)$, but this Lipschitz estimate is an upper bound and the actual change can be much smaller. Therefore, any function with a finite or bounded first derivative is Lipschitz continuous. For example, $\sin(x)$, $\cos(x)$, $x$ and $x^2$ are all Lipschitz, while the binary step function $H(x) = 1$ if $x \geq 0$ (otherwise $H(x) = 0$ if $x < 0$) is not Lipschitz at $x = 0$.

For example, function $f(x) = 3x^2$ is Lipschitz continuous in the domain $Q = [-5, 5]$. For any $x_1, x_2 \in Q$, we have

$$|f(x_1) - f(x_1)| = |3x_1^2 - 3x_2^2| = 3|(x_1^2 - x_2^2)| = 3|x_1 + x_2| \cdot |x_1 - x_2|. \tag{1.49}$$

Since $|x_1 + x_2| \leq |x_1| + |x_2| \leq 5 + 5 = 10$, we have

$$|f(x_1) - f(x_2)| \leq 30|x_1 - x_2|, \tag{1.50}$$

which means that the Lipschitz constant is 30. It is worth pointing out that $L = 30$ is the just one lower value. We can also use $L = 50$ or $L = 100$ such as

$$|f(x_1) - f(x_2) \leq 100|x_1 - x_2|. \tag{1.51}$$

Though Lipschitz constant should not depend on $x_1$ and $x_2$, it may depend on the size of the domain when we try to derive this constant.

### 1.2.5 Integration

Integration is a reverse operation of differentiation. For a univariate function $f(x)$, the meaning of its integral over a finite interval $[a, b]$ is the area enclosed by the curve $f(x)$ and the $x$-axis

$$I = \int_a^b f(x)\mathrm{d}x, \tag{1.52}$$

where $a$ and $b$ are the integration limits, and $f(x)$ is called the integrand. It is worth pointing out that the area above the $x$-axis is considered as positive, while the area under the $x$-axis is negative. If there is a function $F(x)$ whose first derivative is $f(x)$ (i.e. $F'(x) = f(x)$) in the same interval $[a, b]$, we can write

$$\int_a^b f(x)\mathrm{d}x = F(x)\Big|_a^b = F(b) - F(a). \tag{1.53}$$

The above integral is a definite integral with fixed integral limits. If there are no specific integration limits involved in the integral, we can, in general, write it as

$$\int f(x)\mathrm{d}x = F(x) + C, \tag{1.54}$$

which is an indefinite integral and $C$ is the unknown integration constant. This integral constant comes from the fact that a function or curve shifted by a constant will have the same gradient.

One of the integration limits (or both) can be infinite.

**Example 1.9** For example, we have

$$\int_0^\infty \mathrm{e}^{-x}\mathrm{d}x. \tag{1.55}$$

Since $[-\mathrm{e}^{-x}]' = \mathrm{e}^{-x}$, we have $F(x) = -\mathrm{e}^{-x}, f(x) = \mathrm{e}^{-x}$, and $F'(x) = f(x)$. Thus, we get

$$\int_0^\infty \mathrm{e}^{-x}\mathrm{d}x = [-\mathrm{e}^{-x}]\Big|_0^\infty = (-\mathrm{e}^{-\infty}) - (-\mathrm{e}^{-0}) = -0 - (-1) = 1. \tag{1.56}$$

A very useful formula that can be derived from the product rule of differentiation $[u(x)v(x)]' = u'(x)v(x) + u(x)v(x)$ is the integration by parts

$$\int \frac{\mathrm{d}[u(x)v(x)]}{\mathrm{d}x}\mathrm{d}x = \int [u'(x)v(x) + u(x)v'(x)]\mathrm{d}x$$
$$= \int v(x)u'(x)\mathrm{d}x + \int u(x)v'(x)\mathrm{d}x, \tag{1.57}$$

which leads to

$$\int u'(x)v(x)\mathrm{d}x = u(x)v(x) - \int u(x)v'(x)\mathrm{d}x. \tag{1.58}$$

Ignoring the integration constants, the above formula can be written in a more compact form as

$$\int v\mathrm{d}u = uv - \int u\mathrm{d}v. \tag{1.59}$$

Similarly, its corresponding definite integral becomes

$$\int_a^b v\mathrm{d}u = (uv)\Big|_a^b - \int_a^b u\mathrm{d}v. \tag{1.60}$$

Let us look at an example.

**Example 1.10** To evaluate the integral

$$I = \int_0^\infty x\mathrm{e}^{-x}\mathrm{d}x,$$

we have $v(x) = x$ and $u'(x) = \mathrm{e}^{-x}$, which leads to

$$\mathrm{d}v = \mathrm{d}x, \quad v'(x) = 1, \quad u(x) = -\mathrm{e}^{-x}.$$

From the formula of integration by parts, we have

$$I = \int_0^\infty x\mathrm{e}^{-x}\mathrm{d}x = \big[x(-\mathrm{e}^{-x})\big]_0^\infty - \int_0^\infty 1 \cdot (-\mathrm{e}^{-x})\mathrm{d}x$$
$$= -(\infty)\mathrm{e}^{-\infty} - [-0\mathrm{e}^{-0}] + \int_0^\infty \mathrm{e}^{-x} = 0 + 0 + \int_0^\infty \mathrm{e}^{-x}\mathrm{d}x = 1,$$

where we have used the earlier result in Eq. (1.56) and the fact that

$$\lim_{K\to\infty} K\mathrm{e}^{-K} \to 0.$$

The multiple integral of a multivariate function can be defined in a similar manner. For example, for a function of $f(x, y)$ of two independent variables $x$ and $y$, its double integral can be defined as

$$I = \iint f(x,y)\mathrm{d}x\mathrm{d}y. \tag{1.61}$$

In a rectangular domain $D = [a, b] \times [c, d]$ (that is $a \leq x \leq b$ and $c \leq y \leq d$), the following integral means the volume enclosed by the surface $f(x, y)$ over the rectangular domain. We have

$$I = \iint_D f(x, y)\mathrm{d}x\mathrm{d}y = \int_c^d \left[\int_a^b f(x, y)\mathrm{d}x\right] \mathrm{d}y, \tag{1.62}$$

which is the same as

$$I = \int_a^b \left[\int_c^d f(x, y)\mathrm{d}y\right] \mathrm{d}x, \tag{1.63}$$

due to Fubini's theorem that is valid when

$$\iint_D |f(x, y)|\mathrm{d}x\mathrm{d}y < \infty. \tag{1.64}$$

As integration is not quite relevant to most optimization techniques, we will not discuss integration any further. We will introduce more whenever needed in later chapters. In the rest of the chapter, we will review the vector algebra and eigenvalues of matrices before we move onto the introduction of optimality conditions.

## 1.3 Vectors

Loosely speaking, a vector is a quantity with a magnitude and a direction in practice. However, mathematically speaking, a vector can be represented by a set of ordered scalars or numbers. For example, a three-dimensional vector in the Cartesian coordinates can be written as

$$\boldsymbol{a} = \begin{pmatrix} a_1 \\ a_2 \\ a_3 \end{pmatrix} = \begin{pmatrix} x \\ y \\ z \end{pmatrix}, \tag{1.65}$$

where $a_1, a_2, a_3$ (or $x, y, z$) are its three components, along $x$-, $y$-, and $z$-axes, respectively. A vector is usually denoted in a lowercase boldface. Here, we write the component as a column vector. Alternatively, a vector can be equally represented by a row vector in the form:

$$\boldsymbol{a} = \begin{pmatrix} a_1 & a_2 & a_3 \end{pmatrix} = \begin{pmatrix} x & y & z \end{pmatrix}. \tag{1.66}$$

A column vector can be converted into a row vector by a simple transpose (using notation T as a superscript to denote this operation) or vice versa. We have

$$\begin{pmatrix} a_1 & a_2 & a_3 \end{pmatrix}^{\mathrm{T}} = \begin{pmatrix} a_1 \\ a_2 \\ a_3 \end{pmatrix} \quad \text{or} \quad \begin{pmatrix} x \\ y \\ z \end{pmatrix}^{\mathrm{T}} = \begin{pmatrix} x & y & z \end{pmatrix}. \tag{1.67}$$

The magnitude or length of a three-dimensional vector is its Cartesian norm

$$|\boldsymbol{a}| = \sqrt{a_1^2 + a_2^2 + a_3^2} = \sqrt{x^2 + y^2 + z^2}. \tag{1.68}$$

### 1.3.1 Vector Algebra

In general, a vector in an $n$-dimensional space ($n \geq 1$) can be written as a column vector

$$\boldsymbol{x} = \begin{pmatrix} x_1 \\ x_2 \\ \vdots \\ x_n \end{pmatrix} \tag{1.69}$$

or a row vector

$$\boldsymbol{x} = \begin{pmatrix} x_1 & x_2 & \dots & x_n \end{pmatrix}. \tag{1.70}$$

Its length can be written as

$$||\boldsymbol{x}|| = \sqrt{x_1^2 + x_2^2 + \cdots + x_n^2}, \tag{1.71}$$

which is the Euclidean norm.

The addition or substraction of two vectors $\boldsymbol{u}$ and $\boldsymbol{v}$ are the addition or substraction of their corresponding components, that is

$$\boldsymbol{u} \pm \boldsymbol{v} = \begin{pmatrix} u_1 \\ u_2 \\ \vdots \\ u_n \end{pmatrix} \pm \begin{pmatrix} v_1 \\ v_2 \\ \vdots \\ v_n \end{pmatrix} = \begin{pmatrix} u_1 \pm v_1 \\ u_2 \pm v_2 \\ \vdots \\ u_n \pm v_n \end{pmatrix}. \tag{1.72}$$

The dot product, also called the inner product, of two vectors $\boldsymbol{u}$ and $\boldsymbol{v}$ is defined as

$$\boldsymbol{u}^{\mathrm{T}}\boldsymbol{v} \equiv \boldsymbol{u} \cdot \boldsymbol{v} = \sum_{i=1}^{n} u_i v_i = u_1 v_1 + u_2 v_2 + \cdots + u_n v_n. \tag{1.73}$$

### 1.3.2 Norms

For an $n$-dimensional vector $\boldsymbol{x}$, we can define a $p$-norm or $L_p$-norm (also $L^p$-norm) as

$$||\boldsymbol{x}||_p \equiv \left(|x_1|^p + |x_2|^p + \cdots + |x_n|^p\right)^{1/p} = \left(\sum_{i=1}^{n} |x_i|^p\right)^{1/p} \quad (p > 0). \tag{1.74}$$

Obviously, the Cartesian norm or length is an $L_2$-norm

$$||\boldsymbol{x}||_2 = \sqrt{|x_1|^2 + |x_2|^2 + \cdots + |x_n|^2} = \sqrt{x_1^2 + x_2^2 + \cdots + x_n^2}. \tag{1.75}$$

Three most widely used norms are $p = 1, 2$, and $\infty$. When $p = 2$, it becomes the Cartesian $L_2$-norm as discussed above. When $p = 1$, the $L_1$-norm is given by

$$||\boldsymbol{x}||_1 = |x_1| + |x_2| + \cdots + |x_n|. \tag{1.76}$$

For $p = \infty$, it becomes

$$||\boldsymbol{x}||_\infty = \max\{|x_1|, |x_2|, \ldots, |x_n|\} = x_{\max}, \tag{1.77}$$

which is the largest absolute component of $\boldsymbol{x}$. This is because

$$\begin{aligned} ||\boldsymbol{x}||_\infty &= \lim_{p\to\infty}\left(\sum_{i=1}^{p}|x_i|^p\right)^{1/p} = \lim_{p\to\infty}\left(|x_{\max}|^p\sum_{i=1}^{n}\left|\frac{x_i}{x_{\max}}\right|^p\right)^{1/p} \\ &= x_{\max}\lim_{p\to\infty}\left(\sum_{i=1}^{n}\left|\frac{x_i}{x_{\max}}\right|\right)^{1/p} = x_{\max}, \end{aligned} \tag{1.78}$$

where we have used the fact that $|x_i/x_{\max}| < 1$ (except for one component, say, $|x_k| = x_{\max}$). Thus, $\lim_{p\to\infty}|x_i/x_{\max}|^p \to 0$ for all $i \neq k$. Thus, the sum of all ratio terms is 1. That is

$$\left(\lim_{p\to\infty}\left|\frac{x_i}{x_{\max}}\right|^p\right)^{1/p} = 1. \tag{1.79}$$

In general, for any two vectors $\boldsymbol{u}$ and $\boldsymbol{v}$ in the same space, we have the following equality:

$$||\boldsymbol{u}||_p + ||\boldsymbol{v}||_p \geq ||\boldsymbol{u} + \boldsymbol{v}||_p \quad (p \geq 0). \tag{1.80}$$

**Example 1.11** For two vectors $\boldsymbol{u} = [1 \quad 2 \quad 3]^{\mathrm{T}}$ and $\boldsymbol{v} = [1 \quad -2 \quad -1]^{\mathrm{T}}$, we have

$$\boldsymbol{u}^{\mathrm{T}}\boldsymbol{v} = 1 \times 1 + 2 \times (-2) + 3 \times (-1) = -6,$$

$$||\boldsymbol{u}||_1 = |1| + |2| + |3| = 6, \quad ||\boldsymbol{v}||_1 = |1| + |-2| + |-1| = 4,$$

$$||\boldsymbol{u}||_2 = \sqrt{1^2 + 2^2 + 3^2} = \sqrt{14}, \quad ||\boldsymbol{v}||_2 = \sqrt{1^2 + (-2)^2 + (-1)^2} = \sqrt{6},$$

$$||\boldsymbol{u}||_\infty = \max\{|1|, |2|, |3|\} = 3, \quad ||\boldsymbol{v}||_\infty = \max\{|1|, |-2|, |-1|\} = 2,$$

and

$$\boldsymbol{w} = \boldsymbol{u} + \boldsymbol{v} = \begin{bmatrix}1+1 & 2+(-2) & 3+(-1)\end{bmatrix}^{\mathrm{T}} = \begin{bmatrix}2 & 0 & 2\end{bmatrix}^{\mathrm{T}}$$

whose norms are

$$||\boldsymbol{w}||_1 = |2| + |0| + |2| = 2, \quad ||\boldsymbol{w}||_\infty = \max\{|2|, |0|, |2|\} = 2,$$

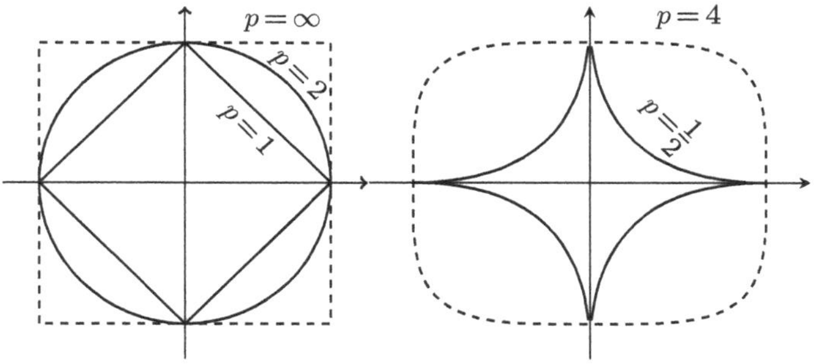


Figure 1.6 Different $p$-norms for $p = 1, 2$, and $\infty$ (left) as well as $p = 1/2$ and $p = 4$ (right).

$$||\boldsymbol{w}||_2 = \sqrt{2^2 + 0^2 + 2^2} = \sqrt{8}.$$

Using the above values, it is straightforward to verify that

$$||\boldsymbol{u}||_p + ||\boldsymbol{v}||_p \geq ||\boldsymbol{u} + \boldsymbol{v}||_p \quad (p = 1, 2, \infty).$$

### 1.3.3 2D Norms

To get a clearer picture about the differences between different norms, we now focus on the vectors in the two-dimensional (2D) Cartesian coordinates $(x, y)$. For $\boldsymbol{u} = (x, y)^{\mathrm{T}}$, we have

$$||\boldsymbol{u}||_p = (|x|^p + |y|^p)^{1/p} \quad (p > 0). \tag{1.81}$$

Obviously, in special cases of $p = 1, 2, \infty$, we have

$$||\boldsymbol{u}||_1 = |x| + |y|, \quad ||\boldsymbol{u}||_2 = \sqrt{x^2 + y^2}, \quad ||\boldsymbol{u}||_\infty = \max\{|x|, |y|\}. \tag{1.82}$$

In order to show the main differences, we can let $-1 \leq x, y \leq 1$ and thus the 2-norm becomes a unit circle $x^2 + y^2 = 1$. All the other norms with different $p$ values can be plotted around this unit circle, as shown in Figure 1.6.

## 1.4 Matrix Algebra

### 1.4.1 Matrices

A matrix is a rectangular array of numbers such as

$$\boldsymbol{A} = \begin{pmatrix} 1 & 2 & 3 \\ 4 & 5 & 6 \end{pmatrix}, \quad \boldsymbol{B} = \begin{pmatrix} 7 & 8 \\ -1 & 3.7 \end{pmatrix}. \tag{1.83}$$

Matrix $\boldsymbol{A}$ has two rows and three columns, thus its size is said to be $2 \times 3$ or 2 by 3. The element in the first row and first column is $a_{11} = 1$, and the element in the second row and third column is $a_{23} = 6$. Similarly, matrix $\boldsymbol{B}$ has a size of $2 \times 2$.

It is customary to use a boldface uppercase letter to represent a matrix, while its element is written in a corresponding lowercase letter. Thus, a matrix $\boldsymbol{A}$ of size $m \times n$ can, in general, be written as

$$\boldsymbol{A} = [a_{ij}] = \begin{pmatrix} a_{11} & a_{12} & \dots & a_{1n} \\ a_{21} & a_{22} & \dots & a_{2n} \\ \vdots & \vdots & \ddots & \vdots \\ a_{m1} & a_{m2} & \dots & a_{mn} \end{pmatrix}, \quad [a_{ij}] \in \mathbb{R}^{m\times n}, \tag{1.84}$$

where $1 \leq i \leq m, 1 \leq j \leq n$. It is worth pointing out that a vector can be considered as a special case of matrices, and thus matrices are the natural extension of vectors. Here, we assume that all the numbers are real numbers, that is $a_{ij} \in \mathbb{R}$, which is most relevant to the contents in this book. In general, the entries in a matrix can be complex numbers. Here, we have used $\mathbb{R}^{m\times n}$ to denote the fact that all the elements in $\boldsymbol{A}$ span a space with a dimensionality of $m \times n$.

The transpose or transposition of an $m \times n$ matrix $\boldsymbol{A} = [a_{ij}]$ is obtained by turning columns into rows and vice versa. This operation is denoted by T or $\boldsymbol{A}^{\mathrm{T}}$. That is

$$\boldsymbol{A}^{\mathrm{T}} = [a_{ij}]^{\mathrm{T}} = [a_{ji}] = \begin{pmatrix} a_{11} & a_{21} & \dots & a_{m1} \\ a_{12} & a_{22} & \dots & a_{m2} \\ \vdots & \vdots & \ddots & \vdots \\ a_{1n} & a_{2n} & \dots & a_{mn} \end{pmatrix}, \quad [a_{ji}] \in \mathbb{R}^{n\times m}. \tag{1.85}$$

In a special case when $m = n$, the matrix becomes a square matrix. If a transpose of a square matrix $\boldsymbol{A}$ is equal to itself, the matrix is said to be symmetric. That is

$$\boldsymbol{A}^{\mathrm{T}} = \boldsymbol{A} \quad \text{or} \quad a_{ij} = a_{ji}, \tag{1.86}$$

for all $1 \leq i, j \leq n$.

For a square matrix $\boldsymbol{A} = [a_{ij}] \in \mathbb{R}^{n\times n}$, its diagonal elements are $a_{ii} (i = 1, 2, \dots, n)$. The trace of the matrix is the sum of all its diagonal elements

$$\mathrm{tr}(\boldsymbol{A}) = a_{11} + a_{22} + \cdots + a_{nn} = \sum_{i=1}^{n} a_{ii}. \tag{1.87}$$

A special and useful square matrix $[a_{ij}]$ is the identity matrix where the diagonal elements are one ($a_{ii} = 1$) and all other elements are zero ($a_{ij} = 0$ if $i \neq j$). That is

$$\boldsymbol{I} = \begin{pmatrix} 1 & 0 & 0 & \dots & 0 \\ 0 & 1 & 0 & \dots & 0 \\ 0 & 0 & 1 & \dots & 0 \\ \vdots & \vdots & \vdots & \ddots & 0 \\ 0 & 0 & 0 & \dots & 1 \end{pmatrix}. \tag{1.88}$$

Addition or subtraction of two matrices is possible only if they are of the same size. For example, if $\boldsymbol{A} = [a_{ij}]$ and $\boldsymbol{B} = [b_{ij}]$ where $1 \leq i \leq m$ and $1 \leq j \leq n$, we have

$$\boldsymbol{A} \pm \boldsymbol{B} = [a_{ij}] \pm [b_{ij}] = [a_{ij} \pm b_{ij}]$$
$$= \begin{pmatrix} a_{11} \pm b_{11} & a_{12} \pm b_{12} & \dots & a_{1n} \pm b_{1n} \\ a_{21} \pm b_{21} & a_{22} \pm b_{22} & \dots & a_{2n} \pm b_{2n} \\ \vdots & \vdots & \ddots & \vdots \\ a_{m1} \pm b_{m1} & a_{m2} \pm b_{m2} & \dots & a_{mn} \pm b_{mn} \end{pmatrix}. \tag{1.89}$$

**Example 1.12** For matrices $\boldsymbol{A} = \begin{pmatrix} 2 & 3 & -1 \\ 1 & 2 & 5 \end{pmatrix}$, $\boldsymbol{B} = \begin{pmatrix} 1 & 2 \\ 2 & 3 \end{pmatrix}$, and $\boldsymbol{D} = \begin{pmatrix} 1 & 1 & 1 \\ 2 & -1 & 0 \end{pmatrix}$ we have

$$\boldsymbol{A} + \boldsymbol{D} = \begin{pmatrix} 2 & 3 & -1 \\ 1 & 2 & 5 \end{pmatrix} + \begin{pmatrix} 1 & 1 & 1 \\ 2 & -1 & 0 \end{pmatrix}$$
$$= \begin{pmatrix} 2+1 & 3+1 & -1+1 \\ 1+2 & 2+(-1) & 5+0 \end{pmatrix} = \begin{pmatrix} 3 & 4 & 0 \\ 3 & 1 & 5 \end{pmatrix}.$$

The transpose of these matrices are

$$\boldsymbol{A}^{\mathrm{T}} = \begin{pmatrix} 2 & 1 \\ 3 & 2 \\ -1 & 5 \end{pmatrix}, \quad \boldsymbol{D}^{\mathrm{T}} = \begin{pmatrix} 1 & 2 \\ 1 & -1 \\ 1 & 0 \end{pmatrix}, \quad \boldsymbol{B}^{\mathrm{T}} = \begin{pmatrix} 1 & 2 \\ 2 & 3 \end{pmatrix} = \boldsymbol{B},$$

which means that the square matrix $\boldsymbol{B}$ is symmetric. In addition, the trace of $\boldsymbol{B}$ is $\mathrm{tr}(\boldsymbol{B}) = 1 + 3 = 4$.

The multiplication of two matrices requires a special condition that the number of columns of the first matrix must be equal to the number of rows of the second matrix. If $\boldsymbol{A}$ has an $m \times n$ matrix and $\boldsymbol{B}$ is an $n \times p$ matrix, then the production $\boldsymbol{C} = \boldsymbol{AB}$ is an $m \times p$ matrix. The element $c_{ij}$ is obtained by the dot product of the $i$th row of $\boldsymbol{A}$ and the $j$th column of $\boldsymbol{B}$. That is

$$c_{ij} = a_{i1}b_{1j} + a_{i2}b_{2j} + \cdots + a_{in}b_{nj} = \sum_{k=1}^{n} a_{ik}b_{kj}. \tag{1.90}$$

Let us look at an example.

**Example 1.13** For $A = \begin{pmatrix} 1 & 2 \\ 3 & 4 \\ 5 & 6 \end{pmatrix}$ and $B = \begin{pmatrix} 1 & -1 \\ -2 & 5 \end{pmatrix}$, we have $C = AB$ so that

$$c_{11} = \begin{pmatrix} 1 & 2 \end{pmatrix} \begin{pmatrix} 1 \\ -2 \end{pmatrix} = 1 \times 1 + 2 \times (-2) = -3,$$

$$c_{12} = \begin{pmatrix} 1 & 2 \end{pmatrix} \begin{pmatrix} -1 \\ 5 \end{pmatrix} = 1 \times (-1) + 2 \times 5 = 9,$$

$$c_{21} = \begin{pmatrix} 3 & 4 \end{pmatrix} \begin{pmatrix} 1 \\ -2 \end{pmatrix} = -5, \quad c_{22} = \begin{pmatrix} 3 & 4 \end{pmatrix} \begin{pmatrix} -1 \\ 5 \end{pmatrix} = 17,$$

$$c_{31} = \begin{pmatrix} 5 & 6 \end{pmatrix} \begin{pmatrix} 1 \\ -2 \end{pmatrix} = -7, \quad c_{32} = \begin{pmatrix} 5 & 6 \end{pmatrix} \begin{pmatrix} -1 \\ 5 \end{pmatrix} = 25.$$

Thus, we have

$$C = AB = \begin{pmatrix} -3 & 9 \\ -5 & 17 \\ -7 & 25 \end{pmatrix}.$$

However, $BA$ does not exist because it is not possible to carry out the multiplication. In general, $AB \neq BA$ even if they exist, which means that matrix multiplication is not commutative.

It is straightforward to show that

$$(AB)^{\mathrm{T}} = B^{\mathrm{T}} A^{\mathrm{T}}. \tag{1.91}$$

**Example 1.14** Let us revisit the previous example. We know that

$$A^{\mathrm{T}} = \begin{pmatrix} 1 & 3 & 5 \\ 2 & 4 & 6 \end{pmatrix}, \quad B^{\mathrm{T}} = \begin{pmatrix} 1 & -2 \\ -1 & 5 \end{pmatrix}.$$

Thus, we have

$$B^{\mathrm{T}} A^{\mathrm{T}} = \begin{pmatrix} 1 & -2 \\ -1 & 5 \end{pmatrix} \begin{pmatrix} 1 & 3 & 5 \\ 2 & 4 & 6 \end{pmatrix}$$
$$= \begin{pmatrix} -3 & -5 & -7 \\ 9 & 17 & 25 \end{pmatrix} = \begin{pmatrix} -3 & 9 \\ -5 & 17 \\ -7 & 25 \end{pmatrix} = (AB)^{\mathrm{T}}.$$

For a square matrix $A$ and an identity matrix $I$ of the same size, it is straightforward to check that

$$AI = IA = A. \tag{1.92}$$

For a square matrix of size $n \times n$, if there exists another unique matrix $\boldsymbol{B}$ of the same size satisfying

$$\boldsymbol{AB} = \boldsymbol{BA} = \boldsymbol{I}, \tag{1.93}$$

then $\boldsymbol{B}$ is called the inverse matrix of $\boldsymbol{A}$. Here, $\boldsymbol{I}$ is an $n \times n$ identity matrix. In this case, we often denote $\boldsymbol{B} = \boldsymbol{A}^{-1}$, which means

$$\boldsymbol{AA}^{-1} = \boldsymbol{A}^{-1}\boldsymbol{A} = \boldsymbol{I}. \tag{1.94}$$

In general, $\boldsymbol{A}^{-1}$ may not exist or be unique. One useful test condition is the determinant to be introduced next.

In addition, in a special case when the inverse of $\boldsymbol{A}$ is the same as its transpose $\boldsymbol{A}^{\mathrm{T}}$ (i.e. $\boldsymbol{A}^{-1} = \boldsymbol{A}^{\mathrm{T}}$), then $\boldsymbol{A}$ is said to be orthogonal, which means

$$\boldsymbol{AA}^{\mathrm{T}} = \boldsymbol{AA}^{-1} = \boldsymbol{I}, \quad \boldsymbol{A}^{-1} = \boldsymbol{A}^{\mathrm{T}}. \tag{1.95}$$

It is easy to check that the rotation matrix

$$\boldsymbol{R} = \begin{pmatrix} \cos\theta & -\sin\theta \\ \sin\theta & \cos\theta \end{pmatrix} \tag{1.96}$$

is orthogonal because $\boldsymbol{R}^{-1} = \begin{pmatrix} \cos\theta & \sin\theta \\ -\sin\theta & \cos\theta \end{pmatrix} = \boldsymbol{R}^{\mathrm{T}}$.

### 1.4.2 Determinant

The determinant of a square matrix $\boldsymbol{A}$ is just a number, denoted by $\det(\boldsymbol{A})$. In the simplest case, for a $2 \times 2$ matrix, we have

$$\det(\boldsymbol{A}) = \det\begin{vmatrix} a & b \\ c & d \end{vmatrix} = ad - bc. \tag{1.97}$$

For a $3 \times 3$ matrix, we have

$$\begin{aligned}\det(\boldsymbol{A}) &= \det\begin{vmatrix} a_{11} & a_{12} & a_{13} \\ a_{21} & a_{22} & a_{23} \\ a_{31} & a_{32} & a_{33} \end{vmatrix} \\ &= a_{11}\det\begin{vmatrix} a_{22} & a_{23} \\ a_{32} & a_{33} \end{vmatrix} - a_{12}\det\begin{vmatrix} a_{21} & a_{23} \\ a_{31} & a_{33} \end{vmatrix} + a_{13}\det\begin{vmatrix} a_{21} & a_{22} \\ a_{31} & a_{32} \end{vmatrix} \\ &= a_{11}(a_{22}a_{33} - a_{32}a_{23}) - a_{12}(a_{21}a_{33} - a_{31}a_{23}) + a_{13}(a_{21}a_{32} - a_{31}a_{22}).\end{aligned} \tag{1.98}$$

In general, the determinant of a matrix can be calculated using a recursive formula such as the Leibniz formula or the Laplace expansion with the adjugate matrices. Interested readers can refer to more advanced literature on this topic.

A square matrix $\boldsymbol{A}$ can have a unique inverse matrix if $\det(\boldsymbol{A}) \neq 0$. Otherwise, the matrix is called singular and not invertible. For an invertible $2 \times 2$ matrix

$$\boldsymbol{A} = \begin{pmatrix} a & b \\ c & d \end{pmatrix}, \tag{1.99}$$

its inverse can be conveniently calculated by

$$\boldsymbol{A}^{-1} = \frac{1}{\det(\boldsymbol{A})} \begin{pmatrix} d & -b \\ -c & a \end{pmatrix} = \frac{1}{ad - bc} \begin{pmatrix} d & -b \\ -c & a \end{pmatrix}, \quad ad - bc \neq 0. \tag{1.100}$$

As an exercise, we leave the reader to show that this is true.

### 1.4.3 Rank of a Matrix

The rank of a matrix $\boldsymbol{A}$ is a useful concept, and it is the maximum number of linearly independent columns or rows. For example, the rank of matrix

$$\boldsymbol{A} = \begin{pmatrix} 1 & 2 & 3 \\ 1 & 1 & 0 \\ 2 & 3 & 3 \end{pmatrix} \tag{1.101}$$

is 2 because there first two rows are independent, and the third row is the sum of the first two rows. We can write it as

$$\text{rank}(\boldsymbol{A}) = 2. \tag{1.102}$$

In some specialized literature, the number of linearly independent rows of $\boldsymbol{A}$ is called row rank, while the number of linearly independent columns if called column rank. It can be proved that the row rank is always equal to the column rank for the same matrix. In general, for an $m \times n$ matrix $\boldsymbol{A}$, we have

$$\text{rank}(\boldsymbol{A}) \leq \min\{m, n\}. \tag{1.103}$$

When the equality holds (i.e. $\text{rank}(\boldsymbol{A}) = \min\{m, m\}$), the matrix is said to be full rank. Thus, the following matrices

$$\boldsymbol{C} = \begin{pmatrix} 1 & 2 & 3 & 4 \\ 1 & 1 & 2 & 2 \\ 2 & 3 & 4 & 5 \end{pmatrix}, \quad \boldsymbol{D} = \begin{pmatrix} 1 & 2 \\ 3 & 4 \\ 3 & 7 \end{pmatrix} \tag{1.104}$$

are all full rank matrices because their ranks are 3 and 2, respectively.

For an $n \times n$ matrix, it becomes a full rank matrix if its rank is $n$. A useful full rank test is that the determinant of $\boldsymbol{A}$ is not zero. That is $\det(\boldsymbol{A}) \neq 0$. For example, matrix

$$\boldsymbol{B} = \begin{pmatrix} 1 & 2 & 3 \\ 1 & 1 & 0 \\ 2 & 3 & 7 \end{pmatrix} \tag{1.105}$$

is a full rank matrix because $\text{rank}(\boldsymbol{B}) = 3$ and $\det(\boldsymbol{B}) = -4$.

There are a few methods such as the Gauss elimination can be used to compute the rank of a matrix. Readers can refer to more advanced literature on this topic.

### 1.4.4 Frobenius Norm

Similar to the norms for vectors, there are also various ways to define norms for a matrix. For a matrix of size $m \times n$, the Frobenius norm is defined by

$$||\boldsymbol{A}||_{\mathrm{F}} = \sqrt{\sum_{i=1}^{m} \sum_{j=1}^{n} |a_{ij}|^2}, \tag{1.106}$$

which is equivalent to

$$||\boldsymbol{A}||_{\mathrm{F}} = \sqrt{\mathrm{tr}(A^{\mathrm{T}}A)} = \sqrt{\mathrm{diag}(A^{\mathrm{T}}A)}. \tag{1.107}$$

The maximum absolute column sum norm is defined by

$$||\boldsymbol{A}||_1 = \max_{1 \le j \le n} \sum_{i=1}^{m} |a_{ij}|. \tag{1.108}$$

Similarly, the maximum absolute row sum norm is defined by

$$||\boldsymbol{A}||_\infty = \max_{1 \le i \le m} \sum_{j=1}^{n} |a_{ij}|. \tag{1.109}$$

**Example 1.15** For $A = \begin{pmatrix} 1 & -2 & 3 \\ -5 & 0 & 7 \end{pmatrix}$, we have

$$||\boldsymbol{A}||_1 = \max\{|1| + |-5|, |-2| + |0|, |3| + |7|\} = \max\{6, 2, 10\} = 10,$$

$$||\boldsymbol{A}||_\infty = \max\{|1| + |-2| + |3|, |-5| + |0| + |7|\} = \max\{6, 12\} = 12,$$

and

$$||\boldsymbol{A}||_{\mathrm{F}} = \sqrt{|1|^2 + |-2|^2 + |3|^2 + |-5|^2 + |0|^2 + |7|^2} = \sqrt{88}.$$

Other norms can also be defined for different applications.

## 1.5 Eigenvalues and Eigenvectors

An eigenvalue $\lambda$ of a square matrix $\boldsymbol{A}$ is defined by

$$\boldsymbol{A}\boldsymbol{u} = \lambda \boldsymbol{u}, \tag{1.110}$$

where a nonzero eigenvector $\boldsymbol{u}$ exists for a corresponding $\lambda$. An $n \times n$ matrix can have at most $n$ different eigenvalues and thus $n$ corresponding eigenvectors.

Multiplying the above Eq. (1.110) by an identity $n \times n$ matrix $\boldsymbol{I}$, we have

$$\boldsymbol{IAu} = \boldsymbol{I}\lambda\boldsymbol{u} = (\lambda\boldsymbol{I})\boldsymbol{u}, \tag{1.111}$$

which becomes

$$\boldsymbol{IAu} - (\lambda\boldsymbol{I})\boldsymbol{u} = (\boldsymbol{A} - \lambda\boldsymbol{I})\boldsymbol{u} = 0, \tag{1.112}$$

where we have used $\boldsymbol{IA} = \boldsymbol{A}$. In order to obtain a non-trial solution $\boldsymbol{u} \neq 0$, it is thus required that the matrix $\boldsymbol{A} - \lambda\boldsymbol{I}$ is not invertible. In other words, its determinant should be zero. That is

$$\det(\boldsymbol{A} - \lambda\boldsymbol{I}) = 0, \tag{1.113}$$

which is equivalent to a polynomial of order $n$. Such a polynomial is called the characteristic polynomial of $\boldsymbol{A}$. All the eigenvalues form a set, called the spectrum of matrix $\boldsymbol{A}$.

**Example 1.16** The eigenvalue of $\boldsymbol{A} = \begin{pmatrix} a & b \\ b & a \end{pmatrix}$ can be calculated by

$$\det(\boldsymbol{A} - \lambda\boldsymbol{I}) = \det\begin{vmatrix} a-\lambda & b \\ b & a-\lambda \end{vmatrix} = 0,$$

which leads to

$$(a - \lambda)^2 - b^2 = 0,$$

or

$$a - \lambda = \pm b.$$

Thus, we have

$$\lambda_1 = a + b, \quad \lambda_2 = a - b.$$

For example, if $\boldsymbol{A} = \begin{pmatrix} 2 & 3 \\ 3 & 2 \end{pmatrix}$, we have

$$\lambda_1 = 5, \quad \lambda_2 = -1.$$

The trace of $\boldsymbol{A}$ is $\text{tr}(\boldsymbol{A}) = 2 + 2 = 4$. Here, the sum of both eigenvalues are $\lambda_1 + \lambda_2 = 5 + (-1) = 4$, which means that $\text{tr}(\boldsymbol{A}) = \lambda_1 + \lambda_2$.

In general, if a square matrix has $n$ different eigenvalues $\lambda_i (i = 1, 2, \ldots, n)$, we have

$$\text{tr}(\boldsymbol{A}) = \sum_{i=1}^{n} a_{ii} = \sum_{i=1}^{n} \lambda_i. \tag{1.114}$$

Let us look at another example.

**Example 1.17** For $A = \begin{pmatrix} 1 & 2 \\ -3 & 8 \end{pmatrix}$, its eigenvalues can be obtained by

$$\det(\boldsymbol{A} - \lambda \boldsymbol{I}) = \det \begin{vmatrix} 1-\lambda & 2 \\ -3 & 8-\lambda \end{vmatrix} = 0,$$

which gives

$$(1-\lambda)(8-\lambda) - 2 \times (-3) = 0,$$

or

$$\lambda^2 - 9\lambda + 14 = (\lambda - 2)(\lambda - 7) = 0.$$

Thus, the two eigenvalues are

$$\lambda_1 = 2, \quad \lambda_2 = 7.$$

The eigenvector $\boldsymbol{u} = (a \ \ b)^T$ corresponding to $\lambda_1 = 2$ can be obtained by the original definition

$$\boldsymbol{A}\boldsymbol{u} = \lambda_1 \boldsymbol{u},$$

or

$$\begin{pmatrix} 1 & 2 \\ -3 & 8 \end{pmatrix} \begin{pmatrix} a \\ b \end{pmatrix} = 2 \begin{pmatrix} a \\ b \end{pmatrix}.$$

This is equivalent to two equations

$$\begin{cases} a + 2b = 2a, \\ -3a + 8b = 2b. \end{cases}$$

Both equations give $a = 2b$, which means that they are not linear independent because one can be obtained by the other via some minor algebraic manipulations. This means that we can determine the direction of the eigenvector, not the magnitude uniquely. In some textbooks, it is assumed that the magnitude of an eigenvector should be one. That is $||\boldsymbol{u}||_2 = 1 = \sqrt{a^2 + b^2}$. However, in many textbooks and many software packages, they usually assume that the first component is 1, which makes subsequent calculations much easier. Thus, we can set $a = 1$, thus $b = 1/2$. The eigenvector for eigenvalue 2 becomes

$$\boldsymbol{u}_1 = \begin{pmatrix} 1 \\ \frac{1}{2} \end{pmatrix}.$$

Similarly, the eigenvector $\boldsymbol{u}_2 = [c \ \ d]^T$ for $\lambda_2 = 7$ can be obtained by

$$\begin{pmatrix} 1 & 2 \\ -3 & 8 \end{pmatrix} \begin{pmatrix} c \\ d \end{pmatrix} = 7 \begin{pmatrix} c \\ d \end{pmatrix}.$$

As both equations lead to $d = 3c$, we impose $c = 1$, which means $d = 3$. Thus, the eigenvector $\boldsymbol{u}_2$ for $\lambda_2 = 7$ is

$$\boldsymbol{u}_2 = \begin{pmatrix} 1 \\ 3 \end{pmatrix}.$$

It is worth pointing out that the eigenvectors $-\boldsymbol{u}_1$ and $-\boldsymbol{u}_2$ for $\lambda_1$ and $\lambda_2$, respectively, are equally valid.

In addition, in some textbooks and many software packages, the unity of the eigenvectors is used, instead of setting the first component as 1. In this case, the above vectors are multiplied by a normalization or scaling constant (usually the length or magnitude). Therefore, in the above example, the eigenvectors can also be written equivalently as

$$\boldsymbol{u}_1 = \frac{1}{\sqrt{5}} \begin{pmatrix} 2 \\ 1 \end{pmatrix}, \quad \boldsymbol{u}_2 = \frac{1}{\sqrt{10}} \begin{pmatrix} 1 \\ 3 \end{pmatrix}. \tag{1.115}$$

A useful theorem is that if a square matrix is real and symmetric $\boldsymbol{A}^T = \boldsymbol{A} \in \mathbb{R}^{n \times n}$, all its eigenvalues are real, and eigenvectors corresponding to distinct eigenvalues are orthogonal. That is, if $\boldsymbol{v}_1$ and $\boldsymbol{v}_2$ are two eigenvectors for $\lambda_1 \neq \lambda_2$, respectively, we have $\boldsymbol{v}_1^{\mathrm{T}} \boldsymbol{v}_2 = \boldsymbol{v}_1 \cdot \boldsymbol{v}_2 = 0$.

However, eigenvalues for real symmetric matrices may not be distinct. For example, both eigenvalues of $\boldsymbol{I} = \begin{pmatrix} 1 & 0 \\ 0 & 1 \end{pmatrix}$ are 1, and they are not distinct.

### 1.5.1 Definiteness

A square symmetric matrix $\boldsymbol{A}$ (i.e. $\boldsymbol{A}^{\mathrm{T}} = \boldsymbol{A}$) is said to be positive definite if all its eigenvalues are strictly positive ($\lambda_i > 0$, where $i = 1, 2, \ldots, n$). By multiplying both sides of $\boldsymbol{A}\boldsymbol{u} = \lambda \boldsymbol{u}$ by $\boldsymbol{u}^{\mathrm{T}}$, we have

$$\boldsymbol{u}^{\mathrm{T}} \boldsymbol{A} \boldsymbol{u} = \boldsymbol{u}^{\mathrm{T}} \lambda \boldsymbol{u} = \lambda \boldsymbol{u}^{\mathrm{T}} \boldsymbol{u}, \tag{1.116}$$

which leads to

$$\lambda = \frac{\boldsymbol{u}^{\mathrm{T}} \boldsymbol{A} \boldsymbol{u}}{\boldsymbol{u}^{\mathrm{T}} \boldsymbol{u}}. \tag{1.117}$$

Since $\boldsymbol{u}^{\mathrm{T}} \boldsymbol{u} = ||\boldsymbol{u}||_2^2 > 0$, this means that

$$\boldsymbol{u}^{\mathrm{T}} \boldsymbol{A} \boldsymbol{u} > 0, \qquad \text{if } \lambda > 0. \tag{1.118}$$

In fact, for any vector $\boldsymbol{v}$, the following relationship holds:

$$\boldsymbol{v}^{\mathrm{T}} \boldsymbol{A} \boldsymbol{v} > 0. \tag{1.119}$$

For $\boldsymbol{v}$ can be a unit vector, all the diagonal elements of $\boldsymbol{A}$ should be strictly positive as well. If the equal sign is included in the definition, we have semidefiniteness. That is, $\boldsymbol{A}$ is called positive semidefinite if $\boldsymbol{u}^{\mathrm{T}}\boldsymbol{A}\boldsymbol{u} \geq 0$, and negative semidefinite if $\boldsymbol{u}^{\mathrm{T}}\boldsymbol{A}\boldsymbol{u} \leq 0$ for all $\boldsymbol{u}$.

If all the eigenvalues are nonnegative or $\lambda_i \geq 0$, then the matrix is positive semi-definite. If all the eigenvalues are nonpositive or $\lambda_i \leq 0$, then the matrix is negative semidefinite. In general, an indefinite matrix can have both positive and negative eigenvalues. Furthermore, the inverse of a positive definite matrix is also positive definite. For a linear system $\boldsymbol{A}\boldsymbol{u} = \boldsymbol{f}$, if $\boldsymbol{A}$ is positive definite, the system can be solved more efficiently by matrix decomposition methods.

Let us look at an example.

**Example 1.18** For matrices

$$\boldsymbol{A} = \begin{pmatrix} 3 & -2 \\ -2 & 3 \end{pmatrix}, \quad \boldsymbol{B} = \begin{pmatrix} 7 & 3 \\ 3 & 7 \end{pmatrix}, \quad \boldsymbol{C} = \begin{pmatrix} 2 & 3 \\ 3 & 2 \end{pmatrix},$$

the eigenvalues of $\boldsymbol{A}$ are 1 and 5, which are both positive. Thus, $\boldsymbol{A}$ is positive definite. Similarly, $\boldsymbol{B}$ is also positive definite because its two eigenvalues are 4 and 10. However, $\boldsymbol{C}$ is indefinite because one of its eigenvalues is negative (−1) and the other eigenvalue is positive (5).

The definiteness of matrices can be useful to determine if a multivariate function has a local maximum or minimum. It is also useful to see if an expression can be written as a quadratic form.

### 1.5.2 Quadratic Form

Quadratic forms are widely used in optimization, especially in convex optimization and quadratic programming. Loosely speaking, a quadratic form is a homogenous polynomial of degree 2 of $n$ variables. For example, $3x^2 + 10xy + 7y^2$ is a binary quadratic form, while $x^2 + 2xy + y^2 - y$ is not.

For a real $n \times n$ symmetric matrix $\boldsymbol{A}$ and a vector $\boldsymbol{u}$ of $n$ elements, their combination

$$Q = \boldsymbol{u}^{\mathrm{T}}\boldsymbol{A}\boldsymbol{u} \tag{1.120}$$

is called a quadratic form. Since $\boldsymbol{A} = [a_{ij}]$, we have

$$Q = \boldsymbol{u}^{\mathrm{T}}\boldsymbol{A}\boldsymbol{u} = \sum_{i=1}^{n}\sum_{j=1}^{n} u_i a_{ij} u_j = \sum_{i=1}^{n}\sum_{j=1}^{n} a_{ij} u_i u_j$$
$$= \sum_{i=1}^{n} a_{ii} u_i^2 + 2\sum_{i=2}^{n}\sum_{j=1}^{i-1} a_{ij} u_i u_j. \tag{1.121}$$

**Example 1.19** For the symmetric matrix $A = \begin{pmatrix} 1 & 2 \\ 2 & 5 \end{pmatrix}$ and $u = \begin{pmatrix} u_1 & u_2 \end{pmatrix}^T$, we have

$$u^T A u = \begin{pmatrix} u_1 & u_2 \end{pmatrix} \begin{pmatrix} 1 & 2 \\ 2 & 5 \end{pmatrix} \begin{pmatrix} u_1 \\ u_2 \end{pmatrix}$$
$$= \begin{pmatrix} u_1 & u_2 \end{pmatrix} \begin{pmatrix} u_1 + 2u_2 \\ \\ 2u_1 + 5u_2 \end{pmatrix} = 3u_1^2 + 10u_1u_2 + 7u_2^2.$$

In fact, for a binary quadratic form $Q(x, y) = ax^2 + bxy + cy^2$, we have

$$\begin{pmatrix} x & y \end{pmatrix} \begin{pmatrix} \alpha & \beta \\ \beta & \gamma \end{pmatrix} \begin{pmatrix} x \\ y \end{pmatrix} = (\alpha + \beta)x^2 + (\alpha + 2\beta + \gamma)xy + (\beta + \gamma)y^2.$$

If this is equivalent to $Q(x, y)$, it requires that

$$\alpha + \beta = a, \qquad \alpha + 2\beta + \gamma = b, \qquad \beta + \gamma = c,$$

which leads to $a + c = b$. This means that not all arbitrary quadratic functions $Q(x, y)$ are quadratic form.

If $A$ is real symmetric, its eigenvalues $\lambda_i$ are real, and the eigenvectors $v_i$ of distinct eigenvalues $\lambda_i$ are orthogonal to each other. Therefore, we can write $u$ using the eigenvector basis and we have

$$u = \sum_{i=1}^{n} \alpha_i v_i. \tag{1.122}$$

In addition, $A$ becomes diagonal in this basis. That is

$$A = \begin{pmatrix} \lambda_1 & & \\ & \ddots & \\ & & \lambda_n \end{pmatrix}. \tag{1.123}$$

Subsequently, we have

$$Au = \sum_{i=1}^{n} \alpha_i A v_i = \sum_{i=1}^{n} \lambda_i \alpha_i v_i, \tag{1.124}$$

which means that

$$u^T A u = \sum_{j=1}^{n} \sum_{i=1}^{n} \lambda_i \alpha_j \alpha_i v_i^T v_i = \sum_{i=1}^{n} \lambda_i \alpha_i^2, \tag{1.125}$$

where we have used the fact that $v_i^T v_i = 1$ (by normalizing the eigenvectors so as to have a magnitude of unity).

## 1.6 Optimization and Optimality

Optimization is everywhere, from engineering design and business planning to artificial intelligence and industries. After all, time and resources are limited, and optimal use of such valuable resources is crucial. In addition, designs of products have to maximize the performance, sustainability, energy efficiency, and to minimize the costs and wastage. Therefore, optimization is specially important for engineering applications, business planning, and industries.

### 1.6.1 Minimum and Maximum

One of the simplest optimization problems is to find the minimum of a function such as $f(x) = x^2$ in the real domain. As $x^2$ is always nonnegative, it is easy to guess that the minimum occurs at $x = 0$.

From basic calculus, we know that, for a given curve described by $f(x)$, its gradient $f'(x)$ describes the rate of change. When $f'(x) = 0$, the curve has a horizontal tangent at that particular point. This means that it becomes a point of special interest. In fact, the maximum or minimum of a curve can only occur at

$$f'(x_*) = 0, \tag{1.126}$$

which is a critical condition or stationary condition. The solution $x_*$ to this equation corresponds to a stationary point and there may be multiple stationary points for a given curve.

In order to see if it is a maximum or minimum at $x = x_*$, we have to use the information of its second derivative $f''(x)$. In fact, $f''(x_*) > 0$ corresponds to a minimum, while $f''(x_*) < 0$ corresponds to a maximum. Let us see a concrete example.

**Example 1.20** To find the minimum of $f(x) = x^2 e^{-x^2}$, we have the stationary condition $f'(x) = 0$ or

$$f'(x) = 2x \times e^{-x^2} + x^2 \times (-2x)e^{-x^2} = 2(x - x^3)e^{-x^2} = 0.$$

As $e^{-x^2} > 0$, we have

$$x(1 - x^2) = 0, \quad \text{or} \quad x = 0, \quad \text{and} \quad x = \pm 1.$$

The second derivative is given by

$$f''(x) = 2e^{-x^2}(1 - 5x^2 + 2x^4),$$

which is an even function with respect to $x$.

So at $x = \pm 1$, $f''(\pm 1) = 2[1 - 5(\pm 1)^2 + 2(\pm 1)^4]e^{-(\pm 1)^2} = -4e^{-1} < 0$. Thus, there are two maxima that occur at $x_* = \pm 1$ with $f_{\max} = e^{-1}$. At $x = 0$, we have $f''(0) = 2 > 0$, thus the minimum of $f(x)$ occurs at $x_* = 0$ with $f_{\min}(0) = 0$.

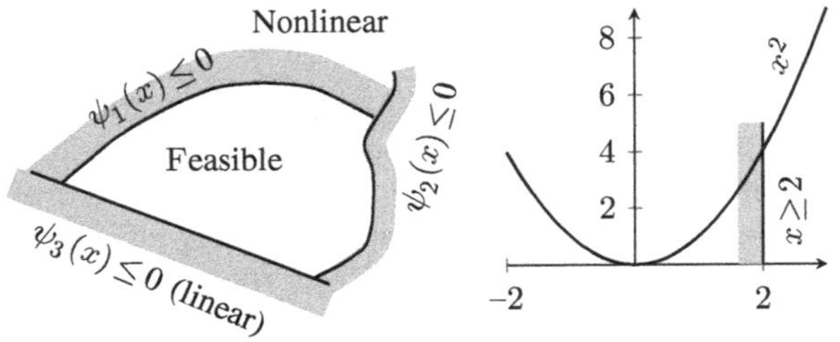


**Figure 1.7** Feasible domain with nonlinear inequality constraints $\psi_1(x)$ and $\psi_2(x)$ (left) as well as a linear inequality constraint $\psi_3(x)$. An example with an objective of $f(x) = x^2$ subject $x \geq 2$ (right).

In mathematical programming, there are many important concepts, and we will first introduce the concepts of feasible solutions, optimality criteria, strong local optima, and weak local optima.

### 1.6.2 Feasible Solution

A point $\boldsymbol{x}$ which satisfies all the constraints is called a feasible point and thus it is a feasible solution to the problem. The set of all feasible points is called the feasible region (see Figure 1.7).

For example, we know that the domain $f(x) = x^2$ consists of all the real numbers. If we want to minimize $f(x)$ without any constraint, all solutions such as $x = -1$, $x = 1$, and $x = 0$ are feasible. In fact, the feasible region is the whole real axis. Obviously, $x = 0$ corresponds to $f(0) = 0$ as the true minimum.

However, if we want to find the minimum of $f(x) = x^2$ subject to $x \geq 2$, it becomes a constrained optimization problem. The points such as $x = 1$ and $x = 0$ are no longer feasible because they do not satisfy $x \geq 2$. In this case, the feasible solutions are all the points that satisfy $x \geq 2$. So $x = 2$, $x = 100$, and $x = 10^8$ are all feasible. It is obvious that the minimum occurs at $x = 2$ with $f(2) = 2^2 = 4$. That is, the optimal solution for this problem occurs at the boundary point $x = 2$ (see Figure 1.7).

### 1.6.3 Gradient and Hessian Matrix

We can extend the optimization procedure for univariate functions to multivariate functions using partial derivatives and relevant conditions. Let us start with an example

$$\text{Minimize} \quad f(x, y) = x^2 + y^2 \quad (x, y \in \mathbb{R}). \tag{1.127}$$

It is obvious that $x = 0$ and $y = 0$ is the minimum solution because $f(0, 0) = 0$. The question is how to solve this problem formally. We can extend the stationary condition to partial derivatives, and we have $\partial f / \partial x = 0$ and $\partial f / \partial y = 0$. In this case, we have

$$\frac{\partial f}{\partial x} = 2x + 0 = 0, \quad \frac{\partial f}{\partial y} = 0 + 2y = 0. \tag{1.128}$$

The solution is obviously $x_* = 0$ and $y_* = 0$.

Now how do we know that it corresponds to a maximum or minimum? If we try to use the second derivatives, we have four different partial derivatives such as $f_{xx}$ and $f_{yy}$ and which one should we use? In fact, we need to define a Hessian matrix from these second partial derivatives and we have

$$\boldsymbol{H} = \begin{pmatrix} f_{xx} & f_{xy} \\ f_{yx} & f_{yy} \end{pmatrix} = \begin{pmatrix} \frac{\partial^2 f}{\partial x^2} & \frac{\partial^2 f}{\partial x \partial y} \\ \frac{\partial^2 f}{\partial y \partial x} & \frac{\partial^2 f}{\partial y^2} \end{pmatrix}. \tag{1.129}$$

Since $\partial x \partial y = \partial y \partial x$ or

$$\frac{\partial^2 f}{\partial x \partial y} = \frac{\partial^2 f}{\partial y \partial x}, \tag{1.130}$$

we can conclude that the Hessian matrix is always symmetric. In the case of $f = x^2 + y^2$, it is easy to check that the Hessian matrix is

$$\boldsymbol{H} = \begin{pmatrix} 2 & 0 \\ 0 & 2 \end{pmatrix}. \tag{1.131}$$

Mathematically speaking, if $\boldsymbol{H}$ is positive definite, then the stationary point $(x_*, y_*)$ corresponds to a local minimum. Similarly, if $\boldsymbol{H}$ is negative definite, the stationary point corresponds to a maximum. Since the Hessian matrix here does not involve any $x$ or $y$, it is always positive definite in the whole search domain $(x, y) \in \mathbb{R}^2$, so we can conclude that the solution at point $(0, 0)$ is the global minimum.

Obviously, this is a special case. In general, the Hessian matrix will depend on the independent variables, but the definiteness test conditions still apply. That is, positive definiteness of a stationary point means a local minimum. Alternatively, for bivariate functions, we can define the determinant of the Hessian matrix in Eq. (1.129) as

$$\Delta = \det(\boldsymbol{H}) = f_{xx} f_{yy} - (f_{xy})^2. \tag{1.132}$$

At the stationary point $(x_*, y_*)$, if $\Delta > 0$ and $f_{xx} > 0$, then $(x_*, y_*)$ is a local minimum. If $\Delta > 0$ but $f_{xx} < 0$, it is a local maximum. If $\Delta = 0$, it is inconclusive and we have to use other information such as higher-order derivatives. However, if $\Delta < 0$, it is a saddle point. A saddle point is a special point where a local minimum occurs along one direction while the maximum occurs along another (orthogonal) direction.

In fact, for a multivariate function $f(x_1, x_2, \dots, x_n)$ in an $n$-dimensional space, the stationary condition can be extended to

$$\boldsymbol{G} = \nabla f = \left( \frac{\partial f}{\partial x_1}, \frac{\partial f}{\partial x_2}, \dots, \frac{\partial f}{\partial x_n} \right)^{\mathrm{T}} = 0, \tag{1.133}$$

where $G$ is called the gradient vector. The second derivative test becomes the definiteness of the Hessian matrix

$$H = \begin{pmatrix} \frac{\partial^2 f}{\partial x_1{}^2} & \frac{\partial^2 f}{\partial x_1 \partial x_2} & \cdots & \frac{\partial^2 f}{\partial x_1 \partial x_n} \\ \frac{\partial^2 f}{\partial x_2 \partial x_1} & \frac{\partial^2 f}{\partial x_2{}^2} & \cdots & \frac{\partial^2 f}{\partial x_2 \partial x_n} \\ \vdots & \vdots & \ddots & \vdots \\ \frac{\partial^2 f}{\partial x_n \partial x_1} & \frac{\partial^2 f}{\partial x_n \partial x_2} & \cdots & \frac{\partial^2 f}{\partial x_n{}^2} \end{pmatrix}. \quad (1.134)$$

At the stationary point defined by $G = \nabla f = 0$, the positive definiteness of $H$ gives a local minimum, while the negative definiteness corresponds to a local maximum. In essence, the eigenvalues of the Hessian matrix $H$ determine the local behavior of the function. As we mentioned before, if $H$ is positive semidefinite, it corresponds to a local minimum.

### 1.6.4 Optimality Conditions

A point $x_*$ is called a strong local maximum of the nonlinearly constrained optimization problem if $f(x)$ is defined in a $\delta$-neigborhood $N(x_*, \delta)$ and satisfies $f(x_*) > f(u)$ for $\forall u \in N(x_*, \delta)$, where $\delta > 0$ and $u \neq x_*$. If $x_*$ is not a strong local maximum, the inclusion of equality in the condition $f(x_*) \geq f(u)$ for $\forall u \in N(x_*, \delta)$ defines the point $x_*$ as a weak local maximum (see Figure 1.8). The local minima can be defined in a similar manner when $>$ and $\geq$ are replaced by $<$ and $\leq$, respectively.

Figure 1.8 shows various local maxima and minima. Point $A$ is a strong local maximum, while point $B$ is a weak local maximum because there are many (in fact infinite) different values of $x$ which will lead to the same value of $f(x_*)$. Point $D$ is the global maximum, and point $E$ is the global minimum. In addition, point $F$ is a strong local minimum.

However, point $C$ is a strong local minimum, but it has a discontinuity in $f'(x_*)$; the stationary condition for this point $f'(x_*) = 0$ is not valid. We will not deal with this type of minima or maxima in detail, though the subgradient method should work well if the function is convex.

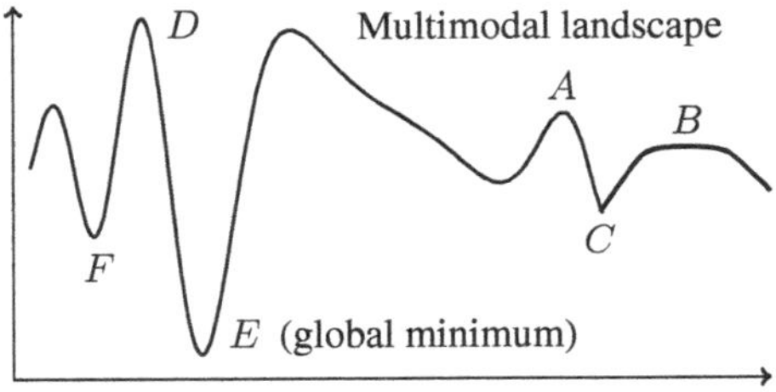


**Figure 1.8** Local optima, weak optima, and global optimality.

As we briefly mentioned before, for a smooth curve $f(x)$, optimal solutions usually occur at stationary points where $f'(x) = 0$. This is not always the case because optimal solutions can also occur at the boundary, as we have seen in the previous example of minimizing $f(x) = x^2$ subject to $x \geq 2$. In our present discussion, we will assume that both $f(\boldsymbol{x})$ and $f'(\boldsymbol{x})$ are always continuous, or $f(\boldsymbol{x})$ is everywhere twice-continuously differentiable. Obviously, the information of $f'(x)$ is not sufficient to determine whether a stationary point is a local maximum or minimum. Thus, higher-order derivatives such as $f''(x)$ are needed, but we do not make any assumption at this stage. We will discuss this further in detail in later chapters.

## 1.7 General Formulation of Optimization Problems

Whatever the real-world applications may be, it is usually possible to formulate an optimization problem in a general mathematical form. All optimization problems with an explicit objective $f(\boldsymbol{x})$ can, in general, be expressed as a nonlinearly constrained optimization problem

$$\text{Maximize/minimize} \quad f(\boldsymbol{x}), \quad \boldsymbol{x} = (x_1, x_2, \dots, x_n)^{\mathrm{T}} \in \mathbb{R}^n,$$

$$\text{Subject to } \phi_j(\boldsymbol{x}) = 0 \quad (j = 1, 2, \dots, M),$$

$$\psi_k(\boldsymbol{x}) \leq 0 \quad (k = 1, \dots, N), \tag{1.135}$$

where $f(\boldsymbol{x})$, $\phi_j(\boldsymbol{x})$ and $\psi_k(\boldsymbol{x})$, are scalar functions of the design vector $\boldsymbol{x}$. Here, the components $x_i$ of $\boldsymbol{x} = (x_1, \dots, x_n)^{\mathrm{T}}$ are called design or decision variables, and they can be either continuous, discrete, or a mixture of these two. The vector $\boldsymbol{x}$ is often called the decision vector, which varies in an $n$-dimensional space $\mathbb{R}^n$. It is worth pointing out that we use a column vector here for $\boldsymbol{x}$ (thus with a transpose T). We can also use a row vector $\boldsymbol{x} = (x_1, \dots, x_n)$ and the results will be the same, though some formulations may be slightly different. Different textbooks may use slightly different formulations. Once we are aware of such minor variations, this causes no difficulty or confusion.

It is worth pointing out that the objectives are explicitly known in all the optimization problems to be discussed in this book. However, in reality, it is often difficult to quantify what we want to achieve, but we still try to optimize certain things such as the degree of enjoyment or the quality of service on holiday. In other cases, it might be impossible to write the objective function in any explicit form mathematically. In any case, we always assume that the values of an objective function are always computable.

## Exercises

**1.1** Find the first and second derivatives of $f(x) = \sin(x)/x$.

**1.2** Find the gradient and Hessian matrix of $f(x, y, z) = x^2 + y^2 + 2xy + 3yz + z^3$. Is the Hessian matrix symmetric?

**1.3** Show that $\int_0^\infty x^3 e^{-x} dx = 6$.

**1.4** Find the eigenvalues and eigenvectors of $\boldsymbol{A} = \begin{pmatrix} 2 & 3 \\ 3 & 4 \end{pmatrix}$.

**1.5** In the second exercise, the Hessian matrix at $z = 1$ becomes

$$\boldsymbol{H} = \begin{pmatrix} 2 & 2 & 0 \\ 2 & 2 & 3 \\ 0 & 3 & 6 \end{pmatrix},$$

what are its eigenvalues? Is this matrix positive definite?

**1.6** Show that $f(x, y) = (x - 1)^2 + x^2y^2$ has a minimum at $(1, 0)$.

## Further Reading

Boyd, S.P. and Vandenberghe, L. (2004). *Convex Optimization*. Cambridge, UK: Cambridge University Press.

Eriksson, K., Estep, D., and Johnson, C. (2004). *Applied Mathematics: Body and Soul, Volume 1: Derivatives and Geometry in IR3*. Berlin: Springer-Verlag.

Gill, P.E., Murray, W., and Wright, M.H. (1982). *Practical Optimization*. Bingley: Emerald Publishing.

Kreyszig, E. (2010). *Advanced Engineering Mathematics*, 10e. Hoboken, NJ: Wiley.

Nocedal, J. and Wright, S.J. (2006). *Numerical Optimization*, 2e. New York: Springer.

Yang, X.S. (2010). *Engineering Optimization: An Introduction with Metaheurisitic Applications*. Hoboken, NJ: Wiley.

Yang, X.S. (2014). *Nature-Inspired Optimization Algorithms*. London: Elsevier.

Yang, X.S. (2017). *Engineering Mathematics with Examples and Applications*. London: Academic Press.

# 2

# Algorithms, Complexity, and Convexity

To find solutions to optimization problems, optimization techniques are used. Though analytical solutions are not only accurate but also can provide insight into the problem; however, analytical solutions are rarely possible in practical applications. Therefore, approximate solutions or sufficiently close solutions are usually sought, and thus a wide range of sophisticated optimization algorithms are widely used for solving different types of problems in practice.

## 2.1 What Is an Algorithm?

An algorithm is a computational, iterative procedure. For example, Newton's method for finding the roots of a polynomial $p(x) = 0$ can be written as

$$x_{t+1} = x_t - \frac{p(x_t)}{p'(x_t)}, \tag{2.1}$$

where $x_t$ is the approximation at iteration $t$ and $p'(x)$ is the first derivative of $p(x)$. This procedure typically starts with an initial guess $x_0$ at $t = 0$.

In most cases, as along as $p' \neq 0$ and $x_0$ is not too far away from the target solution, this algorithm can work very well. As we do not know the target solution $x_* = \lim_{t\to\infty} x_t$ in advance, the initial guess can be an educated guess or a purely random guess. However, if the initial guess is too far away, the algorithm may never reach the final solution or simply fail.

**Example 2.1** For example, we know that the polynomial

$$p(x) = x^2 + 9x - 10 = (x - 1)(x + 10) = 0 \tag{2.2}$$

has two roots $x_* = 1$ and $x_* = -10$. We also have $p'(x) = 2x + 9$ and

$$x_{t+1} = x_t - \frac{(x_t^2 + 9x_t - 10)}{2x_t + 9}. \tag{2.3}$$

*Optimization Techniques and Applications with Examples*, First Edition. Xin-She Yang.

If we start from $x_0 = 10$, we can easily reach $x_* = 1$ in less than five iterations. We have

$$x_1 = x_0 - \frac{x_0^2 + 9x_0 - 10}{2x_0 + 9} = 10 - \frac{10^2 + 9 \times 10 - 10}{2 \times 10 + 9} \approx 3.7931,$$

$$x_2 = x_1 - \frac{x_1^2 + 9x_1 - 10}{2x_1 + 9} \approx 1.4704,$$

$$x_3 \approx 1.0185, \quad x_4 \approx 1.000.$$

If we use $x_0 = 100$, it may take about eight iterations, depending on the accuracy we want. If we start any value $x_0 > 0$, we can only reach $x_* = 1$ and we will never reach the other root $x_* = -10$.

On the other hand, if we start with $x_0 = -5$, we can reach $x_* = -10$ in about seven steps with an accuracy of $10^{-9}$. However, if we start with $x_0 = -4.5$, the algorithm will simply fail because $p'(x_0) = 2x_0 + 9 = 0$.

This has clearly demonstrated that the final solution will usually depend on where the initial starting point is.

This method can be modified to solve optimization problems. For example, for a single objective function $f(x)$, the minimal and maximal values should occur at stationary points $f'(x) = 0$, which becomes a root-finding problem for $f'(x)$. Thus, the maximum or minimum of $f(x)$ can be found by modifying the Newton's method as the following iterative formula:

$$x_{t+1} = x_t - \frac{f'(x_t)}{f''(x_t)}. \tag{2.4}$$

For an $n$-dimensional problem with an objective $f(\boldsymbol{x})$ with independent variables $\boldsymbol{x} = (x_1, x_2, \ldots, x_n)$, the above iteration formula can be generalized to a vector form

$$\boldsymbol{x}^{t+1} = \boldsymbol{x}^t - \frac{\nabla f(\boldsymbol{x}^t)}{\nabla^2 f(\boldsymbol{x}^t)} = \boldsymbol{x}^t - \frac{\nabla f(\boldsymbol{x}^t)}{\boldsymbol{H}}, \tag{2.5}$$

where $\boldsymbol{H}(\boldsymbol{x}^t)$ is the Hessian matrix calculated at $\boldsymbol{x}^t$. Here, we have used the notation convention $\boldsymbol{x}^t$ to denote the current solution vector at iteration $t$ (not to be confused with an exponent). In some textbooks, the notation $\boldsymbol{x}^{(t)}$ is used to avoid any confusion with the exponent notation. However, as $\boldsymbol{x}^t$ is so widely used in the literature on numerical methods and optimization, we will use this notation in this book. Alternatively, we can always rewrite the above equation as

$$\boldsymbol{x}_{t+1} = \boldsymbol{x}_t - \frac{\nabla f(\boldsymbol{x}_t)}{\nabla^2 f(\boldsymbol{x}_t)} = \boldsymbol{x}_t - \boldsymbol{H}^{-1} \nabla f(\boldsymbol{x}_t). \tag{2.6}$$

## 2.2 Order Notations

In the description of algorithmic complexity, we often have to use the order notations, often in terms of big $O$ and small $o$. Loosely speaking, for two functions $f(x)$ and $g(x)$, if

$$\lim_{x \to x_0} \frac{f(x)}{g(x)} \to K, \tag{2.7}$$

where $K$ is a finite, nonzero limit, we write

$$f = O(g). \tag{2.8}$$

The big $O$ notation means that $f$ is asymptotically equivalent to the order of $g(x)$. If the limit is unity or $K = 1$, we say $f(x)$ is order of $g(x)$. In this special case, we write

$$f \sim g, \tag{2.9}$$

which is equivalent to $f/g \to 1$ and $g/f \to 1$ as $x \to x_0$. Obviously, $x_0$ can be any value, including 0 and $\infty$. The notation $\sim$ does not necessarily mean $\approx$ in general, though it might give the same results, especially in the case when $x \to 0$. For example, $\sin x \sim x$ and $\sin x \approx x$ if $x \to 0$.

When we say $f$ is order of 100 (or $f \sim 100$), this does not mean $f \approx 100$, but it can mean that $f$ could be between about 50 and 150. The small $o$ notation is often used if the limit tends to 0. That is

$$\lim_{x \to x_0} \frac{f}{g} \to 0 \tag{2.10}$$

or

$$f = o(g). \tag{2.11}$$

If $g > 0, f = o(g)$ is equivalent to $f \ll g$ (that is, $f$ is much less that $g$).

**Example 2.2** For example, for $\forall x \in \mathbb{R}$, we have

$$e^x = 1 + x + \frac{x^2}{2!} + \frac{x^3}{3!} + \cdots + \frac{x^n}{n!} + \cdots, \tag{2.12}$$

which can be written as

$$e^x \approx 1 + x + O(x^2) \approx 1 + x + \frac{x^2}{2} + o(x), \tag{2.13}$$

depending on the accuracy of the approximation of interest.

A very useful classic example is Stirling's asymptotic series for factorials

$$n! \sim \sqrt{2\pi n} \left(\frac{n}{e}\right)^n \left(1 + \frac{1}{12n} + \frac{1}{288n^2} - \frac{139}{51\,480n^3} - \cdots\right), \tag{2.14}$$

for sufficiently large $n$. This approximation can demonstrate the fundamental difference between asymptotic series and the standard approximate expansions. For standard power expansions, the error $R_k(h^k) \to 0$, but for an asymptotic series, the error of the truncated series $R_k$ decreases compared with the leading term (here $\sqrt{2\pi n}(n/e)^n$). However, $R_n$ does not necessarily tend to zero. In fact, the term

$$R_2 = \frac{1}{12n} \cdot \sqrt{2\pi n}\,(\frac{n}{e})^n \tag{2.15}$$

is still very large as $R_2 \to \infty$ if $n \gg 1$. For example, for $n = 100$, we have

$$n! = 9.3326 \times 10^{157}, \tag{2.16}$$

while the leading approximation is

$$\sqrt{2\pi n}(\frac{n}{e})^n = 9.3248 \times 10^{157}. \tag{2.17}$$

The difference between these two values is $7.7740 \times 10^{154}$, which is still very large, though three orders smaller than the leading approximation. For this reason, order notations are also called asymptotic notations.

It is worth pointing out that the expressions in computational complexity are most concerned with functions such as $f(n)$ of an input of problem size $n$ where $n \in \mathbb{N}$ is an integer in the natural numbers $\mathbb{N} = \{1, 2, 3, \ldots\}$.

For example, for functions $f(n) = 10n^2 + 20n + 100$ and $g(n) = 5n^2$, we have

$$f(n) = O\left(g(n)\right), \tag{2.18}$$

for every sufficiently large $n$. As $n$ is sufficient, $n^2$ will be much larger than $n$ (i.e. $n^2 \gg n$), then $n^2$ terms dominate two expressions. In order to emphasize the input $n$, we can often write

$$f(n) = O\left(g(n)\right) = O(n^2). \tag{2.19}$$

In addition, $f(n)$ is in general a polynomial of $n$, which not only includes terms such as $n^3$ and $n^2$, it also may include $n^{2.5}$ or $\log(n)$. Therefore, $f(n) = 100n^3 + 20n^{2.5} + 25n\log(n) + 123n$ is a valid polynomial in the context of computational complexity. In this case, we have

$$f(n) = 100n^3 + 20n^{2.5} + 25n\log(n) + 123n = O(n^3). \tag{2.20}$$

Here, we always implicitly assume that $n$ is sufficiently large and the base of the logarithm is base 2.

## 2.3 Convergence Rate

Almost all algorithms are iterative and the solutions form a sequence of $s_0$, $s_1$, $s_2$, ..., $s_n$, ...; that is, $s_n (n = 0, 1, 2, \ldots)$. If the sequence $s_n$ converges towards a fixed solution (a fixed point), we have

$$\lim_{n\to\infty} s_n = R. \tag{2.21}$$

The rate of convergence measures how quickly the error $E_n = s_n - R$ reduces to zero, which is defined as

$$\lim_{n\to\infty} \frac{|E_{n+1}|}{|E_n|^q} = \lim_{n\to\infty} \frac{|s_{n+1} - R|}{|s_n - R|^q} = A \quad (A > 0, \quad q \geq 1), \tag{2.22}$$

where $q$ represents the order of convergence of the iterative sequence.

In other words, we say that the sequence converges to $R$ with the order of $q$. Here, $A$ is called the asymptotic error constant or the rate of convergence.

- If $q = 1$ and $0 < A < 1$, we say the convergence is linear. The convergence is said to be superlinear if $A = 0$ and $q = 1$. In case of $A = 1$ and $q = 1$, the convergence is sublinear.
- If $q = 2$, the convergence is quadratic. We can also say that the sequence has a quadratic rate of convergence.
- If $q = 3$, the convergence is cubic.

For example, the following sequence

$$s_n = 2^{-n} \quad (n = 0, 1, 2, \ldots) \tag{2.23}$$

or

$$1, \frac{1}{2}, \frac{1}{4}, \frac{1}{8}, \frac{1}{16}, \ldots, \tag{2.24}$$

converges towards 0. That is $\lim_{n\to\infty} s_n = R = 0$. Let us try

$$\lim_{n\to\infty} \frac{|s_{n+1} - R|}{|s_n - R|} = \lim_{n\to\infty} \frac{|2^{-(n+1)} - 0|}{|2^{-n} - 0|} = \lim_{n\to\infty} \frac{2^{-n}2^{-1}}{2^{-n}} = \frac{1}{2} = A. \tag{2.25}$$

Thus, $q = 1$ and $A = 1/2$, so the sequence converges linearly.

It is straightforward to show that $2^{-n^2}$ converges to 0 superlinearly because

$$\lim_{n\to\infty} \frac{|2^{-(n+1)^2} - 0|}{|2^{-n^2} - 0|} = \lim_{n\to\infty} \frac{2^{-n^2}2^{-(2n+1)}}{2^{-n^2}} = \lim_{n\to\infty} 2^{-(2n+1)} = 0. \tag{2.26}$$

However, the sequence $s_n = 2^{-2^n}$ converges to 0 quadratically because

$$\lim_{n\to\infty} \frac{|2^{-2^{n+1}} - 0|}{|2^{-2^n} - 0|^2} = \lim_{n\to\infty} \frac{2^{-2^n\times 2}}{2^{-2^2\times 2}} = 1, \tag{2.27}$$

which gives $q = 2$ and $A = 1$.

## 2.4 Computational Complexity

To measure how easy or hard that a problem can be solved, we need to estimate its computational complexity. We cannot simply ask how long it takes to solve a particular problem instance because the actual computational time will depend on both the hardware and software used to solve it. Thus, time does not make much sense in this context. A useful measure of complexity should be independent of the hardware and software used. However, such complexity is closely linked to the algorithms used.

### 2.4.1 Time and Space Complexity

To find the maximum (or minimum) among $n$ different numbers, we only need to go through each number once by simply comparing the current number with the highest (or lowest) number once and update the new highest (or lowest) when necessary. Thus, the number of mathematical operations is simply $O(n)$, which is the time complexity of this problem.

In practice, comparing two big numbers may take slightly longer, and different representations of numbers may also affect the speed of this comparison. In addition, multiplication and division usually take more time than simple addition and substraction. However, in computational complexity, we usually ignore such minor differences and simply treat all operations as equal. In this sense, the complexity is about the number or order of mathematical operations, not the actual order of computational time.

On the other hand, space computational complexity estimates the size of computer memory needed to solve the problem. In the above simple problem of finding the maximum or minimum among $n$ different numbers, the memory needed is $O(n)$ because it needs at $n$ different entries in the computer memory to store $n$ different numbers. Though we need one more entry to store the largest or smallest number, this minor change does not affect the order of complexity because we implicitly assume that $n$ is sufficiently large.

In most literature, if there is no time or space explicitly used when talking about computational complexity, it usually means time complexity. In discussing computational complexity, we often use the word "problem" to mean a class of problems of the same type, and an "instance" to mean a specific example of a problem class. Thus, $\boldsymbol{Ax} = \boldsymbol{b}$ is a problem (class) for linear algebra, while

$$\begin{pmatrix} 2 & 3 \\ 1 & 1 \end{pmatrix} \begin{pmatrix} x \\ y \end{pmatrix} = \begin{pmatrix} 8 \\ 3 \end{pmatrix} \tag{2.28}$$

is an instance. In addition, a decision problem is a yes–no problem where an output is binary (0 or 1), even though the inputs can be any values.

Complexity classes are often associated with the computation on a Turing machine. Informally speaking, a Turing machine is an abstract machine that reads an input (a bit or a symbol) and manipulates one symbol at a time, following a set of fixed rules, by shifting a tape to the left or right. It can also write a symbol or bit to the tape (or scratch pad) which has an unlimited memory capacity. This basic concept can be extended to a universal Turing machine where the input and output can be represented as strings (either binary strings or strings in general), the tape or scratch pad can be considered as memory, and the manipulation as a transition function such as a simple Boolean function. It has been proved that a Turing machine is capable of carrying out the computation of any computable function. In this case, a universal Turing machine can be considered as a general-purpose modern computer with the transition function being the central processing unit, the tape being the computer memory and the inputs/outputs being the string representations. As the rules are fixed and at most one action is allowed at any given time, such a Turing machine is called a deterministic Turing machine.

In a more general case, a Turing machine has a set of rules that may allow more than one action, and the manipulation of symbols can have multiple branches, depending on current states. Thus, the transitions on the machine are not deterministic and multiple decision structures exist in this case. Such a Turing machine is called a non-deterministic Turing machine.

The above two types of computational complexity are closely linked to the type of problems. Even for the same type of problem, different algorithms can be used, and the number of basic mathematical operations may be different. In this case, we are concerned about the complexity of an algorithm in terms of arithmetic complexity.

### 2.4.2 Class P

Computational complexity is often divided into different complexity classes. Loosely speaking, a complexity class is a set of decision problems or functions that can be computed by a Turing machine with a fixed resource.

The class P is a set of all functions or problems that can be solved by a deterministic Turing machine using a computation time of a polynomial $f(n)$ in terms of input size $n$, denoted by DTIME $\big(f(n)\big)$ or DTIME($n^k$) where $k \geq 1$. Since class P includes all such problems, we have

$$\text{Class P} = \cup_{k\geq 1}\text{DTIME}\left(n^k\right), \tag{2.29}$$

where $\cup$ is the union of all relevant problems or functions in this class. Here, the notation DTIME highlights the deterministic nature of the computation.

Theoretically speaking, any problem in Class P can be solved efficiently within a polynomial time. However, this does not mean they can always be solved in a practically acceptable time, and actual computation time can be still impractically long in practice. For example, the computation time of DTIME $\left(n^{15}\right)$ when $n = 10\,000$ is still a very long time $O(10^{60})$. Even for all the supercomputers in the world starting to run from the Big Bang, it would take much longer than the age of the universe. To solve a problem in a practically acceptable timescale, we usually try to find algorithms that DTIME $\left(n^k\right)$ with $1 \leq k \leq 3$, otherwise, we may have to satisfy with any sufficient good approximation solution.

On the other hand, a much harder and larger class is the EXPTIME that consists of problems that can only be solved in an exponential time

$$\text{EXPTIME} = \cup_{1 \leq k \in \mathbb{N}} \text{DTIME}\left(2^{n^k}\right). \tag{2.30}$$

Even with a small $k = 2$ and $n = 10$, the computation time is still $O(2^{10^2}) = O(2^{100}) = O(10^{30})$. Thus, a Class P problem can be solved efficiently (in a polynomial time sense), not necessarily quickly in practice.

### 2.4.3 Class NP

In many cases, it is easier to verify a given solution, while it is much harder to find the solution in the first place. This feature can be defined using a non-deterministic Turing machine, which leads to a so-called Class NP that consists of a set of all problems that can be solved by a non-deterministic Turing machine in a polynomial time NTIME($f(n)$), while the verification of a solution can be done in polynomial time by a deterministic Turing machine. Here, we use notation N to show the non-deterministic nature of such computation. Thus, the Class NP can be defined by

$$\text{Class NP} = \cup_{1 \leq k \in \mathbb{N}} \text{NTime}\left(n^k\right), \tag{2.31}$$

where NP means "non-deterministic polynomial-time." An extension of NTIME to a harder and larger class is NEXPTIME:

$$\text{NEXPTIME} = \cup_{1 \leq k \in \mathbb{N}} \text{NTime}\left(2^{n^k}\right). \tag{2.32}$$

From the above definitions, it is straightforward to conclude that

$$\text{Class P} \subseteq \text{Class NP} \subset \text{EXPTIME}. \tag{2.33}$$

### 2.4.4 NP-Completeness

To solve a problem, we can often solve another problem (ideally easier problem or a problem with known solution methods) by reduction or transformation. However, such a reduction may not be always possible.

If a problem $A$ can be reduced to another problem $B$ using a polynomial time, then we say $A$ is reducible to $B$ in polynomial time. The definition of NP-hard is rather vague. An NP-hard problem means that it is at least as hard as any NP-problem. In other words, if a problem $H$ is NP-hard and if there is a polynomial time algorithm to solve $H$, then it implies that every problem in this class can be solved in polynomial time. However, the current literature indicates that NP-hard problems cannot be solved in polynomial time. That is, N $\neq$ NP.

A problem $H$ is called NP-complete if $H$ is both NP-hard and in class NP, and every problems in NP can be reduced to $H$ in polynomial time. As a result, NP-complete problems are the hardest problems in NP.

Examples of NP-complete problems are the traveling salesman problem (visiting $n$ cities exactly once so as to minimize the total distance traveled), Hamiltonian cycle on graphs, knapsack problem, and integer linear programming.

### 2.4.5 Complexity of Algorithms

The computational complexity discussed up to now has focused on the problems, and the algorithms are mainly described as simply in terms of polynomial time or exponential time. From the perspective of algorithm development and analysis, different algorithms will have different complexity, even for the same type of problems. In this case, we have to estimate the arithmetic complexity of an algorithm, or simply algorithmic complexity.

For example, to solve a sorting problem with $n$ different numbers so as to sort them from the smallest to the largest, we can use different algorithms. For example, the selection sort uses two loops for sorting $n$, which has an algorithmic complexity of $O(n^2)$, while the quicksort (or partition and exchange sort) has a complexity of $O(n \log n)$. There are many different sorting algorithms with different complexities.

It is worth pointing out that the algorithmic complexity here is mainly about time complexity because space (memory) complexity is less important. In this case, the space algorithmic complexity is $O(n)$.

**Example 2.3** The multiplication of two $n \times n$ matrices $\boldsymbol{A}$ and $\boldsymbol{B}$ using simple matrix multiplication rules has a complexity of $O(n^3)$. There are $n$ rows and $n$ columns for each matrix, and their product $\boldsymbol{C}$ has $n \times n$ entries. To get each entry, we need to carry out the multiplication of a row of $\boldsymbol{A}$ by a corresponding column of $\boldsymbol{B}$ and calculate their sum, thus the complexity is $O(n)$. As there are $n \times n = n^2$ entries, the overall complexity is $O(n^2) \times O(n) = O(n^3)$.

In the rest of this book, we analyze different algorithms, the complexity to be given is usually the arithmetic complexity of an algorithm under discussion.

To understand different types of optimization problems and thus enable us to select a suitable algorithm, we need to look the problem functions such as linearity, convexity, and nonlinearity.

## 2.5 Convexity

### 2.5.1 Linear and Affine Functions

Generally speaking, a function is a mapping from the independent variables or inputs to a dependent variable or variables/outputs. For example, the following function

$$f(x, y) = x^2 + y^2 + xy, \tag{2.34}$$

is a function which depends on two independent variables. This function maps the domain $\mathbb{R}^2$ (for $-\infty < x < \infty$ and $-\infty < y < \infty$) to $f$ on the real axis as its range. So we use the notation $f\colon \mathbb{R}^2 \to \mathbb{R}$ to denote this.

In general, a function $f(x, y, z, \dots)$ will map $n$ independent variables to $m$ dependent variables, and we use the notation $f\colon \mathbb{R}^n \to \mathbb{R}^m$ to mean that the domain of the function is a subset of $\mathbb{R}^n$ while its range is a subset of $\mathbb{R}^m$. The domain of a function is sometimes denoted by dom($f$) or dom $f$.

The inputs or independent variables can often be written as a vector. For simplicity, we often use a vector $\boldsymbol{x} = (x, y, z, \dots)^{\mathrm{T}} = (x_1, x_2, \dots, x_n)^{\mathrm{T}}$ for multiple variables. Therefore, $f(\boldsymbol{x})$ is often used to mean $f(x, y, z, \dots)$ or $f(x_1, x_2, \dots, x_n)$.

A function $\mathcal{L}(\boldsymbol{x})$ is called linear if it satisfies

$$\mathcal{L}(\boldsymbol{x} + \boldsymbol{y}) = \mathcal{L}(\boldsymbol{x}) + \mathcal{L}(\boldsymbol{y}), \qquad \mathcal{L}(\alpha\boldsymbol{x}) = \alpha\mathcal{L}(\boldsymbol{x}), \tag{2.35}$$

for any vectors $\boldsymbol{x}$ and $\boldsymbol{y}$, and any scalar $\alpha \in \mathbb{R}$.

**Example 2.4** To see if $f(\boldsymbol{x}) = f(x_1, x_2) = 2x_1 + 3x_2$ is linear, we use

$$\begin{aligned} f(x_1 + y_1, x_2 + y_2) &= 2(x_1 + y_1) + 3(x_2 + y_2) = 2x_1 + 2y_1 + 3x_2 + 3y_2 \\ &= [2x_1 + 3x_2] + [2y_1 + 3y_2] = f(x_1, x_2) + f(y_1, y_2). \end{aligned}$$

In addition, for any scalar $\alpha$, we have

$$f(\alpha x_1, \alpha x_2) = 2\alpha x_1 + 3\alpha x_2 = \alpha[2x_1 + 3x_2] = \alpha f(x_1, x_2).$$

Therefore, this function is indeed linear. This function can also be written as a vector form

$$f(\boldsymbol{x}) = \begin{pmatrix} 2 & 3 \end{pmatrix} \begin{pmatrix} x_1 \\ x_2 \end{pmatrix} = \boldsymbol{a} \cdot \boldsymbol{x} = \boldsymbol{a}^{\mathrm{T}}\boldsymbol{x},$$

where $\boldsymbol{a} = (2 \;\; 3)^{\mathrm{T}}$ and $\boldsymbol{x} = (x_1 \;\; x_2)^{\mathrm{T}}$.

Functions can be a multiple-component vector, which can be written as $\boldsymbol{F}$. A function $\boldsymbol{F}$ is called affine if there exists a linear function $\mathcal{L}$ and a vector constant $\boldsymbol{b}$ such that $\boldsymbol{F} = \mathcal{L}(\boldsymbol{x}) + \boldsymbol{b}$. In general, an affine function is a linear function with a translation, which can be written as a matrix form $\boldsymbol{F} = \boldsymbol{A}\boldsymbol{x} + \boldsymbol{b}$, where $\boldsymbol{A}$ is an $m \times n$ matrix, and $\boldsymbol{b}$ is a column vector in $\mathbb{R}^n$.

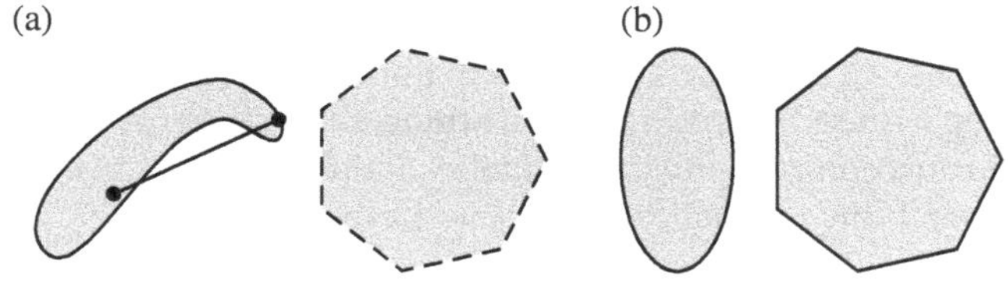


**Figure 2.1** Convexity: (a) non-convex and (b) convex.

Knowing the properties of a function can be useful for finding the maximum or minimum of the function. In fact, in mathematical optimization, nonlinear problems are often classified according to the convexity of the defining function(s). Geometrically speaking, an object is convex if, for any two points within the object, every point on the straight line segment joining them is also within the object. Examples are a solid ball, a cube, or a pyramid. Obviously, a hollow object is not convex. Four examples are given in Figure 2.1.

A set is called an affine set if the set contains the line through any two distinct points $\boldsymbol{x}_1$ and $\boldsymbol{x}_2$ in the set $S$. That is, the whole line as a linear combination

$$\boldsymbol{x} = \theta \boldsymbol{x}_1 + (1 - \theta)\boldsymbol{x}_2 \quad (\theta \in \mathbb{R}), \tag{2.36}$$

is contained in $S$. All the affine functions form an affine set. For example, all the solutions to a linear system $\boldsymbol{Au} = \boldsymbol{b}$ form an affine set $\{\boldsymbol{u}|\boldsymbol{Au} = \boldsymbol{b}\}$.

Mathematically speaking, a set $S \in \mathbb{R}^n$ in a real vector space is called a convex set if

$$\theta \boldsymbol{x} + (1 - \theta)\boldsymbol{y} \in S, \qquad \forall(\boldsymbol{x}, \boldsymbol{y}) \in S, \ \ \theta \in [0, 1]. \tag{2.37}$$

Thus, an affine set is always convex, but a convex set is not necessarily affine.

A convex hull, also called convex envelope, is a set $\boldsymbol{S}_{\text{convex}}$ formed by the convex combination

$$\boldsymbol{x} = \sum_{i=1}^{k} \theta_i \boldsymbol{x}_i = \theta_1 \boldsymbol{x}_1 + \cdots + \theta_k \boldsymbol{x}_k, \qquad \sum_{i=1}^{k} \theta_i = 1, \tag{2.38}$$

where $\theta_i \geq 0$ are nonnegative for all $i = 1, 2, \ldots, k$. It can be considered as the minimal set containing all the points.

On the other hand, a convex cone is a set containing all the conic combinations of the points in the form

$$\boldsymbol{x} = \theta_1 \boldsymbol{x}_1 + \theta_2 \boldsymbol{x}_2 \qquad (\theta \geq 0, \ \ \theta_2 \geq 0), \tag{2.39}$$

for any two points $\boldsymbol{x}_1, \boldsymbol{x}_2$ in the set.

A componentwise inequality for vectors and matrices is defined

$$\boldsymbol{u} \preceq \boldsymbol{v} \Longleftrightarrow u_i \leq v_i \qquad (i = 1, 2, \ldots, n). \tag{2.40}$$

It is a short-hand notation for $n$ inequalities for all $n$ components of the vectors $\boldsymbol{u}$ and $\boldsymbol{v}$. Here $\Longleftrightarrow$ means "if and only if." It is worth pointing out in some literature the notation $\boldsymbol{u} \leq \boldsymbol{v}$ is used to mean the same thing, though $\preceq$ provides a specific reminder to mean a componentwise inequality. If there is no confusion caused, we can use either. If the inequality is strict, we have

$$\boldsymbol{u} \prec \boldsymbol{v} \Longleftrightarrow u_i < v_i \qquad (\forall i = 1, \ldots, n). \tag{2.41}$$

Using this notation, we can now define hyperplanes and halfspaces. A hyperplane is a set satisfying $\{\boldsymbol{x}|\boldsymbol{A}\boldsymbol{x} = \boldsymbol{b}\}$ with a nonzero normal vector $\boldsymbol{A} \neq 0$, while a halfspace is a set formed by $\{\boldsymbol{x}|\boldsymbol{A}\boldsymbol{x} \preceq \boldsymbol{b}\}$ with $\boldsymbol{A} \neq 0$. It is straightforward to verify that a hyperplane is affine and convex, while a halfspace is convex.

### 2.5.2 Convex Functions

A function $f(x)$ defined on a convex set $\Omega$ is called convex if and only if it satisfies

$$f(\alpha x + \beta y) \leq \alpha f(x) + \beta f(y) \qquad (\forall x, y \in \Omega), \tag{2.42}$$

and

$$\alpha \geq 0, \quad \beta \geq 0, \quad \alpha + \beta = 1. \tag{2.43}$$

**Example 2.5** For example, the convexity of $f(x) = x^2 - 1$ requires

$$(\alpha x + \beta y)^2 - 1 \leq \alpha(x^2 - 1) + \beta(y^2 - 1) \qquad \forall(x, y \in \Re),$$

where $\alpha, \beta \geq 0$ and $\alpha + \beta = 1$. This is equivalent to

$$\alpha x^2 + \beta y^2 - (\alpha x + \beta y)^2 \geq 0,$$

where we have used $\alpha + \beta = 1$. We now have

$$\begin{aligned} \alpha x^2 + \beta y^2 - \alpha^2 x^2 - 2\alpha\beta xy - \beta^2 y^2 &= \alpha(1-\alpha)(x-y)^2 \\ &= \alpha\beta(x-y)^2 \geq 0, \end{aligned}$$

which is always true because $\alpha, \beta \geq 0$ and $(x - y)^2 \geq 0$. Therefore, $f(x) = x^2 - 1$ is convex for $\forall x \in \mathbb{R}$.

A function $f(x)$ on $\Omega$ is concave if and only if $g(x) = -f(x)$ is convex. An interesting property of a convex function $f$ is that the vanishing of the gradient $\mathrm{d}f/\mathrm{d}x|_{x_*} = 0$ guarantees that the point $x_*$ is the global minimum of $f$. Similarly, for a concave function, any local maximum is also the global maximum. If a function is not convex or concave, then it is much more difficult to find its global minima or maxima.

The test of convexity using the above definition is tedious. A more quick and efficient way is to use the definiteness of the Hessian matrix. A function is convex if its Hessian matrix is positive semidefinite for every point in the domain.

Conversely, a function becomes concave if its Hessian is negative semidefinite for every point in its domain. This is a very powerful method, and let us revisit the above example $f(x) = x^2 - 1$. We know its second derivative or Hessian is simply $f''(x) = 2$, which is always positive for every point in the domain $\mathfrak{R}$. We can quickly draw the conclusion that $f(x) = x^2 - 1$ is convex in the domain.

In fact, for a univariate function $f(x)$, if $f''(x) \geq 0$ for all $x$, it is convex. If $f''(x) > 0$ for all $x$ (without the equality), then $f$ is strictly convex. Thus, $x^2 - 1$ is not only convex, but also strictly convex.

**Example 2.6** To see if $f(x, y) = x^2 + 2y^2$ is convex, let us calculate its Hessian first, and we have

$$H = \begin{pmatrix} 2 & 0 \\ 0 & 4 \end{pmatrix},$$

which is obviously positive definite because it is a diagonal matrix and its eigenvalues 2 and 4 are both positive. Therefore, $f(x, y) = x^2 + 2y^2$ is convex on the whole domain $(x, y) \in \mathbb{R}^2$.

Examples of convex functions defined on $\mathbb{R}$ are $\alpha x + \beta$ for $\alpha, \beta \in \mathbb{R}$, $\exp[\alpha x]$ for $\alpha \in \mathbb{R}$, and $|x|^\alpha$ for $\alpha \geq 1$. Examples of concave functions are $\alpha x + \beta$ for $\alpha, \beta \in \mathbb{R}$, $x^\alpha$ for $x > 0$ and $0 \leq p \leq 1$, and $\log x$ for $x > 0$. It is worth pointing out that a linear function is both convex and concave.

There are some important mathematical operations that still preserve the convexity such as nonnegative weighted sum, composition using affine functions, and maximization or minimization. For example, if $f$ is convex, then $\beta f$ is also convex for $\beta \geq 0$. The nonnegative sum $\alpha f_1 + \beta f_2$ is convex if $f_1, f_2$ are convex and $\alpha, \beta \geq 0$.

The composition using an affine function also holds. For example, $f(\boldsymbol{A}\boldsymbol{x} + \boldsymbol{b})$ is convex if $f$ is convex. In addition, if $f_1, f_2, \ldots, f_n$ are convex, then $\max\{f_1, f_2, \ldots, f_n\}$ is also convex. Similarly, the piecewise-linear function $\max_{i=1}^n(\boldsymbol{A}_i\boldsymbol{x} + \boldsymbol{b}_i)$ is also convex.

If both $f$ and $g$ are convex, then $\psi(\boldsymbol{x}) = f[g(\boldsymbol{x})]$ can also be convex under certain non-decreasing conditions. For example, $\exp[f(\boldsymbol{x})]$ is convex if $f(\boldsymbol{x})$ is convex. This can be extended to the vector composition, and most interestingly, the log–sum–exp function

$$f(\boldsymbol{x}) = \log\left[\sum_{k=1}^{n} e^{x_k}\right] \tag{2.44}$$

is convex. For a more comprehensive introduction of convex functions, readers can refer to more advanced literature such as the book by Boyd and Vandenberghe listed at the end of this chapter.

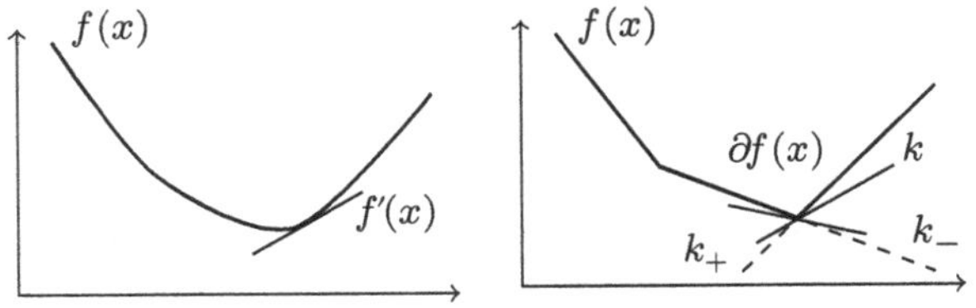


**Figure 2.2** Gradient (left) and subgradients (right) of a function $f(x)$ at $x_0$.

For the applications discussed later in this book, two important examples are quadratic functions and least-squares functions. If $\boldsymbol{A}$ is positive definite, then the quadratic function

$$f(\boldsymbol{x}) = \frac{1}{2}\boldsymbol{u}^{\mathrm{T}}\boldsymbol{A}\boldsymbol{u} + \boldsymbol{b}^{\mathrm{T}}\boldsymbol{u} + \boldsymbol{c} \tag{2.45}$$

is convex. In addition, the least-squares minimization

$$f(\boldsymbol{x}) = ||\boldsymbol{A}\boldsymbol{u} - \boldsymbol{b}||_2^2 \tag{2.46}$$

is convex for any $\boldsymbol{A}$. This is because $\nabla f = 2\boldsymbol{A}^{\mathrm{T}}(\boldsymbol{A}\boldsymbol{u} - \boldsymbol{b})$ and $\boldsymbol{H} = \nabla^2 f(\boldsymbol{x}) = 2\boldsymbol{A}^{\mathrm{T}}\boldsymbol{A}$ is positive definite for any $\boldsymbol{A}$.

### 2.5.3 Subgradients

Convex functions may be non-smooth. For example, from the definition of Eq. (2.42), we can prove that $f(x) = |x|$ is convex, but its derivatives are not well defined at $x = 0$. In this case, subgradients can be defined.

For a convex function $f(x)$ that is not differentiable at a point $x = x_0 \in [a, b]$, a subgradient $k$ can be defined by the inequality

$$f(x) - f(x_0) \geq k(x - x_0), \quad \forall x \in [a, b], \tag{2.47}$$

which can usually lead to more than one value of $k \in \mathbb{R}$. Any $k$ satisfying this condition is a subgradient or subderivative of $f(x)$ at $x_0$ (see Figure 2.2). In fact, the above definition is equivalent to the one-sided limits as given by

$$k_- = \lim_{x \to x_0^-} \frac{f(x) - f(x_0)}{(x - x_0)}, \quad k_+ = \lim_{x \to x_0^+} \frac{f(x) - f(x_0)}{(x - x_0)}, \tag{2.48}$$

such that

$$k_- \leq k_+. \tag{2.49}$$

As there are more than one value of $k$, the whole set in $[k_-, k_+]$, which is always no empty, is referred to as the subdifferential of $f(x)$ at $x_0$, often denoted by $\partial f(x)$ or simply $\partial f$. In the special case when $k_- = k_+$, there is only one single element of subgradients, then the subgradient becomes the standard gradient $f'(x)$.

As a simple example, the function $f(x) = |x - x_0|$ is not differentiable at $x = x_0$, but its subgradients are the set in $[-1, +1]$ because we have $k_- = -1$ and $k_+ = +1$ for this function.

It is worth pointing out that the above definition of subgradients for univariate convex functions can be extended to a non-differentiable, multivariate, convex function $f(\boldsymbol{x})$ where $\boldsymbol{x} = (x_1, x_2, \dots, x_n)^{\mathrm{T}}$. In general, we have a subgradient vector $\boldsymbol{k} = (k_1, k_2, \dots, k_n)^{\mathrm{T}}$ that satisfies the inequality

$$f(\boldsymbol{x}) - f(\boldsymbol{x}_0) \geq \boldsymbol{k}^{\mathrm{T}}(\boldsymbol{x} - \boldsymbol{x}_0). \tag{2.50}$$

Obviously, when the function becomes differentiable at a point, the subgradients become the standard gradient vector $\nabla f$.

## 2.6 Stochastic Nature in Algorithms

### 2.6.1 Algorithms with Randomization

Algorithms such as Newton's method are deterministic algorithms in the sense that the results are exactly repeatable if started with the same initial starting points. The advantage of such algorithms is that they are often effective; however, the final results will largely depend on the initial starting points.

For problems, especially hard problems, such dependence on the initial configuration can be problematic. As the search space can be large, it may be advantageous if different solutions can be obtained in different runs. To achieve this, some randomness can be used. In fact, there is a class of algorithms, called heuristics, that use trials and errors to search for feasible solutions to problems. Such heuristics can be very effective and often optimal or suboptimal solutions can be found quite quickly in practice.

From an earlier example about Newton's root-finding method of Eq. (2.2), the actual root found by the iterations will depend on the starting point as pointed in that example. If we can randomly initialize the starting point, then both roots can be found in the end. This is in fact the Newton method with random restart. The probability of finding each of two roots (1 and $-10$) is not equal, and the actual probability will depend on the initialization probability distribution and the structure of the algorithm.

### 2.6.2 Random Variables

Randomness such as roulette-rolling and noise arises from the lack of information, or incomplete knowledge of reality. It can also come from the intrinsic complexity, diversity, and perturbations of the system. Probability $P$ is a number or an expected frequency assigned to an event $A$ that indicates how likely it is that the event will occur when a random experiment is performed. This probability is often written as $P(A)$ to show that the probability $P$ is associated with event $A$.

For a discrete random variable $X$ with distinct values such as the number of cars passing through a junction, each value $x_i$ may occur with a certain probability $p(x_i)$. In other words, the probability varies and is associated with its corresponding random variable. Traditionally, an uppercase letter such as $X$ is used to denote a random variable, while a lowercase letter such as $x_i$ to represent its values. For example, if $X$ means a coin-flipping event, then $x_i = 0$ (tail) or 1 (head). A probability function $p(x_i)$ is a function that assigns probabilities to all the discrete values $x_i$ of the random variable $X$.

As an event must occur inside a sample space, the requirement that all the probabilities must be summed to one, which leads to

$$\sum_{i=1}^{n} p(x_i) = 1. \tag{2.51}$$

For example, the outcomes of tossing a fair coin form a sample space. The outcome of a head (H) is an event with a probability of $P(\mathrm{H}) = 1/2$, and the outcome of a tail (T) is also an event with a probability of $P(\mathrm{T}) = 1/2$. The sum of both probabilities should be one, that is

$$P(\mathrm{H}) + P(\mathrm{T}) = \frac{1}{2} + \frac{1}{2} = 1. \tag{2.52}$$

The cumulative probability function (CPF) of $X$ is defined by

$$P(X \le x) = \sum_{x_i < x} p(x_i). \tag{2.53}$$

For a continuous random variable $X$ that takes a continuous range of values (such as the level of noise), its distribution is continuous and the probability density function (PDF) $p(x)$ is defined for a range of values $x \in [a, b]$ for given limits $a$ and $b$ (or even over the whole real axis $x \in (-\infty, \infty)$). In this case, we always use the interval $(x, x + \mathrm{d}x]$ so that $p(x)$ is the probability that the random variable $X$ takes the value $x < X \le x + \mathrm{d}x$ is

$$\Phi(x) = P(x < X \le x + \mathrm{d}x) = p(x)\mathrm{d}x. \tag{2.54}$$

As all the probabilities of the distribution shall be added to unity, we have

$$\int_a^b p(x)\mathrm{d}x = 1. \tag{2.55}$$

The CPF becomes

$$\Phi(x) = P(X \le x) = \int_a^x p(x)\mathrm{d}x, \tag{2.56}$$

which is the definite integral of the PDF between the lower limit $a$ up to the present value $X = x$.

Two main measures for a random variable $X$ with a given probability distribution $p(x)$ are its mean and variance. The mean $\mu$ or expectation of $E[X]$ is defined by

$$\mu \equiv E[X] \equiv <X> = \int xp(x)\mathrm{d}x \tag{2.57}$$

for a continuous distribution and the integration is within the integration limits. If the random variable is discrete, then the integration becomes the weighted sum

$$E[X] = \sum_i x_i p(x_i). \tag{2.58}$$

The variance $\mathrm{var}[X] = \sigma^2$ is the expectation value of the deviation squared $(X - \mu)^2$. That is

$$\sigma^2 \equiv \mathrm{var}[X] = E[(X-\mu)^2] = \int (x-\mu)^2 p(x)\mathrm{d}x. \tag{2.59}$$

The square root of the variance $\sigma = \sqrt{\mathrm{var}[X]}$ is called the standard deviation, which is simply $\sigma$.

**Example 2.7** For example, for a simple uniform distribution, its probability density can be written as

$$p(x) = \frac{1}{b-a} \quad (a \le x \le b, \ \ b > a > 0),$$

which means that $p(x)$ is simply a constant.

The mean can be calculated by

$$\begin{aligned} \mu &= \int_a^b xp(x)\mathrm{d}x = \int_a^b \frac{x}{(b-a)}\mathrm{d}x = \frac{1}{(b-a)}\int_a^b x\mathrm{d}x \\ &= \frac{1}{(b-a)}\left[\frac{x^2}{2}\right]_a^b = \frac{1}{(b-a)} \cdot \left[\frac{b^2}{2} - \frac{a^2}{2}\right] = \frac{a+b}{2}. \end{aligned}$$

In addition, the variance is

$$\begin{aligned} \sigma^2 &= E[(X-\mu)^2] \\ &= \int_a^b (x-\mu)^2 p(x)\mathrm{d}x = \int_a^b \left[x - \frac{(a+b)}{2}\right]^2 \cdot \frac{1}{(b-a)}\mathrm{d}x \\ &= \frac{1}{(b-a)}\int_a^b \left[x^2 - (a+b)x + \frac{(a+b)^2}{4}\right]\mathrm{d}x \\ &= \frac{1}{(b-a)}\left[\frac{x^3}{3}\Big|_a^b - \frac{(a+b)}{2}x^2\Big|_a^b + \frac{(a+b)^2}{4}x\Big|_a^b\right] \\ &= \frac{1}{12(b-a)}(b^3 + 3a^2b - 3ab^3 - a^3) = \frac{1}{12(b-a)}(b-a)^3 = \frac{(b-a)^2}{12}. \end{aligned}$$

For a discrete distribution, the variance simply becomes the following sum:

$$\sigma^2 = \sum_i (x - \mu)^2 p(x_i). \tag{2.60}$$

In addition, any other formula for a continuous distribution can be converted to their counterparts for a discrete distribution if the integration is replaced by the sum. Therefore, we will mainly focus on the continuous distribution in the rest of the section.

From the above definitions, it is straightforward to prove

$$E[\alpha x + \beta] = \alpha E[X] + \beta, \qquad E[X^2] = \mu^2 + \sigma^2, \tag{2.61}$$

and

$$\mathrm{var}[\alpha x + \beta] = \alpha^2 \mathrm{var}[X], \tag{2.62}$$

where $\alpha$ and $\beta$ are constants.

Other frequently used measures are the mode and median. The mode of a distribution is defined by the value at which the PDF $p(x)$ is the maximum. For an even number of data sets, the mode may have two values. The median $m$ of a distribution corresponds to the value at which the CPF $\Phi(m) = 1/2$. The upper and lower quartiles $Q_U$ and $Q_L$ are defined by $\Phi(Q_U) = 3/4$ and $\Phi(Q_L) = 1/4$.

### 2.6.3 Poisson Distribution and Gaussian Distribution

The Poisson distribution is the distribution for small-probability discrete events. Typically, it is concerned with the number of events that occur in a certain time interval (e.g. number of telephone calls in an hour) or spatial area.

The PDF of the Poisson distribution is

$$P(X = x) = \frac{\lambda^x e^{-\lambda}}{x!} \qquad (\lambda > 0), \tag{2.63}$$

where $x = 0, 1, 2, \ldots, n$ and $\lambda$ is the mean of the distribution.

Obviously, the sum of all the probabilities must be equal to one. That is

$$\sum_{x=0}^{\infty} \frac{\lambda^x e^{-\lambda}}{x!} = \frac{\lambda^0 e^{-\lambda}}{0!} + \frac{\lambda^1 e^{-\lambda}}{1!} + \frac{\lambda^2 e^{-\lambda}}{2!} + \frac{\lambda^3 e^{-\lambda}}{3!} + \cdots$$
$$= e^{-\lambda}\left[1 + \lambda + \frac{\lambda^2}{2!} + \frac{\lambda^3}{3!} + \cdots\right] = e^{-\lambda} e^{\lambda} = e^{-\lambda+\lambda} = e^0 = 1. \tag{2.64}$$

Many stochastic processes such as the number of phone calls in a call center and the number of cars passing through a junction obey the Poisson distribution. If we are concerned with a Poisson distribution with a time interval $t$, $\lambda$ will be the arrival rate per unit time. However, in general, we should use $x = \lambda t$

to replace $x$ when dealing with the arrivals in a fixed period $t$. Thus, the Poisson distribution becomes

$$P(X = n) = \frac{(\lambda t)^n e^{-\lambda t}}{n!}. \tag{2.65}$$

**Example 2.8** A manufacturer produces a product that has a 1.5% probability of defects. In a batch of 200 same products, what is the probability of this batch containing exactly two defective products?

The parameter $\lambda$ can be calculated as

$$\lambda = 200 \times 1.5\% = 200 \times 0.015 = 3.$$

So the probability of getting exactly two defective products is

$$P(X = 2) = \frac{3^2 e^{-3}}{2!} \approx 0.224.$$

The probability of at least one defective product is

$$P(X \geq 1) = 1 - P(X = 0) = 1 - \frac{3^0 e^{-3}}{0!} \approx 0.95.$$

Using the definitions of mean and variance, it is straightforward to prove that $\mu = \lambda$ and $\sigma^2 = \lambda$ for the Poisson distribution.

The Gaussian distribution or normal distribution is the most important continuous distribution in probability and it has a wide range of applications. For a continuous random variable $X$, the PDF of the Gaussian distribution is given by

$$p(x) = \frac{1}{\sigma\sqrt{2\pi}} e^{-(x-\mu)^2/(2\sigma^2)}, \tag{2.66}$$

where $\sigma^2 = \text{var}[X]$ is the variance and $\mu = E[X]$ is the mean of the Gaussian distribution. It is straightforward to verify (using the integral in Chapter 1) that

$$\int_{-\infty}^{\infty} p(x) dx = 1, \tag{2.67}$$

and this is exactly the reason why the factor $1/\sqrt{2\pi}$ is required in the normalization of all the probabilities.

The probability function reaches a peak at $x = \mu$ and the variance $\sigma^2$ controls the width of the peak (see Figure 2.3) where $\mu = 0$ is used.

The CPF for a normal distribution is the integral of $p(x)$, which is defined by

$$\Phi(x) = P(X < x) = \frac{1}{\sqrt{2\pi\sigma^2}} \int_{-\infty}^{x} e^{-\frac{(\zeta-\mu)^2}{2\sigma^2}} d\zeta, \tag{2.68}$$

which can be written as

$$\Phi(x) = \frac{1}{\sqrt{2}} \left[1 + \text{erf}\left(\frac{x-\mu}{\sqrt{2}\sigma}\right)\right], \tag{2.69}$$

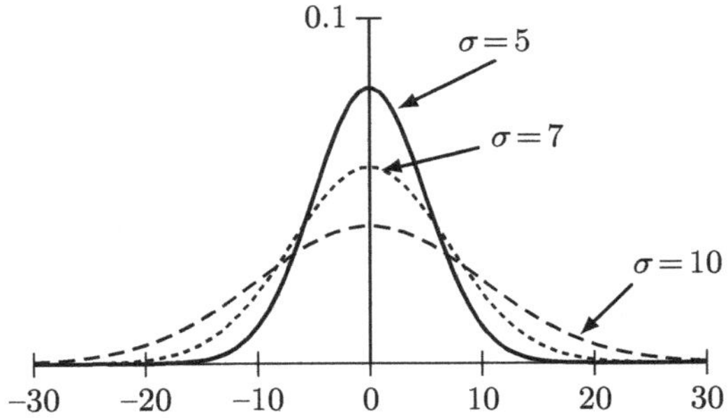


**Figure 2.3** Gaussian distributions for $\sigma = 5, 7, 10$.

where the error function is defined as

$$\mathrm{erf}(x) = \frac{2}{\sqrt{\pi}} \int_0^x e^{-\zeta^2} d\zeta. \tag{2.70}$$

### 2.6.4 Monte Carlo

In many applications, the number of possible combinations and states is so astronomical that it is impossible to carry out evaluations over all possible combinations systematically. In this case, the Monte Carlo method is one of the best alternatives. Monte Carlo is in fact a class of methods now widely used in computer simulations, machine learning, and weather forecasting. Since the pioneer studies in 1940s and 1950s, especially the work by Ulam, von Newmann, and Metropolis, it has been applied in almost all area of simulations, from numerical integration and Internet routing to financial market and climate simulations.

The fundamental idea of Monte Carlo integration is to randomly sample the domain of integration inside a control volume (often a regular region), and the integral of interest is estimated using the fraction of the random sampling points and the volume of the control region. Mathematically speaking, that is to say,

$$I = \int_\Omega f dV \approx V \left[ \frac{1}{N} \sum_{i=1}^{N} f_i \right] + O(\epsilon), \tag{2.71}$$

where $V$ is the volume of the domain $\Omega$, and $f_i$ is the evaluation of $f(x, y, z, \dots)$ at the sampling point $(x_i, y_i, z_i, \dots)$. The error estimate $\epsilon$ is given by

$$\epsilon = VS = V \sqrt{\frac{\mu_2 - \mu^2}{N}}, \tag{2.72}$$

where

$$\mu = \frac{1}{N} \sum_{i=1}^{N} f_i, \qquad \mu_2 = \frac{1}{N} \sum_{i=1}^{n} f_i^2. \tag{2.73}$$

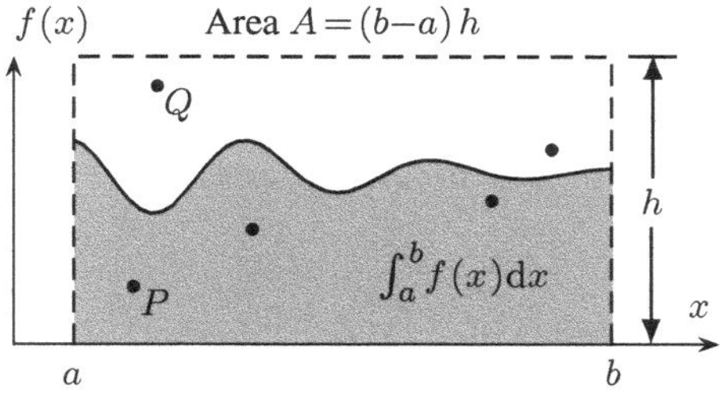


**Figure 2.4** Representation of Monte Carlo integration.

Here, the sample variance $S^2$ can be estimated by

$$S^2 = \frac{1}{N-1}\sum_{i=1}^{N}(f_i - \mu)^2 \approx \frac{1}{N}\sum_{i=1}^{N} f_i^2 - \mu^2, \tag{2.74}$$

which is the approximation of the variance $\sigma_f^2$:

$$\sigma_f^2 = \frac{1}{V}\int (f - \mu)^2 \mathrm{d}V. \tag{2.75}$$

The law of large number asymptotics implies that $\epsilon \to 0$ as $N \to \infty$. That is to say

$$V\mu = \lim_{N\to\infty} \frac{V}{N}\sum_{i=1}^{N} \to I. \tag{2.76}$$

In the simplest case in the domain [0, 1], we have

$$\int_0^1 f(x)\mathrm{d}x = \lim_{N\to\infty}\left[\frac{1}{N}\sum_{i=1}^{N} f(x_i)\right]. \tag{2.77}$$

We can see that the error of Monte Carlo integration decreases with $N$ in a manner of $1/\sqrt{N}$, which is independent of the number ($D$) of dimensions. This becomes advantageous over other conventional methods for multiple integrals in higher dimensions.

The basic procedure is to generate the random points so that they distribute uniformly inside the domain. In order to calculate the volume in higher dimensions, it is better to use a regular control domain to enclose the domain $\Omega$ of the integration.

For simplicity of discussion, we now use the integral of a univariate function $f(x)$ over the interval $[a, b]$ (see Figure 2.4). Let us estimate the integral

$$I = \int_a^b f(x)\mathrm{d}x = (b-a)\left[\frac{1}{N}\sum_{i=1}^{N} f(x_i)\right], \tag{2.78}$$

where $a$ and $b$ are finite integration limits. Now, we first use a regular control volume or a bounding box so that it encloses the interval $[a, b]$ and the curve $f(x)$ itself. As the length of the domain is $(b - a)$ and the height of the box is $h$, the area $A$ (or more generally the volume in higher dimension) is simply

$$A = (b - a)h. \tag{2.79}$$

We know the integral $I$ is the shaded area under the curve inside the bounding box, then the fraction or ratio $I/A$ of the integral (area) to the total area of the bounding box is statistically equivalent to the fraction or probability of uniformly-distributed random sampling points falling in the shaded area inside the box. Suppose we generate $N$ sampling points which are uniformly distributed inside the box. If there are $K$ points that are under the curve inside the box (shaded region), then the integral $I$ can be estimated by

$$I \approx A\frac{K}{N} = \frac{(b - a)hK}{N}. \tag{2.80}$$

The errors in Monte Carlo integration decrease in the form of $1/\sqrt{N}$ as $N$ increases. As the true randomness of sampling points are not essential as long as the sampling points can be distributed as uniformly as possible. In fact, studies show that it is possible to sample the points in a certain deterministic way so as to minimize the error of the Monte Carlo integration. In this case, the error may decrease in terms of $(\ln N)^D/N$ where $D$ is the dimension if appropriate methods such as Halton sequences are used. In addition, quasi-Monte Carlo methods using low discrepancy pseudo-random numbers are more efficient than standard Monte Carlo methods. Readers interested in such topics can refer to more advanced literature.

### 2.6.5 Common Probability Distributions

There are a number of other important distributions such as the exponential distribution, binomial distribution, Cauchy distribution, Lévy distribution, and Student $t$-distribution.

A Bernoulli distribution is a distribution of outcomes of a binary random variable $X$ where the random variable can only take two values: either 1 (success or yes) or 0 (failure or no). The probability of taking 1 is $0 \le p \le 1$, while the probability of taking 0 is $q = 1 - p$. Then, the probability mass function can be written as

$$B(m, p) = \begin{cases} p & \text{if} \quad m = 1, \\ 1 - p, & \text{if} \quad m = 0, \end{cases} \tag{2.81}$$

which can be written more compactly as

$$B(m, p) = p^m (1 - p)^{1-m}, \quad m \in \{0, 1\}. \tag{2.82}$$

It is straightforward to show that its mean and variance are

$$E[X] = p, \quad \mathrm{var}[X] = pq = p(1-p). \tag{2.83}$$

This is the probability of a single experiment with two distinct outcomes. In case of multiple experiments or trials ($n$), the probability distribution of exactly $m$ successes becomes the binomial distribution

$$B_n(m, n, p) = \binom{n}{m} p^m (1-p)^{n-m}, \tag{2.84}$$

where

$$\binom{n}{m} = \frac{n!}{m!(n-m)!} \tag{2.85}$$

is the binomial coefficient. Here, $n!$ is the factorial and $n = n(n-1)(n-2)\ldots 1$. For example, $5! = 5 \times 4 \times 3 \times 2 \times 1 = 120$. Conventionally, we set $0! = 1$.

It is also straightforward to verify that

$$E[X] = np, \quad \mathrm{var}[X] = np(1-p), \tag{2.86}$$

for $n$ trials.

The exponential distribution has the following PDF:

$$f(x) = \lambda e^{-\lambda x} \quad (\lambda > 0, \quad x > 0) \tag{2.87}$$

and $f(x) = 0$ for $x \leq 0$. Its mean and variance are

$$\mu = \frac{1}{\lambda}, \qquad \sigma^2 = \frac{1}{\lambda^2}. \tag{2.88}$$

**Example 2.9** The expectation $E(X)$ of an exponential distribution is

$$\begin{aligned}\mu = E(X) &= \int_{-\infty}^{\infty} x\lambda e^{-\lambda x} dx = \int_0^{\infty} x\lambda e^{-\lambda x} dx \\ &= \left[-xe^{-\lambda x} - \frac{1}{\lambda}e^{-\lambda x}\right]_0^{\infty} = \frac{1}{\lambda}.\end{aligned}$$

For $E(X^2)$, we have

$$\begin{aligned}E(X^2) &= \int_0^{\infty} x^2\lambda e^{-\lambda x} dx = \left[-x^2 e^{-\lambda x}\right]_0^{\infty} + 2\int_0^{\infty} xe^{-\lambda x} dx \\ &= \left[-x^2 e^{-\lambda x}\right]_0^{\infty} + \left[-\frac{2x}{\lambda}e^{-\lambda x} - \frac{2}{\lambda^2}e^{-\lambda x}\right]_0^{\infty} = \frac{2}{\lambda^2}.\end{aligned}$$

Here, we have used the fact that $x$ and $x^2$ grow slower than $\exp(-\lambda x)$ decreases. That is, $x\exp(-\lambda x) \to 0$ and $x^2\exp(-\lambda x) \to 0$ when $x \to \infty$.

Since $E(X^2) = \mu^2 + \sigma^2 = \mu^2 + \text{Var}(X)$ from Eq. (2.61), we have

$$\text{Var}(X) = \frac{2}{\lambda^2} - \left(\frac{1}{\lambda}\right)^2 = \frac{1}{\lambda^2}.$$

Cauchy probability distribution can be written as

$$p(x, \mu, \gamma) = \frac{1}{\pi\gamma}\left[\frac{\gamma^2}{(x-\mu)^2 + \gamma^2}\right] \quad (-\infty < x < \infty), \tag{2.89}$$

its mean and variance are undefined or infinite, which is a true indication that this distribution is heavy-tailed. The cumulative distribution function of the Cauchy distribution is

$$F(x) = \frac{1}{\pi}\tan^{-1}\left(\frac{x-\mu}{\gamma}\right) + \frac{1}{2}. \tag{2.90}$$

It is worth pointing out that this distribution can have a heavy tail or a fat tail where probability can be still significantly nonzero at the tails as $x \to \infty$. Thus, such a distribution belongs to the heavy-tailed or fat-tailed distributions.

Other heavy-tailed distributions include Pareto distribution, power-law distributions, and Lévy distribution.

A Pareto distribution is defined by

$$p(x) = \begin{cases} \frac{\alpha x_0^\alpha}{x^{\alpha+1}} & \text{if } x \geq x_0, \\ 1 & \text{if } x < x_0, \end{cases} \tag{2.91}$$

where $\alpha > 0$ and $x_0 > 0$ is the minimum value of $x$. The CPF is

$$F(x) = \begin{cases} 1 - \left(\frac{x_0}{x}\right)^\alpha & \text{for } x \geq x_0, \\ 0 & \text{for } x < x_0. \end{cases} \tag{2.92}$$

The power-law probability distribution can be written as

$$p(x) = Ax^{-\alpha}, \quad x \geq x_0 > 0, \tag{2.93}$$

where $\alpha > 1$ is an exponent and $A = (\alpha - 1)x_0^{\alpha-1}$ is the normalization constant. Alteratively, we can write the above as

$$p(x) = \frac{\alpha - 1}{x_0}\left(\frac{x}{x_0}\right)^{-\alpha}. \tag{2.94}$$

Lévy probability distribution is given by

$$p(x, \mu, \gamma) = \frac{\sqrt{\gamma/(2\pi)}}{(x-\mu)^{3/2}}e^{-\gamma/2(x-\mu)} \quad (x \geq \mu), \tag{2.95}$$

where $\mu > 0$ controls its location and $\gamma$ controls its scale. For a more general case with an exponent $\beta$, we have to use the integral to define Lévy distribution

$$p(x) = \frac{1}{\pi} \int_0^{\infty} \cos(kx) e^{-\alpha |k|^{\beta}} dk \quad (0 < \beta \leq 2), \tag{2.96}$$

where $\alpha > 0$. The special case of $\beta = 1$ becomes a Cauchy distribution, while $\beta = 2$ corresponds to a normal distribution.

The Student's $t$-distribution is given by

$$p(t) = K \left(1 + \frac{t^2}{n}\right)^{-(n+1)/2} \quad (-\infty < t < +\infty), \tag{2.97}$$

where $n$ is the number of degrees of freedom, and

$$K = \frac{\Gamma[(n+1)/2]}{\sqrt{n\pi}\,\Gamma(n/2)}, \tag{2.98}$$

where $\Gamma$ is the gamma function defined by

$$\Gamma(z) = \int_0^{\infty} x^{z-1} e^{-x} dx. \tag{2.99}$$

When $z = n$ is an integer, we have $\Gamma(n) = (n-1)!$.

It is worth pointing out that Pareto distribution, power-law distribution, and Lévy distributions are one-tailed because they are valid for $x > x_{\min}$. On the other hand, both the Cauchy distribution and Student's $t$-distribution are two-tailed as they are valid for the whole real domain.

With all these fundamentals, we are now ready to introduce optimization techniques more formally in the next chapters.

## Exercises

**2.1** Simplify the following order expressions:
- $f(n) = 20n^3 + 20\log(n) + n\log(n)$.
- $g(n) = 5n^{2.5} + n^2 + 10n$.
- $f(n) + 2g(n)$.

**2.2** Test if the following functions are convex.
- $x^4$ and $|x|$.
- $\exp(-x)$.
- $1/x$ and $1/x^4$.
- $f(x, y, z) = x^2 + y^2 + z^2$.

**2.3** For the Poisson distribution $p(x) = \lambda^x e^{-\lambda}/x!$ where $\lambda > 0$, find its mean.

**2.4** The Rayleigh distribution is given by $p(x) = (x/\sigma^2)e^{-x^2/(2\sigma^2)}$ for $x \geq 0$. Find its cumulative probability distribution and mean.

**2.5** Use a Monte Carlo method to estimate $\pi$.

## Bibliography


Arara, S. and Barak, B. (2009). *Computational Complexity: A Modern Approach*. Cambridge, UK: Cambridge University Press.

Bertsekas, D.P., Nedic, A., and Ozdaglar, A. 2003. *Convex Analysis and Optimization*, 2e. Belmont, MA: Athena Scientific.

Boyd, S.P. and Vandenberghe, L. (2004). *Convex Optimization*. Cambridge, UK: Cambridge University Press.

Cook, S. (1983). An overview of computational complexity. *Communications of the ACM* 26 (6): 400–408.

Fishman, G.S. (1995). *Monte Carlo: Concepts, Algorithms and Applications*. New York: Springer.

Garey, M.R. and Johnson, D.S. (1981). *Computers and Intractability: A Guide to the Theory of NP-Completeness*. San Francisco, CA: W.H. Freeman.

Goldreich, O. (2008). *Computational Complexity: A Conceptual Perspective*. Cambridge, UK: Cambridge University Press.

Grindstead, C.M. and Snell, J.L. (1997). *Introduction to Probability*, 2e. Providence, RI: American Mathematical Society.

Kiwiel, K. (1985). *Methods of Descent for Nondifferentiable Optimization*. Berlin: Springer-Verlag.

Metropolis, N. and Ulam, S. (1949). The Monte Carlo method. *Journal of the American Statistical Association*, 44, 335–341.

Papadimitriou, C.H. and Steiglitz, K. (1998). *Combinatorial Optimization: Algorithms and Complexity*. Mineola, NY: Dover Publication.

Schrijver, A. (1998). *Theory of Linear and Integer Programming*. New York: Wiley.

Shor, N.Z. (1985). *Minimization Methods for Non-differential Functions*. Berlin: Springer-Verlag.

Tovey, C.A. (2002). Tutorial on computational complexity. *Interfaces* 32 (3): 30–61.

Turing, A.M. (1948). *Intelligent Machinery*. London: National Physical Laboratory Report.

# Part II

# Optimization Techniques and Algorithms

# 3

# Optimization

From the brief introduction in Chapter 1, we know that the formulation of optimization in general will have an objective and multiple constraints, which forms constrained optimization. If there is no constraint at all, the problem becomes an unconstrained problem. In this chapter, we will introduce some commonly used techniques for solving optimization problems, including gradient-based algorithms and gradient-free algorithms.

## 3.1 Unconstrained Optimization

### 3.1.1 Univariate Functions

The simplest optimization problem without any constraints is probably the search for the maxima or minima of a univariate function $f(x)$ for $-\infty < x < +\infty$ (or in the whole real domain $\mathbb{R}$), we simply write

$$\text{maximize or minimize } f(x), \quad x \in \mathbb{R} \tag{3.1}$$

or simply

$$\max \text{ or } \min f(x), \quad x \in \mathbb{R}. \tag{3.2}$$

For unconstrained optimization problems, the optimality occurs at the critical points given by the stationary condition $f'(x) = 0$. However, this stationary condition is just a necessary condition, but it is not a sufficient condition. If $f'(x_*) = 0$ and $f''(x_*) > 0$, it is a local minimum. Conversely, if $f'(x_*) = 0$ and $f''(x_*) < 0$, then it is a local maximum.

However, if $f'(x_*) = 0$ but $f''(x)$ is indefinite (both positive and negative) when $x \to x_*$, then $x_*$ corresponds to a saddle point. For example, $f(x) = x^3$ has a saddle point $x_* = 0$ because $f'(0) = 0$ but $f''$ changes sign from $f''(0+) > 0$ to $f''(0-) < 0$.

However, in general, a function can have multiple stationary points. In order to find the global maximum or minimum, we may have to go through every stationary point, unless the objective function is convex.

*Optimization Techniques and Applications with Examples*, First Edition. Xin-She Yang.

It is worth pointing out that the notation argmin or argmax is used in some textbooks, thus the above optimization can be written as

$$\text{argmax}_{x \in \mathbb{R}} f(x) \tag{3.3}$$

or

$$\text{argmin}_{x \in \mathbb{R}} f(x). \tag{3.4}$$

This notation puts its emphasis on the argument $x$ so that the optimization task is to find a point (or points) in the domain of $f(x)$ that maximizes (or minimizes) the function values. On the other hand, the notation we used in Eq. (3.1) emphasizes the maximum or minimum value of the objective function. Both types of notations are used in the literature.

**Example 3.1** For $f(x) = 3x^4 - 20x^3 - 24x^2 + 240x + 400$ where $-\infty < x < +\infty$, its stationary or optimality condition is

$$f'(x) = 12x^3 - 60x^2 - 48x + 240 = 0,$$

which seems not easy to solve anlytically. However, we can rewrite it as

$$f'(x) = 12(x + 2)(x - 2)(x - 5) = 0.$$

Thus, there are three solutions $x* = -2, +2$, and $+5$. The second derivative is

$$f''(x) = 36x^2 - 120x - 48.$$

At $x = 2$, we have $f(2) = 672$ and $f''(2) = -144$, thus this point is a local maximum.

At $x = 5$, we have $f(5) = 375$ and $f''(5) = 252$, which means that this point is a local minimum.

On the other hand, at $x = -2$, we have $f''(-2) = 336$ and $f(-2) = 32$, thus this point is a local minimum. Comparing the two minima at $x* = -2$ and $x* = 5$, we can conclude that the global minimum occur at $x = -2$ with $f_{\min} = 32$.

As there is one local maximum at $x* = 2$, can we conclude that the global maximum is $f_{\max} = 672$? The answer is no. If we look any other points such as $x = 7$ or $x = -5$, we have $f(7) = 1247$ and $f(-5) = 2975$, which are much larger than 672, thus 672 cannot be a global maximum. In fact, this function is unbounded and thus its maximum is $+\infty$.

All these properties become clear if we look at the plot of $f(x)$ as shown in Figure 3.1. This kind of function is often referred to be as multimodal.

However, if we impose a simple interval such that $x \in [-3, 7]$, then we can conclude that the global maximum occurs at $x = 7$ (the right boundary) with $f(7) = 1247$. However, in this case, the maximum does not occur at a stationary point.

This example clearly shows that care should be taken when dealing with multimodal functions.

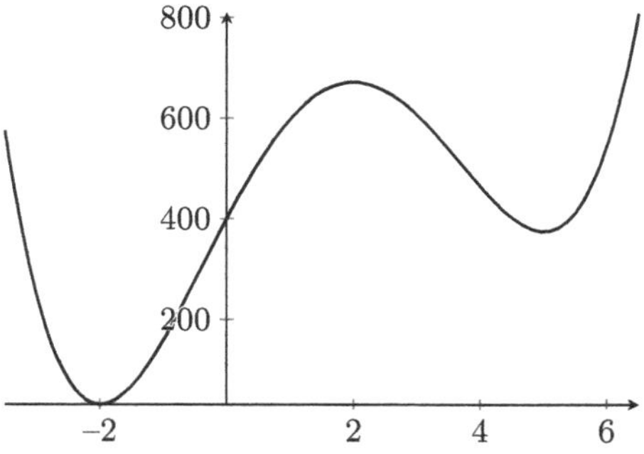


**Figure 3.1** Plot of $f(x)$ where $x = -2$ corresponds to the global minimum.

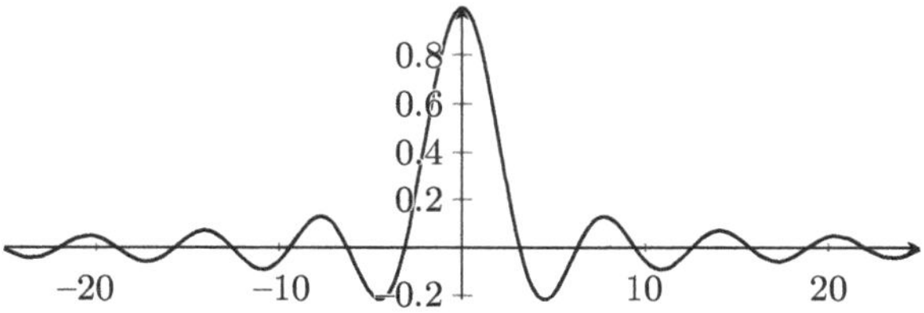


**Figure 3.2** Function $\sin(x)/x$ has an infinite number of local minima and maxima.

A maximum of a function $f(x)$ can be converted into a minimum of $A - f(x)$, where $A$ is usually a positive number (though $A = 0$ will do). For example, we know that the maximum of $f(x) = e^{-x^2}$, for $x \in (-\infty, \infty)$, is 1 at $x_* = 0$. This problem can be converted to a minimization problem $-f(x)$. For this reason, the optimization problems can be expressed as either minimization or maximization, depending on the context and convenience of formulations.

For a simple function optimization problem with one independent variable, the mathematical principle may be easy to understand, and the optimality occurs either at $f'(x) = 0$ (stationary points) or at boundaries (limits of simple intervals). However, it may not be so easy to find the actual optimal solution, even for seemingly simple functions such as $\text{sinc}(x) = \sin(x)/x$ (see Figure 3.2).

**Example 3.2** For $f(x) = \sin(x)/x$, it has an infinite number of local minima and maximum with the global maximum $f_{\max} = 1$ occurring at $x = 0$ as seen in Figure 3.2. However, the analytical method may not be so straightforward.

We know that

$$f'(x) = \frac{\cos(x)}{x} - \frac{\sin(x)}{x^2} = 0, \quad (x \neq 0),$$

which leads to

$$\tan(x) = x, \quad (x \neq 0).$$

But there is no explicit formula for the solutions, except for some approximations. Even we can solve it numerically to find some roots, but we cannot find all the roots (infinitely many).

In addition, these roots do not give any clear indication that $x = 0$ corresponds to the global maximum. In fact, the maximum at $x = 0$ can only be obtained by other methods such as taking limit of an alternative form or by some complex integrals.

As we have seen from this example, numerical methods have to be used to find the actual optimal points. This highlights a key issue: even we know the basic theory of optimization, it may not directly help much in solving certain classes of problems such as highly nonlinear, multimodal optimization problems. In fact, analytical methods can solve only a small set of problems. For a vast majority of problems, numerical algorithms become essential.

### 3.1.2 Multivariate Functions

For a multivariate function $f(\boldsymbol{x})$ where $\boldsymbol{x} = (x_1, \dots, x_n)^{\mathrm{T}}$, its optimization can be expressed in a similar way to a univariate optimization problem.

$$\text{minimize/maximize} \;\; f(\boldsymbol{x}), \quad \boldsymbol{x} \in \mathbb{R}^n. \tag{3.5}$$

Here, we have used the notation $\mathbb{R}^n$ to denote that the vector $\boldsymbol{x}$ is in an $n$-dimensional space where each component $x_i$ is a real number. That is, $-\infty < x_i < +\infty$ for $i = 1, 2, \dots, n$.

For a function $f(\boldsymbol{x})$, we can expand it locally using Taylor series about a point $\boldsymbol{x} = \boldsymbol{x}_*$ so that $\boldsymbol{x} = \boldsymbol{x}_* + \epsilon \boldsymbol{u}$,

$$f(\boldsymbol{x} + \epsilon \boldsymbol{u}) = f(\boldsymbol{x}_*) + \epsilon \boldsymbol{u} \boldsymbol{G}(\boldsymbol{x}_*) + \frac{1}{2}\epsilon^2 \boldsymbol{u}^{\mathrm{T}} \boldsymbol{H}(\boldsymbol{x}_*) \boldsymbol{u} + \cdots, \tag{3.6}$$

where $\boldsymbol{G}$ and $\boldsymbol{H}$ are its gradient vector and Hessian matrix, respectively. $\epsilon$ is a small parameter and $\boldsymbol{u}$ is a vector. For example, for a generic quadratic function

$$f(\boldsymbol{x}) = \frac{1}{2}\boldsymbol{x}^{\mathrm{T}} \boldsymbol{A} \boldsymbol{x} + \boldsymbol{k}^{\mathrm{T}} \boldsymbol{x} + \boldsymbol{b},$$

where $\boldsymbol{A}$ is a constant square matrix, $\boldsymbol{k}$ is the gradient vector and $\boldsymbol{b}$ is a vector constant, we have

$$f(\boldsymbol{x}_* + \epsilon \boldsymbol{u}) = f(\boldsymbol{x}_*) + \epsilon \boldsymbol{u}^{\mathrm{T}} \boldsymbol{k} + \frac{1}{2}\epsilon^2 \boldsymbol{u}^{\mathrm{T}} \boldsymbol{A} \boldsymbol{u} + \cdots, \tag{3.7}$$

where

$$f(\boldsymbol{x}_*) = \frac{1}{2}\boldsymbol{x}_*^{\mathrm{T}} \boldsymbol{A} \boldsymbol{x}_* + \boldsymbol{k}^{\mathrm{T}} \boldsymbol{x}_* + \boldsymbol{b}. \tag{3.8}$$

Thus, in order to study the local behavior of a quadratic function, we mainly need to study $\boldsymbol{G}$ and $\boldsymbol{H}$. In addition, for simplicity, we can take $\boldsymbol{b} = \boldsymbol{0}$ as it is a constant vector anyway.

At a stationary point $\boldsymbol{x}_*$, the first derivatives are zero or $\boldsymbol{G}(\boldsymbol{x}_*) = 0$, therefore, Eq. (3.6) becomes

$$f(\boldsymbol{x}_* + \epsilon \boldsymbol{u}) \approx f(\boldsymbol{x}_*) + \frac{1}{2}\epsilon^2 \boldsymbol{u}^{\mathrm{T}} \boldsymbol{H} \boldsymbol{u}. \tag{3.9}$$

If $\boldsymbol{H} = \boldsymbol{A}$, then

$$\boldsymbol{A}\boldsymbol{v} = \lambda \boldsymbol{v} \tag{3.10}$$

forms an eigenvalue problem. For an $n \times n$ matrix $\boldsymbol{A}$, there will be $n$ eigenvalues $\lambda_j (j = 1, \dots, n)$ with $n$ corresponding eigenvectors $\boldsymbol{v}$. As we have seen earlier that $\boldsymbol{A}$ is symmetric, these eigenvectors are either orthonormal or can be converted to be orthonormal. That is,

$$\boldsymbol{v}_i^{\mathrm{T}} \boldsymbol{v}_j = \delta_{ij}. \tag{3.11}$$

Near any stationary point $\boldsymbol{x}_*$, if we take $\boldsymbol{u}_j = \boldsymbol{v}_j$ as the local coordinate systems, we then have

$$f(\boldsymbol{x}_* + \epsilon \boldsymbol{v}_j) = f(\boldsymbol{x}_*) + \frac{1}{2}\epsilon^2 \lambda_j, \tag{3.12}$$

which means that the variations of $f(\boldsymbol{x})$, when $\boldsymbol{x}$ moves away from the stationary point $\boldsymbol{x}_*$ along the direction $\boldsymbol{v}_j$, are characterized by the eigenvalues. If $\lambda_j > 0$, $|\epsilon| > 0$ will lead to $|\Delta f| = |f(\boldsymbol{x}) - f(\boldsymbol{x}_*)| > 0$. In other words, $f(\boldsymbol{x})$ will increase as $|\epsilon|$ increases. Conversely, if $\lambda_j < 0$, $f(\boldsymbol{x})$ will decrease as $|\epsilon| > 0$ increases. Obviously, in the special case $\lambda_j = 0$, the function $f(\boldsymbol{x})$ will remain constant along the corresponding direction of $\boldsymbol{v}_j$.

**Example 3.3** We know that the function

$$f(x, y) = xy$$

has a saddle point at $(0, 0)$. It increases along the $x = y$ direction and decreases along $x = -y$ direction. From the above analysis, we know that $\boldsymbol{x}_* = (x_*, y_*)^{\mathrm{T}} = (0, 0)^{\mathrm{T}}$ and $f(x_*, y_*) = 0$. We now have

$$f(\boldsymbol{x}_* + \epsilon \boldsymbol{u}) \approx f(\boldsymbol{x}_*) + \frac{1}{2}\epsilon^2 \boldsymbol{u}^{\mathrm{T}} \boldsymbol{A} \boldsymbol{u},$$

where

$$\boldsymbol{A} = \nabla^2 f(\boldsymbol{x}_*) = \begin{pmatrix} \frac{\partial^2 f}{\partial x^2} & \frac{\partial^2 f}{\partial x \partial y} \\ \frac{\partial^2 f}{\partial x \partial y} & \frac{\partial^2 f}{\partial y^2} \end{pmatrix} = \begin{pmatrix} 0 & 1 \\ 1 & 0 \end{pmatrix}.$$

The eigenvalue problem is simply

$$\boldsymbol{A}\boldsymbol{v}_j = \lambda_j \boldsymbol{v}_j, \quad (j = 1, 2)$$

or

$$\begin{vmatrix} 0-\lambda_j & 1 \\ 1 & 0-\lambda_j \end{vmatrix} = 0,$$

whose solutions are

$$\lambda_j = \pm 1.$$

For $\lambda_1 = 1$, the corresponding eigenvector is

$$\boldsymbol{v}_1 = \begin{pmatrix} 1 \\ 1 \end{pmatrix}.$$

Similarly, for $\lambda_2 = -1$, the eigenvector is

$$\boldsymbol{v}_2 = \begin{pmatrix} 1 \\ -1 \end{pmatrix}.$$

Since $\boldsymbol{A}$ is symmetric, $\boldsymbol{v}_1$ and $\boldsymbol{v}_2$ are orthonormal because

$$\boldsymbol{v}_1^{\mathrm{T}}\boldsymbol{v}_2 = 1 \times 1 + 1 \times (-1) = 0.$$

Thus, we have

$$f(\boldsymbol{x}_* + \epsilon \boldsymbol{v}_j) = \frac{1}{2}\epsilon^2 \lambda_j, \qquad (j = 1, 2). \tag{3.13}$$

As $\lambda_1 = 1$ is positive, $f$ increases along the direction $\boldsymbol{v}_1 = (1\ 1)^{\mathrm{T}}$, which is indeed along the line $x = y$.

Similarly, for $\lambda_2 = -1$, $f$ will decrease along $\boldsymbol{v}_2 = (1\ -1)^{\mathrm{T}}$, which is exactly along the line $x = -y$. As there is no zero eigenvalue, the function will not remain constant in the region around $(0, 0)$.

This clearly shows that the properties of local Hessian matrices can indicate how the function may vary in that neighborhood, and thus provides enough information about its local optimality.

## 3.2 Gradient-Based Methods

The gradient-based methods are iterative methods that extensively use the information of the gradient of the objective function during iterations. For the minimization of a function $f(\boldsymbol{x})$, the essence of this method is

$$\boldsymbol{x}^{(k+1)} = \boldsymbol{x}^{(k)} + \alpha g(\nabla f, \boldsymbol{x}^{(k)}), \tag{3.14}$$

where $\alpha$ is the step size which can vary during iterations, and $k$ is the iteration counter. In addition, $g(\nabla f, \boldsymbol{x}^{(k)})$ is a function of the gradient $\nabla f$ and the current location $\boldsymbol{x}^{(k)}$, and the search direction is usually along the negative gradient direction (or $-\nabla f$) for minimization problems. Different methods use different forms of $g(\nabla f, \boldsymbol{x}^{(k)})$.

### 3.2.1 Newton's Method

Newton's method is a root-finding algorithm, but it can be modified for solving optimization problems. This is because optimization is equivalent to finding the root of the first derivative $f'(\boldsymbol{x})$ based on the stationary conditions once the objective function $f(\boldsymbol{x})$ is given. For a continuously differentiable function $f(\boldsymbol{x})$, we have the Taylor expansion in terms of $\Delta x = \boldsymbol{x} - \boldsymbol{x}_k$ about a fixed point $\boldsymbol{x}_k$,

$$f(\boldsymbol{x}) = f(\boldsymbol{x}_k) + [\nabla f(\boldsymbol{x}_k)]^{\mathrm{T}} \Delta \boldsymbol{x} + \frac{1}{2} \Delta \boldsymbol{x}^{\mathrm{T}} \nabla^2 f(\boldsymbol{x}_k) \Delta \boldsymbol{x} + \cdots,$$

whose third term is a quadratic form. Hence, $f(\boldsymbol{x})$ is minimized if $\Delta \boldsymbol{x}$ is the solution of the following linear equation (after taking the first derivative with respect to the increment vector $\Delta \boldsymbol{x} = \boldsymbol{x} - \boldsymbol{x}_k$)

$$\nabla f(\boldsymbol{x}_k) + \nabla^2 f(\boldsymbol{x}_k) \Delta \boldsymbol{x} = 0, \tag{3.15}$$

which gives

$$\Delta \boldsymbol{x} = -\frac{\nabla f(\boldsymbol{x}_k)}{\nabla^2 f(\boldsymbol{x}_k)}. \tag{3.16}$$

This can be rewritten as

$$\boldsymbol{x} = \boldsymbol{x}_k - \boldsymbol{H}^{-1} \nabla f(\boldsymbol{x}_k), \tag{3.17}$$

where $\boldsymbol{H}^{-1}(\boldsymbol{x}_k)$ is the inverse of the Hessian matrix $\boldsymbol{H} = \nabla^2 f(\boldsymbol{x}_k)$, which is defined as

$$\boldsymbol{H}(\boldsymbol{x}) \equiv \nabla^2 f(\boldsymbol{x}) \equiv \begin{pmatrix} \frac{\partial^2 f}{\partial x_1^2} & \cdots & \frac{\partial^2 f}{\partial x_1 \partial x_n} \\ \vdots & \ddots & \vdots \\ \frac{\partial^2 f}{\partial x_n \partial x_1} & \cdots & \frac{\partial^2 f}{\partial {x_n}^2} \end{pmatrix}. \tag{3.18}$$

This matrix is symmetric due to the fact that

$$\frac{\partial^2 f}{\partial x_i \partial x_j} = \frac{\partial^2 f}{\partial x_j \partial x_i}. \tag{3.19}$$

As each step is an approximation, the next solution $\boldsymbol{x}_{k+1}$ is approximately

$$\boldsymbol{x}_{k+1} = \boldsymbol{x}_k - \boldsymbol{H}^{-1} \nabla f(\boldsymbol{x}_k), \tag{3.20}$$

or in different notations as

$$\boldsymbol{x}^{(k+1)} = \boldsymbol{x}^{(k)} - \boldsymbol{H}^{-1} \nabla f(\boldsymbol{x}^{(k)}). \tag{3.21}$$

If the iteration procedure starts from the initial vector $\boldsymbol{x}^{(0)}$, usually a guessed point in the feasible region, then Newton's formula for the $k$th iteration becomes

$$\boldsymbol{x}^{(k+1)} = \boldsymbol{x}^{(k)} - \boldsymbol{H}^{-1}(\boldsymbol{x}^{(k)}) \nabla f(\boldsymbol{x}^{(k)}). \tag{3.22}$$

It is worth pointing out that if $f(\boldsymbol{x})$ is quadratic, then the solution can be found exactly in a single step.

In order to speed up the convergence, we can use a smaller step size $\alpha \in (0, 1]$ and we have the modified Newton's method

$$\boldsymbol{x}^{(k+1)} = \boldsymbol{x}^{(k)} - \alpha \boldsymbol{H}^{-1}(\boldsymbol{x}^{(k)})\nabla f(\boldsymbol{x}^{(k)}). \tag{3.23}$$

Sometimes, it might be time-consuming to calculate the Hessian matrix for second derivatives, especially when the dimensionality is high. A good alternative is to use an $n \times n$ identity matrix $\boldsymbol{I}$ to approximate $\boldsymbol{H}$ so that $\boldsymbol{H}^{-1} = \boldsymbol{I}$, and we have the quasi-Newton method

$$\boldsymbol{x}^{(k+1)} = \boldsymbol{x}^{(k)} - \alpha \boldsymbol{I}\nabla f(\boldsymbol{x}^{(k)}) = \boldsymbol{x}^{(k)} - \alpha \nabla f(\boldsymbol{x}^{(k)}), \tag{3.24}$$

which is usually called the steepest descent method for minimization problems. Here, the step size $\alpha$ is also called the learning rate in the literature.

### 3.2.2 Convergence Analysis

Newton's method for optimization can be analyzed from the point of view of root-finding algorithms. In essence, the iteration process is to find the root of the gradient $g(x) = f'(x) = 0$. Let us focus on the case of univariate function, so the iterative algorithm can be written as

$$x_{k+1} = x_k - \frac{g(x_k)}{g'(x_k)}. \tag{3.25}$$

If we define

$$\phi(x) = x - \frac{g(x)}{g'(x)}, \tag{3.26}$$

we have

$$x_{k+1} = \phi(x_k). \tag{3.27}$$

Let $R$ be the final solution as $k \to \infty$. Then we can expand $\phi$ around $R$ and we have

$$\phi(x) = \phi(R) + \phi'(R)(x - R) + \frac{\phi''(R)}{2}(x - R)^2 + \cdots. \tag{3.28}$$

If we only consider the expansion up to the second order and using $\phi(R) = R$ (a fixed point), we have

$$x_{k+1} \approx R + \phi'(R)(x_k - R) + \frac{\phi''(R)}{2}(x_k - R)^2, \tag{3.29}$$

where we have used $x_{k+1} = \phi(x_k)$.

Since

$$\phi'(x) = \frac{g(x)g''(x)}{[g'(x)]^2}, \tag{3.30}$$

we know that there are two cases:

1) If $\phi'(R) \neq 0$, we know that $x_k \to R$, so $(x_k - R)^2 \to 0$ or

$$\frac{\phi''(R)}{2}(x_k - R)^2 \to 0, \quad k \to \infty,$$

which means

$$\lim_{k\to\infty} \frac{|x_{k+1} - R|}{|x_k - R|} = |g'(R)| \geq 0. \tag{3.31}$$

The sequence thus converges linearly.

2) If $\phi'(R) = 0$, the second-order term dominates and we have

$$\lim_{k\to\infty} \frac{|x_{k+1} - R|}{|x_k - R|^2} = \frac{|\phi''(R)|}{2}. \tag{3.32}$$

This means that the convergence in this case is quadratic.

By the same philosophy, if $\phi'(R) = \phi''(R) = 0$, then the rate of convergence can be cubic if $\phi'''(R) \neq 0$.

For example, for $g(x) = x^2$, we have $g'(x) = 2x$ and $g''(x) = 2$, so

$$\phi'(x) = \frac{g(x)g''(x)}{[g'(x)]^2} = \frac{x^2 \times 2}{(2x)^2} = \frac{1}{2} \neq 0, \tag{3.33}$$

which means that the convergence is linear.

Interestingly, for $g(x) = x^2 - 1$, we know that its roots are $\pm 1$. We know that $g'(x) = 2x$ and $g''(x) = 2$, thus

$$\phi'(x) = \frac{g(x)g''(x)}{[g'(x)]^2} = \frac{(x^2 - 1) \times 2}{(2x)^2} = 0, \tag{3.34}$$

where we have used $x^2 - 1 = 0$. Therefore, the iterative sequence converges to $\pm 1$ quadratically.

### 3.2.3 Steepest Descent Method

The essence of this method is to find the lowest possible value of the objective function $f(\boldsymbol{x})$ from the current point $\boldsymbol{x}^{(k)}$. From the Taylor expansion of $f(\boldsymbol{x})$ about $\boldsymbol{x}^{(k)}$, we have

$$f(\boldsymbol{x}^{(k+1)}) = f(\boldsymbol{x}^{(k)} + \Delta \boldsymbol{s}) \approx f(\boldsymbol{x}^{(k)}) + [\nabla f(\boldsymbol{x}^{(k)})]^{\mathrm{T}} \Delta \boldsymbol{s}, \tag{3.35}$$

where $\Delta \boldsymbol{s} = \boldsymbol{x}^{(k+1)} - \boldsymbol{x}^{(k)}$ is the increment vector. Since we try to find a lower (better) approximation to the objective function, it requires that the second term on the right-hand side is negative. That is

$$f(\boldsymbol{x}^{(k)} + \Delta \boldsymbol{s}) - f(\boldsymbol{x}^{(k)}) = (\nabla f)^{\mathrm{T}} \Delta \boldsymbol{s} < 0. \tag{3.36}$$

From vector analysis, we know the inner product $\boldsymbol{u}^T\boldsymbol{v}$ of two vectors $\boldsymbol{u}$ and $\boldsymbol{v}$ is largest when they are parallel (but in opposite directions if larger negative is sought). Therefore, $(\nabla f)^T \Delta \boldsymbol{s}$ becomes the largest descent when

$$\Delta \boldsymbol{s} = -\alpha \nabla f(\boldsymbol{x}^{(k)}), \tag{3.37}$$

where $\alpha > 0$ is the step size. This is the case when the direction $\Delta \boldsymbol{s}$ is along the steepest descent in the negative gradient direction. As we have seen earlier, this method is a quasi-Newton method.

The choice of the step size $\alpha$ is very important. A very small step size means slow movement towards the local minimum, while a large step may overshoot and subsequently makes it move far away from the local minimum. Therefore, the step size $\alpha = \alpha^{(k)}$ should be different at each iteration step and should be chosen so that it minimizes the objective function $f(\boldsymbol{x}^{(k+1)}) = f(\boldsymbol{x}^{(k)}, \alpha^{(k)})$. In each iteration, the gradient and step size will be calculated. Again, a good initial guess of both the starting point and the step size is useful.

**Example 3.4** Let us minimize the function

$$f(x_1, x_2) = 10x_1^2 + 5x_1x_2 + 10(x_2 - 3)^2,$$

where

$$(x_1, x_2) = [-10, 10] \times [-15, 15],$$

using the steepest descent method starting with the initial $\boldsymbol{x}^{(0)} = (10, 15)^T$. We know that the gradient

$$\nabla f = (20x_1 + 5x_2, \;\; 5x_1 + 20x_2 - 60)^T,$$

therefore

$$\nabla f(\boldsymbol{x}^{(0)}) = (275, \;\; 290)^T.$$

In the first iteration, we have

$$\boldsymbol{x}^{(1)} = \boldsymbol{x}^{(0)} - \alpha_0 \begin{pmatrix} 275 \\ 290 \end{pmatrix}.$$

The step size $\alpha_0$ should be chosen such that $f(\boldsymbol{x}^{(1)})$ is at the minimum, which means that

$$f(\alpha_0) = 10(10 - 275\alpha_0)^2 \\ + 5(10 - 275\alpha_0)(15 - 290\alpha_0) + 10(12 - 290\alpha_0)^2,$$

should be minimized. This becomes an optimization problem for a single independent variable $\alpha_0$. All the techniques for univariate optimization problems such as Newton's method can be used to find $\alpha_0$. We can also obtain the solution by setting

$$\frac{\mathrm{d}f}{\mathrm{d}\alpha_0} = -159\,725 + 3\,992\,000\alpha_0 = 0,$$

whose solution is $\alpha_0 \approx 0.040\,01$.

At the second step, we have

$$\nabla f(\boldsymbol{x}^{(1)}) = (-3.078, 2.919)^{\mathrm{T}}, \quad \boldsymbol{x}^{(2)} = \boldsymbol{x}^{(1)} - \alpha_1 \begin{pmatrix} -3.078 \\ 2.919 \end{pmatrix}.$$

The minimization of $f(\alpha_1)$ gives $\alpha_1 \approx 0.066$, and the new location of the steepest descent is

$$\boldsymbol{x}^{(2)} \approx (-0.797, 3.202)^{\mathrm{T}}.$$

At the third iteration, we have

$$\nabla f(\boldsymbol{x}^{(2)}) = (0.060, 0.064)^{\mathrm{T}}, \quad \boldsymbol{x}^{(3)} = \boldsymbol{x}^{(2)} - \alpha_2 \begin{pmatrix} 0.060 \\ 0.064 \end{pmatrix}.$$

The minimization of $f(\alpha_2)$ leads to $\alpha_2 \approx 0.040$, and we have

$$\boldsymbol{x}^{(3)} \approx (-0.800\,029\,9, 3.200\,29)^{\mathrm{T}}.$$

Then, the iterations continue until a prescribed tolerance is met.

From calculus, we know that we can set the first partial derivatives equal to zero

$$\frac{\partial f}{\partial x_1} = 20x_1 + 5x_2 = 0, \quad \frac{\partial f}{\partial x_2} = 5x_1 + 20x_2 - 60 = 0,$$

and we also know that the minimum occurs exactly at

$$\boldsymbol{x}_* = (-\frac{4}{5}, \frac{16}{5})^{\mathrm{T}} = (-0.8, 3.2)^{\mathrm{T}}.$$

The steepest descent method gives almost the exact solution after only three iterations.

In finding the step size $\alpha_k$ in the above steepest descent method, we have used the stationary condition $\mathrm{d}f(\alpha_k)/\mathrm{d}\alpha_k = 0$. Well, you may say that if we use this stationary condition for $f(\alpha_0)$, why not use the same method to get the minimum point of $f(\boldsymbol{x})$ in the first place. There are two reasons here. The first reason is that this is a simple example for demonstrating how the steepest descent method works. The second reason is that even for complicated multiple variables $f(x_1, \ldots, x_n)$ (say $n = 500$), then $f(\alpha_k)$ at any step $k$ is still a univariate function, and the optimization of such $f(\alpha_k)$ is much simpler compared with the original multivariate problem.

It is worth pointing out that in many cases, we do not need to explicitly calculate the step size $\alpha_k$, and we can instead use a sufficient small value or an adaptive scheme, which can work sufficiently well in practice. In fact, $\alpha_k$ can be considered as a hyper-parameter, which can be either tuned or set adaptively. This point will become clearer when we discuss the line search and stochastic gradient method later in this book.

From our example, we know that the convergence from the second iteration to the third iteration is slow. In fact, the steepest descent is typically slow once the local minimization is near. This is because near the local minima, the gradient is nearly zero, and thus the rate of descent is also slow. If high accuracy is needed near the local minimum, other local search methods should be used.

There are many variations of the steepest descent methods. If the optimization is to find the maximum, then this method becomes the *hill-climbing* method because the aim is to climb up the hill (along the gradient direction) to the highest peak.

The standard steepest descent method works well for convex functions and near a local peak (valley) of most smooth multimodal functions, though this local peak is not necessarily the global best. However, for some tough functions, it is not a good method. This is better demonstrated by the following example:

**Example 3.5** Let us minimize the so-called banana function introduced by Rosenbrock

$$f(x_1, x_2) = (1 - x_1)^2 + 100(x_2 - x_1^2)^2,$$

where

$$(x_1, x_2) \in [-5, 5] \times [-5, 5].$$

This function has a global minimum $f_{\min} = 0$ at $(1, 1)$ which can be determined by

$$\frac{\partial f}{\partial x_1} = -2(1 - x_1) - 400x_1(x_2 - x_1^2) = 0$$

and

$$\frac{\partial f}{\partial x_2} = 200(x_2 - x_1)^2 = 0,$$

whose unique solutions are

$$x_1^* = 1, \quad x_2^* = 1.$$

Now we try to find its minimum by using the steepest descent method with the initial guess $\boldsymbol{x}^{(0)} = (5, 5)$. We know that the gradient is

$$\nabla f = \left[ -2(1 - x_1) - 400x_1(x_2 - x_1^2),\ 200(x_2 - x_1^2) \right]^{\mathrm{T}}.$$

Initially, we have

$$\nabla f(\boldsymbol{x}^{(0)}) = \left(40\,008,\ -4000\right)^{\mathrm{T}}.$$

In the first iteration, we have

$$\boldsymbol{x}^{(1)} = \boldsymbol{x}^{(0)} - \alpha_0 \begin{pmatrix} 40\,008 \\ -4000 \end{pmatrix}.$$

The step size $\alpha_0$ should be chosen such that $f(\boldsymbol{x}^{(1)})$ reaches its minimum, which means that

$$f(\alpha_0) = [1 - (5 - 40\,008\alpha_0)]^2 + 100[(5 + 4000\alpha_0) - (5 - 40\,008\alpha_0)^2]^2$$

should be minimized. The stationary condition becomes

$$\frac{\mathrm{d}f}{\mathrm{d}\alpha_0} = 1.0248 \times 10^{21}\alpha_0^3 - 3.8807 \times 10^{17}\alpha_0^2 + 0.4546 \times 10^{14}\alpha_0 - 1.6166 \times 10^9 = 0,$$

which has three solutions

$$\alpha_0 \approx 0.000\,067\,61, \quad 0.000\,126\,2, \quad 0.000\,184\,8.$$

Whichever these values we use, the new iteration $x_2^{(1)} = x_2^{(0)} + 4000\alpha_0$ is always greater that $x_2^{(0)} = 5$, which moves away from the best solution $(1, 1)$. In this case, the simple steepest descent method does not work well. We have to use other more elaborate methods such as the conjugate gradient method.

In fact, Rosenbrock's banana function is a very tough test function for optimization algorithms. Its solution requires more elaborate methods, and we will discuss some of these methods in later chapters.

### 3.2.4 Line Search

In the steepest descent method, there are two important parts: the descent direction and the step size (or how far to descend). The calculations of the exact step size may be very time consuming. In reality, we intend to find the right descent direction. Then a reasonable amount of descent, not necessarily the exact amount, during each iteration will usually be sufficient. For this, we essentially use a line search method.

To find the local minimum of the objective function $f(\boldsymbol{x})$, we try to search along a descent direction $\boldsymbol{s}_k$ with an adjustable step size $\alpha_k$ so that

$$\psi(\alpha_k) = f(\boldsymbol{x}_k + \alpha_k \boldsymbol{s}_k) \tag{3.38}$$

decreases as much as possible, depending on the value of $\alpha_k$. Loosely speaking, a reasonably right step size should satisfy the Wolfe's conditions:

$$f(\boldsymbol{x}_k + \alpha_k \boldsymbol{s}_k) \leq f(\boldsymbol{x}_k) + \gamma_1 \alpha_k \boldsymbol{s}_k^{\mathrm{T}} \nabla f(\boldsymbol{x}_k) \tag{3.39}$$

and

$$s_k^{\mathrm{T}} \nabla f(\boldsymbol{x}_k + \alpha_k \boldsymbol{s}_k) \geq \gamma_2 \boldsymbol{s}_k^{\mathrm{T}} \nabla f(\boldsymbol{x}_k), \tag{3.40}$$

where $0 < \gamma_1 < \gamma_2 < 1$ are algorithm-dependent parameters. The first condition is a sufficient decrease condition for $\alpha_k$, often called the Armijo condition or rule, while the second inequality is often referred to as the curvature condition. For most functions, we can use $\gamma_1 = 10^{-4}$–$10^{-2}$ and $\gamma_2 = 0.1$–$0.9$. These conditions are usually sufficient to ensure that the algorithm converges in most cases; however, more strong conditions may be needed for some tough functions. The basic steps of the line search method can be summarized in Algorithm 3.1.

**Algorithm 3.1** The basic steps of a line search method.

Initial guess $\boldsymbol{x}_0$ at $k = 0$
**while** ($\|\nabla f(\boldsymbol{x}_k)\| >$ accuracy)
Find the search direction $\boldsymbol{s}_k = -\nabla f(\boldsymbol{x}_k)$
Solve for $\alpha_k$ by decreasing $f(\boldsymbol{x}_k + \alpha \boldsymbol{s}_k)$ significantly
    while satisfying the Wolfe's conditions
Update the result $\boldsymbol{x}_{k+1} = \boldsymbol{x}_k + \alpha_k \boldsymbol{s}_k$
$k = k + 1$
**end**

### 3.2.5 Conjugate Gradient Method

The method of conjugate gradient belongs to a wider class of the so-called Krylov subspace iteration methods. The conjugate gradient method was pioneered by Magnus Hestenes, Eduard Stiefel, and Cornelius Lanczos in the 1950s. It was named as one of the top 10 algorithms of the twentieth century.

The conjugate gradient method can be used to solve the following linear system:

$$\boldsymbol{A}\boldsymbol{u} = \boldsymbol{b}, \tag{3.41}$$

where $\boldsymbol{A}$ is often a symmetric positive definite matrix. The above system is equivalent to minimizing the following function $f(\boldsymbol{u})$:

$$f(\boldsymbol{u}) = \frac{1}{2}\boldsymbol{u}^{\mathrm{T}}\boldsymbol{A}\boldsymbol{u} - \boldsymbol{b}^{\mathrm{T}}\boldsymbol{u} + \boldsymbol{v}, \tag{3.42}$$

where $\boldsymbol{v}$ is a vector constant and can be taken to be zero. We can easily see that $\nabla f(\boldsymbol{u}) = 0$ leads to $\boldsymbol{A}\boldsymbol{u} = \boldsymbol{b}$.

In general, the size of $\boldsymbol{A}$ can be very large and sparse with $n > 100\,000$, but it is not required that $\boldsymbol{A}$ is strictly symmetric positive definite. In fact, the main condition is that $\boldsymbol{A}$ should be a normal matrix. A square matrix $\boldsymbol{A}$ is called normal if $\boldsymbol{A}^{\mathrm{T}}\boldsymbol{A} = \boldsymbol{A}\boldsymbol{A}^{\mathrm{T}}$. Therefore, a symmetric matrix is a normal matrix, so it is an orthogonal matrix because an orthogonal matrix $\boldsymbol{Q}$ satisfying $\boldsymbol{Q}\boldsymbol{Q}^{\mathrm{T}} = \boldsymbol{Q}^{\mathrm{T}}\boldsymbol{Q} = \boldsymbol{I}$.

The theory behind these iterative methods is closely related to the Krylov subspace $\mathcal{K}_k$ spanned by $\boldsymbol{A}$ and $\boldsymbol{b}$ as defined by

$$\mathcal{K}_k(\boldsymbol{A}, \boldsymbol{b}) = \{\boldsymbol{I}\boldsymbol{b}, \boldsymbol{A}\boldsymbol{b}, \boldsymbol{A}^2\boldsymbol{b}, \dots, \boldsymbol{A}^{n-1}\boldsymbol{b}\}, \tag{3.43}$$

where $\boldsymbol{A}^0 = \boldsymbol{I}$.

If we use an iterative procedure to obtain the approximate solution $\boldsymbol{u}_k$ to $\boldsymbol{A}\boldsymbol{u} = \boldsymbol{b}$ at $k$th iteration, the residual is given by

$$\boldsymbol{r}_k = \boldsymbol{b} - \boldsymbol{A}\boldsymbol{u}_k, \tag{3.44}$$

which is essentially the negative gradient $\nabla f(\boldsymbol{u}_k)$. The search direction vector in the conjugate gradient method is subsequently determined by

$$\boldsymbol{d}_{k+1} = \boldsymbol{r}_k - \frac{\boldsymbol{d}_k^{\mathrm{T}} \boldsymbol{A} \boldsymbol{r}_k}{\boldsymbol{d}_k^{\mathrm{T}} \boldsymbol{A} \boldsymbol{d}_k} \boldsymbol{d}_k. \tag{3.45}$$

The solution often starts with an initial guess $\boldsymbol{u}_0$ at $k = 0$, and proceeds iteratively. The above steps can compactly be written as

$$\boldsymbol{u}_{k+1} = \boldsymbol{u}_k + \alpha_k \boldsymbol{d}_k, \quad \boldsymbol{r}_{k+1} = \boldsymbol{r}_k - \alpha_k \boldsymbol{A} \boldsymbol{d}_k, \tag{3.46}$$

and

$$\boldsymbol{d}_{k+1} = \boldsymbol{r}_{k+1} + \beta_k \boldsymbol{d}_k \tag{3.47}$$

where

$$\alpha_k = \frac{\boldsymbol{r}_k^{\mathrm{T}} \boldsymbol{r}_k}{\boldsymbol{d}_k^{\mathrm{T}} \boldsymbol{A} \boldsymbol{d}_k}, \qquad \beta_k = \frac{\boldsymbol{r}_{k+1}^{\mathrm{T}} \boldsymbol{r}_{k+1}}{\boldsymbol{r}_k^{\mathrm{T}} \boldsymbol{r}_k}. \tag{3.48}$$

Iterations stop when a prescribed accuracy is reached. This can easily be programmed in any programming language, especially Matlab.

It is worth pointing out that the initial guess $\boldsymbol{r}_0$ can be any educated guess; however, $\boldsymbol{d}_0$ should be taken as $\boldsymbol{d}_0 = \boldsymbol{r}_0$, otherwise, the algorithm may not converge. In the case when $\boldsymbol{A}$ is not symmetric, we can use the generalized minimal residual (GMRES) algorithm developed by Y. Saad and M.H. Schultz in 1986.

### 3.2.6 Stochastic Gradient Descent

In many optimization problems, especially in deep learning, the objective function or risk function to be minimized can be written in the following form:

$$E(\boldsymbol{w}) = \frac{1}{m} \sum_{i=1}^{m} f_i(\boldsymbol{x}_i, \boldsymbol{w}) = \frac{1}{m} \sum_{i=1}^{m} \left[u_i(\boldsymbol{x}_i, \boldsymbol{w}) - \bar{y}_i\right]^2, \tag{3.49}$$

where

$$f_i(\boldsymbol{x}_i, \boldsymbol{w}) = \left[u_i(\boldsymbol{x}_i, \boldsymbol{w}) - \bar{y}_i\right]^2. \tag{3.50}$$

Here, $\boldsymbol{w} = (w_1, w_2, \ldots, w_K)^T$ is a parameter vector such as the weights in an neural network. In addition, $\bar{y}_i (i = 1, 2, \ldots, m)$ are the target or real data (data points or data sets), while $u_i(\boldsymbol{x}_i)$ are the predicted values based on the inputs $\boldsymbol{x}_i$ by a model such as the models based on trained neural networks.

The standard gradient descent for finding new weight parameters in terms of iterative formula can be written as

$$\boldsymbol{w}^{t+1} = \boldsymbol{w}^t - \frac{\eta}{m} \sum_{i=1}^{m} \nabla f_i, \tag{3.51}$$

where $0 < \eta \leq 1$ is the learning rate or step size. Here, the gradient $\nabla f_i$ is with respect to $\boldsymbol{w}$. This requires the calculations of $m$ gradients. When $m$ is large and the number of iteration $t$ is large, this can be very expensive.

In order to save computation, the true gradient can be approximated by the gradient at a single value at $f_i$ instead of all $m$ values. That is

$$\boldsymbol{w}^{t+1} = \boldsymbol{w}^t - \eta_t \nabla f_i, \tag{3.52}$$

where $\eta_t$ is the learning rate at iteration $t$, which can be varying with iterations. Though this is a crude estimate at a randomly selected point $i$ at iteration $t$ to the true gradient, the computation costs have dramatically reduced by a factor of $1/m$. Due to the random nature of the selection of a sample $i$ (which can be very different at each iteration), this way of calculating gradient is called stochastic gradient. The method based on such crude estimation of gradients is called stochastic gradient descent (SGD) for minimization or stochastic gradient ascent (SGA) for maximization.

The learning rate $\eta_t$ should be reduced gradually. For example, a commonly used reduction of learning rates is

$$\eta_t = \frac{1}{1 + \beta t}, \quad t = 1, 2, \ldots, \tag{3.53}$$

where $\beta > 0$ is a hyper-parameter. Bottou showed that SGD will almost surely converge if

$$\sum_t \eta_t = \infty, \quad \sum_t \eta_t^2 < \infty. \tag{3.54}$$

The best convergence rate is $\eta_t \sim 1/t$ with the averaged residual error decreasing in the manner of $E \sim 1/t$.

It is worth pointing out the SGD is not the direct descent in the true gradient sense, but the descent is in terms of average or expectation. Thus, the paths can still be zig-zag, sometimes, it may be up the gradient, not necessarily all the way down the gradient directions, but the overall computation efficiency is usually much higher than the true gradient descent for large-scale problems. Therefore, it is widely used for deep learning problems and large-scale problems.

### 3.2.7 Subgradient Method

All the above gradient-based methods assume implicitly that the functions are differentiable. In the case of non-differentiable functions, we have to use the subgradient method for non-differential convex functions or more generally gradient-free methods for nonlinear functions to be introduced later in this chapter.

For non-differentiable convex functions, subgradient vectors $\boldsymbol{v}_k$ can be defined by

$$f(\boldsymbol{x}) - f(\boldsymbol{x}_k) \geq \boldsymbol{v}_k^{\mathrm{T}}(\boldsymbol{x} - \boldsymbol{x}_k) \tag{3.55}$$

and Newton's iteration formula (3.24) can be replaced by

$$\boldsymbol{x}^{k+1} = \boldsymbol{x}^k - \alpha_k \boldsymbol{v}_k, \tag{3.56}$$

where $\alpha_k$ is the step size at iteration $k$. As the iteration formula involves the subgradient $\boldsymbol{v}_k = \partial f(\boldsymbol{x}_k)$ calculated at iteration $k$, the method is called the subgradient method.

It is worth pointing out that since there are many arbitrary subgradients, the subgradient calculated at $\boldsymbol{x}_k$ may not be in the desirable direction. Some choices such as choosing the larger values of the norm $\boldsymbol{v}$ can be expected.

In addition, though a constant step size $\alpha_k = \alpha$ where $0 < \alpha < 1$ can work well in many cases, it is desirable that the step size $\alpha_k$ should vary and be scaled when appropriate. For example, a commonly used scheme for varying step sizes is $\alpha_k \geq 0$, $\sum_{k=1}^{\infty} \alpha_k^2 < \infty$, and $\lim_{k \to \infty} \alpha_k = 0$.

The subgradient method is still in used in practice, and it can be very effective in combination with the stochastic gradient method, which leads to a class of so-called stochastic subgradient methods. Convergence can be proved and interested readers can refer to more advanced literature such as Bertsekas et al. (2003).

The limitation of the subgradient method is that it is mainly for convex functions. In case of general nonlinear, non-differentiable, non-convex functions, we can use gradient-free methods and we will introduce some of these methods in the next section.

## 3.3 Gradient-Free Nelder–Mead Method

### 3.3.1 A Simplex

In the $n$-dimensional space, a simplex, which is a generalization of a triangle on a plane, is a convex hull with $n + 1$ distinct points. For simplicity, a simplex in the $n$-dimensional space is referred to as an $n$-simplex. Therefore, 1-simplex is a line segment, 2-simplex is a triangle, a 3-simplex is a tetrahedron (see Figure 3.3), and so on.

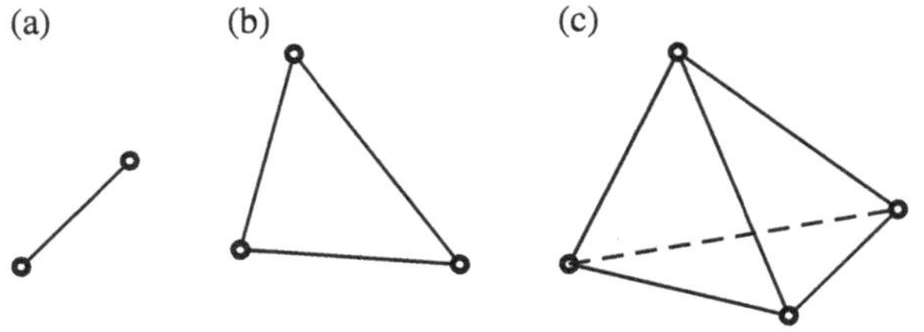


**Figure 3.3** The concept of a simplex: (a) 1-simplex, (b) 2-simplex, and (c) 3-simplex.

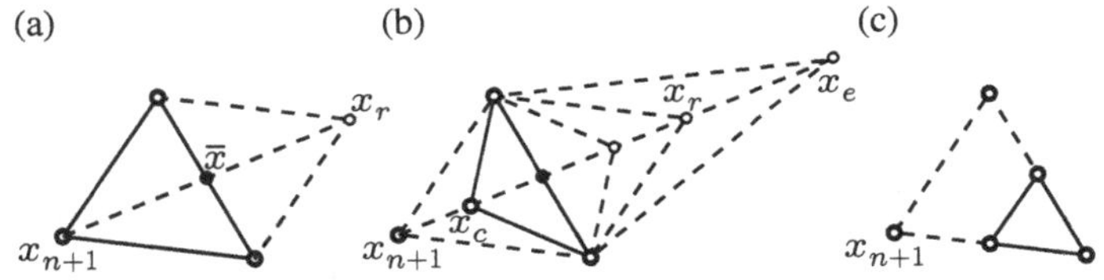


**Figure 3.4** Simplex manipulations: (a) reflection with fixed volume (area), (b) expansion or contraction along the line of reflection, and (c) reduction.

### 3.3.2 Nelder–Mead Downhill Simplex Method

The Nelder–Mead method is a downhill simplex algorithm for unconstrained optimization without using derivatives, and it was first developed by J.A. Nelder and R. Mead in 1965. This is one of the widely used traditional methods since its computational effort is relatively small and is something to get a quick grasp of the optimization problem. The basic idea of this method is to use the flexibility of the constructed simplex via amoeba-style manipulations by reflection, expansion, contraction, and reduction (see Figure 3.4). In some books such as the best known *Numerical Recipes*, it is also called the "Amoeba Algorithm." It is worth pointing out that this downhill simplex method has nothing to do with the simplex method for linear programming.

There are a few variants of the algorithm that use slightly different ways of constructing initial simplex and convergence criteria. However, the fundamental procedure is the same (see Algorithm 3.2).

The first step is to construct an initial $n$-simplex with $n + 1$ vertices and to evaluate the objective function at the vertices. Then, by ranking the objective values and re-ordering the vertices, we have an ordered set so that

$$f(\boldsymbol{x}_1) \leq f(\boldsymbol{x}_2) \leq \cdots \leq f(\boldsymbol{x}_{n+1}), \tag{3.57}$$

at $\boldsymbol{x}_1, \boldsymbol{x}_2, \ldots$, and $\boldsymbol{x}_{n+1}$, respectively. As the downhill simplex method is for minimization, we use the convention that $\boldsymbol{x}_{n+1}$ is the worse point (solution), and $\boldsymbol{x}_1$ is the best solution. Then, at each iteration, similar ranking manipulations are carried out.

**Algorithm 3.2** Pseudocode of Nelder–Mead's downhill simplex method.

Initialize a simplex with $n+1$ vertices in $n$ dimension.
**while** (stop criterion is not true)
(1) Reorder the points such that $f(\boldsymbol{x}_1) \leq f(\boldsymbol{x}_2) \leq \cdots \leq f(\boldsymbol{x}_{n+1})$
  with $\boldsymbol{x}_1$ being the best and $\boldsymbol{x}_{n+1}$ being the worse (highest value)
(2) Find the centroid $\bar{\boldsymbol{x}}$ using $\bar{\boldsymbol{x}} = \sum_{i=1}^{n} \boldsymbol{x}_i/n$ excluding $\boldsymbol{x}_{n+1}$.
(3) Generate a trial point via the reflection of the worse vertex
    $\boldsymbol{x}_r = \bar{\boldsymbol{x}} + \alpha(\bar{\boldsymbol{x}} - \boldsymbol{x}_{n+1})$ where $\alpha > 0$ (typically $\alpha = 1$)
  (a) **if** $f(\boldsymbol{x}_1) \leq f(\boldsymbol{x}_r) < f(\boldsymbol{x}_n)$, $\boldsymbol{x}_{n+1} \leftarrow \boldsymbol{x}_r$; **end**
  (b) **if** $f(\boldsymbol{x}_r) < f(\boldsymbol{x}_1)$,
      Expand in the direction of reflection $\boldsymbol{x}_e = \boldsymbol{x}_r + \beta(\boldsymbol{x}_r - \bar{\boldsymbol{x}})$
      **if** $f(\boldsymbol{x}_e) < f(\boldsymbol{x}_r)$, $\boldsymbol{x}_{n+1} \leftarrow \boldsymbol{x}_e$; **else** $\boldsymbol{x}_{n+1} \leftarrow \boldsymbol{x}_r$; **end**
    **end**
  (c) **if** $f(\boldsymbol{x}_r) > f(\boldsymbol{x}_n)$, Contract by $\boldsymbol{x}_c = \boldsymbol{x}_{n+1} + \gamma(\bar{\boldsymbol{x}} - \boldsymbol{x}_{n+1})$;
      **if** $f(\boldsymbol{x}_c) < f(\boldsymbol{x}_{n+1})$, $\boldsymbol{x}_{n+1} \leftarrow \boldsymbol{x}_c$;
      **else** Reduction by $\boldsymbol{x}_i = \boldsymbol{x}_1 + \delta(\boldsymbol{x}_i - \boldsymbol{x}_1)$, $(i = 2, \ldots, n+1)$; **end**
    **end**
**end**

Then, we have to calculate the centroid $\boldsymbol{x}$ of simplex excluding the worst vertex $\boldsymbol{x}_{n+1}$:

$$\bar{\boldsymbol{x}} = \frac{1}{n} \sum_{i=1}^{n} \boldsymbol{x}_i. \tag{3.58}$$

Using the centroid as the basis point, we try to find the reflection of the worse point $\boldsymbol{x}_{n+1}$ by

$$\boldsymbol{x}_r = \bar{\boldsymbol{x}} + \alpha(\bar{\boldsymbol{x}} - \boldsymbol{x}_{n+1}), \qquad (\alpha > 0), \tag{3.59}$$

though the typical value of $\alpha = 1$ is often used.

Whether the new trial solution is accepted or not and how to update the new vertex, depends on the objective function at $\boldsymbol{x}_r$. There are three possibilities:

1) If $f(\boldsymbol{x}_1) \leq f(\boldsymbol{x}_r) < f(\boldsymbol{x}_n)$, then replace the worst vertex $\boldsymbol{x}_{n+1}$ by $\boldsymbol{x}_r$, that is $\boldsymbol{x}_{n+1} \leftarrow \boldsymbol{x}_r$.
2) If $f(\boldsymbol{x}_r) < f(\boldsymbol{x}_1)$ which means the objective improves, we then seek a more bold move to see if we can improve the objective even further by moving or expanding the vertex further along the line of reflection to a new trial solution

$$\boldsymbol{x}_e = \boldsymbol{x}_r + \beta(\boldsymbol{x}_r - \bar{\boldsymbol{x}}), \tag{3.60}$$

where $\beta = 2$. Now we have to test if $f(\boldsymbol{x}_e)$ improves even better. If $f(\boldsymbol{x}_e) < f(\boldsymbol{x}_r)$, we accept it and update $\boldsymbol{x}_{n+1} \leftarrow \boldsymbol{x}_e$; otherwise, we can use the result of the reflection, that is, $\boldsymbol{x}_{n+1} \leftarrow \boldsymbol{x}_r$.

3) If there is no improvement or $f(\boldsymbol{x}_r) > f(\boldsymbol{x}_n)$, we have to reduce the size of the simplex while maintaining the best sides. This is contraction

$$\boldsymbol{x}_c = \boldsymbol{x}_{n+1} + \gamma(\overline{\boldsymbol{x}} - \boldsymbol{x}_{n+1}), \tag{3.61}$$

where $0 < \gamma < 1$, though $\gamma = 1/2$ is usually used. If $f(\boldsymbol{x}_c) < f(\boldsymbol{x}_{n+1})$ is true, we then update $\boldsymbol{x}_{n+1} \leftarrow \boldsymbol{x}_c$.

If all the above steps fail, we should reduce the size of the simplex towards the best vertex $\boldsymbol{x}_1$. This is the reduction step

$$\boldsymbol{x}_i = \boldsymbol{x}_1 + \delta(\boldsymbol{x}_i - \boldsymbol{x}_1), \qquad (i = 2, 3, \dots, n+1). \tag{3.62}$$

Then, we go to the first step of the iteration process and start over again.

All the methods discussed in this chapter have been implemented in all major software packages such as Matlab, R, Python, Mathematica, and others.

## Exercises

**3.1** Find the maxima and minima of $f(x) = x^3 - (15/2)x^2 + 12x + 7$ in the interval $[-10, 10]$.

**3.2** What is the global minimum of $f(x, y) = x^4 + y^2 + 2x^2y$ in $\mathbb{R}^2$. Check the obtained optimality using the definiteness of the local Hessian.

**3.3** Use any programming language or software to show that the minimum of $f(x, y) = x^4 + y^2 + 2x^2y$ is zero.

## Bibliography

Antoniou, A. and Lu, W.S. (2007). *Practical Optimization: Algorithms and Engineering Applications*. New York: Springer.

Bartholomew-Biggs, M. (2008). *Nonlinear Optimization with Engineering Applications*. New York: Springer.

Bertsekas, D.P. (2010). *Incremental Gradient, Subgradient, and Proximal Methods for Convex Optimization: A Survey*. Report LIDS-2848. Cambridge, MA: Laboratory for Information and Decision Systems, MIT.

Bertsekas, D.P., Nedic, A., and Ozdaglar, A. (2003). *Convex Analysis and Optimization*, 2e. Belmont, MA: Athena Scientific.

Bottou, L. (1998). Online algorithms and stochastic approximations. In: *Online Learning and Neural Networks*. Cambridge, UK: Cambridge University Press.

Bottou, L. (2004). Stochastic learning. In: *Advanced Lectures on Machine Learning* (ed. O. Bousquet and U. von Luxburg), Lecture Notes in Artificial Intelligence. Berlin: Springer-Verlag.

Bottou, L. (2010). Large-scale machine learning with stochastic gradient descent. In: *Proceedings of COMPSTAT'2010* (ed. Y. Lechevallier and G. Saporta), 177–186.

Boyd, S.P. and Vandenberghe, L. (2004). *Convex Optimization*. Cambridge, UK: Cambridge University Press.

Celis, M., Dennis, J.E., and Tapia, R.A. (1985). A trust region strategy for nonlinear equality constrained optimization. In: *Numerical Optimization 1994* (ed. P. Boggs, R. Byrd, and R. Schnabel), 71–82. Philadelphia: SIAM.

Fiacco, A.V. and McCormick, G.P. (1969). *Nonlinear Porgramming: Sequential Unconstrained Minimization Techniques*. New York: Wiley.

Fletcher, R. (2000). *Practical Methods of Optimization*, 2e. New York: Wiley.

Hestenes, M.R. and Stiefel, E. (1952). Methods of conjugate gradients for solving linear systems. *Journal of Research of the National Bureaus of Standards* 49 (6): 409–436.

Jeffrey, A. (2002). *Advanced Engineering Mathematics*. San Diego: Academic Press.

Kiwiel, K. (1985). *Methods of Descent for Nondifferentiable Optimization*. Berlin: Springer-Verlag.

Nelder, J.A. and Mead, R. (1965). A simplex method for function optimization. *Computer Journal* 7: 308–313.

Press, W.H., Teukolsky, S.A., Vetterling, W.T. et al. (2007). *Numerical Recipes: The Art of Scientific Computing*, 3e. Cambridge, UK: Cambridge University Press.

Saad, Y. and Schultz, M.H. (1986). GMRES: a generalized minimal residual algorithm for solving nonsymmetric linear systems. *SIAM Journal on Scientific and Statistical Computing* 7: 856–869.

Shor, N.Z. (1985). *Minimization Methods for Non-differential Functions*. Berlin: Springer-Verlag.

Yang, X.S. (2014). *Introduction to Computational Mathematics*, 2e. Singapore: World Scientific.

Yang, X.S. (2017). *Engineering Mathematics with Examples and Applications*. London: Academic Press.

# 4

# Constrained Optimization

The unconstrained optimization discussed in Chapter 3 mainly concerns the maximization and minimization of an objective function, and the search domain is usually quite regular. In reality, almost all problems have constraints, which can make the search domain quite complicated and irregular. In this case, we have to deal with constrained optimization.

## 4.1 Mathematical Formulation

Whatever the real world problem is, it is usually possible to formulate any constrained optimization problem in a general form:

$$\begin{aligned} &\underset{\boldsymbol{x}\in\mathbb{R}^n}{\text{maximize/minimize}} \quad f(\boldsymbol{x}), \quad \boldsymbol{x} = (x_1, x_2, \ldots, x_n)^{\mathrm{T}} \in \mathbb{R}^n, \\ &\text{subject to} \quad h_i(\boldsymbol{x}) = 0 \quad (i = 1, 2, \ldots, M), \\ &\qquad\qquad\quad g_j(\boldsymbol{x}) \leq 0 \quad (j = 1, \ldots, N), \end{aligned} \tag{4.1}$$

where $f(\boldsymbol{x})$, $h_i(\boldsymbol{x})$, and $g_j(\boldsymbol{x})$ are scalar functions of the real vector $\boldsymbol{x}$. There are $M$ equality constraints and $N$ inequality constraints. In general, all problem functions [$f(\boldsymbol{x})$, $h(\boldsymbol{x})$, and $g(\boldsymbol{x})$] are nonlinear, which requires sophisticated optimization techniques. In a special case when all these functions are linear, they become linear programming (LP) problem, or simple linear programs, that can be solved using a simplex method to be introduced later in this book.

Now the question is how to use the methods discussed in Chapter 3 to solve constrained optimization problems? Alternatively, can we somehow convert a constrained optimization problem into an unconstrained one by taking care of the constraints properly? These will be the main topics for this chapter.

## 4.2 Lagrange Multipliers

The method of Lagrange multipliers is a method of handling equality constraints.

If we want to minimize a function

$$\underset{\boldsymbol{x}\in\mathbb{R}^n}{\text{minimize}}\; f(\boldsymbol{x}), \qquad \boldsymbol{x} = (x_1, \dots, x_n)^{\mathrm{T}} \in \mathbb{R}^n, \tag{4.2}$$

subject to the following nonlinear equality constraint

$$h(\boldsymbol{x}) = 0, \tag{4.3}$$

then we can combine the objective function $f(\boldsymbol{x})$ with the equality to form a new function, called the Lagrangian

$$L(\boldsymbol{x}, \lambda) = f(\boldsymbol{x}) + \lambda h(\boldsymbol{x}), \tag{4.4}$$

where $\lambda$ is the Lagrange multiplier, which is an unknown scalar to be determined. This essentially converts the constrained optimization into an unconstrained problem for $L(\boldsymbol{x}, \lambda)$, which is the beauty of this method.

Now the question is what value of $\lambda$ should be used? To demonstrate the importance of $\lambda$, we now look at a simple example.

**Example 4.1** To solve the nonlinear constrained problem

$$\text{maximize} f(x, y) = 10 - x^2 - (y-2)^2,$$

$$\text{subject to } x + 2y = 5,$$

we can reformulate it using a Lagrange multiplier $\lambda$. Obviously, if there is no constraint at all, the optimal solution is simply $x = 0$ and $y = 2$. To incorporate the equality constraint, we can write

$$L(x, y, \lambda) = 10 - x^2 - (y-2)^2 + \lambda(x + 2y - 5),$$

which becomes an unconstrained optimization problem for $L(x, y, \lambda)$. Obviously, if $\lambda = 0$, then the constraint has no effect at all. If $|\lambda| \gg 1$, then we may have put more weight on the constraint. For various values of $\lambda$, we can try to maximize $L$ using the standard stationary conditions

$$\frac{\partial L}{\partial x} = -2x + \lambda = 0, \qquad \frac{\partial L}{\partial y} = -2(y-2) + 2\lambda = 0,$$

or

$$x = \frac{\lambda}{2}, \qquad y = \lambda + 2.$$

Now we have

- $\lambda = -1$, we have $x = -1/2$ and $y = 1$;
- $\lambda = 0$, we have $x = 0$ and $y = 2$ (no constraint at all);
- $\lambda = 1$, we have $x = 1/2$ and $y = 3$;
- $\lambda = 100$, we get $x = 50$ and $y = 102$.

We can see that none of these solutions can satisfy the equality $x + 2y = 5$. However, if we try $\lambda = 2/5$, we got $x = 1/5$ and $y = 12/5$ and equality constraint is indeed satisfied. Therefore, $\lambda = 2/5$ is the right value for $\lambda$.

Now the question is how to calculate $\lambda$? If we take the derivative of $L$ with respect to $\lambda$, we get

$$\frac{\partial L}{\partial \lambda} = x + 2y - 5,$$

which becomes the original constraint if we set it to zero. This means that a constrained problem can be converted into an unconstrained one if we consider $\lambda$ as an additional parameter or independent variable. Then, the standard stationary conditions apply.

This method can be extended to nonlinear optimization with multiple equality constraints. If we have $M$ equalities,

$$h_j(\boldsymbol{x}) = 0 \qquad (j = 1, \dots, M), \tag{4.5}$$

then we need $M$ Lagrange multipliers $\lambda_j (j = 1, \dots, M)$. We thus have

$$L(x, \lambda_j) = f(\boldsymbol{x}) + \sum_{j=1}^{M} \lambda_j h_j(\boldsymbol{x}). \tag{4.6}$$

The requirement of stationary conditions leads to

$$\frac{\partial L}{\partial x_i} = \frac{\partial f}{\partial x_i} + \sum_{j=1}^{M} \lambda_j \frac{\partial h_j}{\partial x_i} \quad (i = 1, \dots, n) \tag{4.7}$$

and

$$\frac{\partial L}{\partial \lambda_j} = h_j = 0 \;\; (j = 1, \dots, M). \tag{4.8}$$

These $M + n$ equations will determine the $n$ components of $\boldsymbol{x}$ and $M$ Lagrange multipliers.

From

$$\frac{\partial L}{\partial h_j} = \lambda_j, \tag{4.9}$$

we can consider $\lambda_j$ as the rates of the change of the quantity $L(\boldsymbol{x}, \lambda_j)$ as a functional of $h_j$.

**Example 4.2** To solve the optimization problem

$$\underset{(x,y)\in\mathbb{R}^2}{\text{minimize}} f(x, y) = x + y^2,$$

subject to the equality constraints

$$h_1(x, y) = x^2 + 2y^2 - 1 = 0, \qquad h_2(x, y) = x - y + 1 = 0,$$

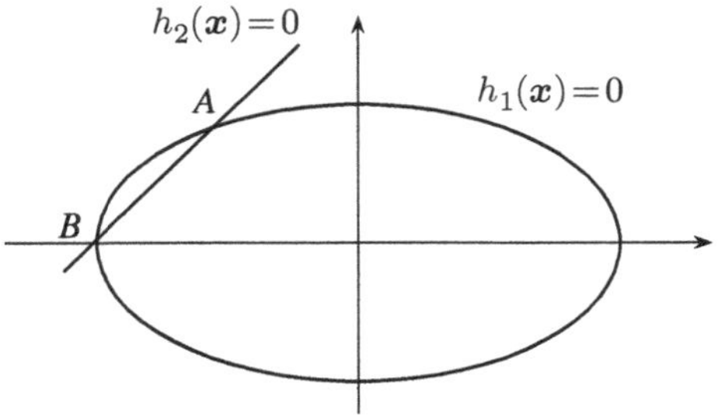


**Figure 4.1** Minimization of a function with the two equality constraints.

we can now define

$$L(x, y, \lambda_1, \lambda_2) = x + y^2 + \lambda_1(x^2 + 2y^2 - 1) + \lambda_2(x - y + 1).$$

The stationary conditions are

$$\frac{\partial L}{\partial x} = 1 + 2\lambda_1 x + \lambda_2 = 0,$$

$$\frac{\partial L}{\partial y} = 2y + 4y\lambda_1 - \lambda_2 = 0,$$

$$\frac{\partial L}{\partial \lambda_1} = x^2 + 2y^2 - 1 = 0,$$

$$\frac{\partial L}{\partial \lambda_2} = x - y + 1 = 0.$$

The first two conditions give

$$x = \frac{-(\lambda_2 + 1)}{2\lambda_1}, \qquad y = \frac{\lambda_2}{2 + 4\lambda_1}.$$

Substituting them into the last two conditions, we have

$$\frac{(\lambda_2 + 1)^2}{4\lambda_1^2} + \frac{2\lambda_2^2}{(2 + 4\lambda_1)^2} = 1, \quad \frac{(\lambda_2 + 1)}{2\lambda_1} + \frac{\lambda_2}{2 + 4\lambda_1} = 1.$$

After some straightforward algebraic manipulations, we have the solutions

$$\lambda_1 = -\frac{7}{6}, \ \lambda_2 = -\frac{16}{9}, \quad \text{or} \quad \lambda_1 = \frac{1}{2}, \ \lambda_2 = 0.$$

These correspond to two points $A(-1/3, 2/3)$ and $B(-1, 0)$. The objective function takes the value $1/9$ and $-1$ at $A$ and $B$, respectively. So the the best solution is $f_{\min} = -1$ at $x_* = -1$ and $y_* = 0$ (see point B in Figure 4.1).

We have a method to deal with equality constraints. It seems that this method cannot be directly used to cope with inequality constraints. Now let us see how inequality constraints $g_j(\boldsymbol{x}) \leq 0$ can be handled.

## 4.3 Slack Variables

A tradition method of converting an inequality such as $g(\boldsymbol{x}) \leq 0$ is to use a slack variable $s$ so that

$$g(\boldsymbol{x}) \leq 0 \tag{4.10}$$

becomes

$$g(\boldsymbol{x}) + s = 0 \quad (s \geq 0). \tag{4.11}$$

The nonnegativeness of $s$ can be achieved by setting $s = t^2$ for $t \in \mathbb{R}$ and we can also write the above equation as

$$g(\boldsymbol{x}) + t^2 = 0. \tag{4.12}$$

Obviously, if there are $m$ inequalities, we need to introduce $m$ additional slack variables. Once all the inequality constraints have been converted to equalities, we can use the standard method of Lagrange multipliers.

In essence, the introduction of slack variables is to reformulate the original optimization problem in a higher-dimensional space such that a boundary imposed by an equality becomes a hyper-surface as viewed from the additional dimension (that is $s$). For example, the inequality $x^2 + y^2 \leq 1$ is a circular domain in the two-dimensional (2D) plane; when it is changed to an equality $x^2 + y^2 + s = 1$, this circular domain becomes a cone in the three-dimensional (3D) space for $s \in [0, 1]$. Alternatively, we can view the circular domain on the 2D plane as the projection of the 3D cone onto the plane of $s = 0$.

Therefore, for $m$ inequalities with $m$ slack variables, the dimensionality of the optimization problem increases by $m$. Thus, a constrained optimization problem in an $n$-dimensional space with $m$ inequalities becomes a corresponding unconstrained problem in an $n + m$-dimensional space.

Let us look at an example.

**Example 4.3** For the objective $f(x) = x^2$, we know its minimum occurs at $x = 0$. By adding an inequality constraint $x \geq 2$, we have the following optimization problem:

$$\text{minimize } f(x) = x^2, \qquad \text{subject to } \quad x \geq 2.$$

For this special case, we know that the minimum should occur at $x = 2$ with $f_{\min} = 4$. Let us use a slack variable $s = t^2$ to convert $2 - x \leq 0$, we have

$$\text{minimize } f(x) = x^2, \qquad \text{subject to } \quad 2 - x + t^2 = 0.$$

Defining

$$L = x^2 + \lambda(2 - x + t^2),$$

we have

$$\frac{\partial L}{\partial x} = 2x - \lambda = 0, \quad \frac{\partial L}{\partial t} = 2t\lambda = 0,$$

and

$$\frac{\partial L}{\partial \lambda} = 2 - x + t^2 = 0.$$

The first condition gives $x = \lambda/2$. Substituting it into the third condition, we have

$$2 - \frac{\lambda}{2} + t^2 = 0.$$

In addition, the second condition means either $\lambda = 0$ or $t = 0$.

If $\lambda = 0$, $2 - \lambda/2 + t^2 = 0$ gives $2 + t^2 = 0$, which is impossible for any real numbers. Thus, the other case $t = 0$ must be true. This means that

$$2 - \frac{\lambda}{2} = 0,$$

which gives $\lambda = 4$ and $x = 2$. This indeed corresponds to the minimum $f_{\min} = 4$ at $x = 2$.

This example is too simple, but it demonstrates how the method works. Now let us look at an example with two decision variables.

**Example 4.4** For the minimization of $f(x, y) = x^2 + y^2$ subject to $y \geq x^2 + 1$, we have $g(x) = x^2 + 1 - y \leq 0$. Now we can use a slack variable $s = t^2$ and we get

$$x^2 + 1 - y + t^2 = 0.$$

We can write the Lagrangian as

$$L = x^2 + y^2 + \lambda(x^2 + 1 - y + t^2),$$

whose stationary conditions become

$$\frac{\partial L}{\partial x} = 2x + 2x\lambda = 0, \quad \frac{\partial L}{\partial y} = 2y - \lambda = 0,$$

$$\frac{\partial L}{\partial t} = 2t\lambda = 0, \quad \frac{\partial L}{\partial \lambda} = x^2 + 1 - y + t^2 = 0.$$

The first condition gives either $x = 0$ or $\lambda = -1$. The second condition gives $y = \lambda/2$. The third condition gives either $t = 0$ or $\lambda = 0$. The final condition gives

$$x^2 + 1 - \frac{\lambda}{2} + t^2 = 0.$$

Case A: $x = 0$ and $t = 0$. The above equation gives that $\lambda = 2$, thus $y = 1$. This corresponds to the point $x = 0$ and $y = 1$, which is the minimum with $f_{\min} = 0^2 + 1^2 = 1$.

Case B: $x = 0$ and $\lambda = 0$. This means that $1 + t^2 = 0$, which is impossible in the real domain.

Case C: $\lambda = -1$ and $t = 0$. We have $x^2 + 1 + 1/2 = 0$, which has no real solution.

Case D: $\lambda = -1$ and $\lambda = 0$, these two conditions are contradicting each other.

Therefore, the only optimal solution is Case A with $x = 0$ and $y = 1$.

We have seen that $\lambda t$ appears as a product, which means that the product of the slack variable $t$ and its corresponding Lagrange multiplier (i.e. $t\lambda$) is zero. This condition is the so-called complementary slackness.

If more inequalities are involved, the solution procedure may become tedious, but the principle is the same. Let us look at another example.

**Example 4.5** Let us solve

$$\text{minimize} f(x, y) = x^2 + y^2 + xy, \tag{4.13}$$

subject to

$$x^2 + 2y \geq 3, \quad x + y \leq 7. \tag{4.14}$$

To convert the above two inequalities into equalities, we use $s^2$ and $t^2$ as the slack variables. We thus have $3 - x^2 - 2y + s^2 = 0$ and $x + y + t^2 - 7 = 0$. So we can rewrite the above problem as

$$\text{minimize} f(x, y, s, t) = x^2 + y^2 + xy, \tag{4.15}$$

subject to

$$3 - x^2 - 2y + s^2 = 0, \quad x + y + t^2 - 7 = 0. \tag{4.16}$$

Using two Lagrange multipliers $\lambda_1$ and $\lambda_2$, we have

$$L = x^2 + y^2 + xy + \lambda_1(3 - x^2 - 2y + s^2) + \lambda_2(x + y + t^2 - 7),$$

whose stationary conditions become

$$\frac{\partial L}{\partial x} = 2x + y - 2x\lambda_1 + \lambda_2 = 0, \quad \frac{\partial L}{\partial y} = 2y + x - 2\lambda_1 + \lambda_2 = 0,$$

$$\frac{\partial L}{\partial \lambda_1} = 3 - x^2 - 2y + s^2 = 0, \quad \frac{\partial L}{\partial \lambda_2} = x + y + t^2 - 7 = 0,$$

and

$$\frac{\partial L}{\partial s} = 2\lambda_1 s = 0, \quad \frac{\partial L}{\partial t} = 2\lambda_2 t = 0.$$

The last two conditions give either $\lambda_1 = 0$ or $s = 0$ as well as either $\lambda_2 = 0$ or $t = 0$. The first two equations give

$$x = \frac{2\lambda_1 + \lambda_2}{4\lambda_1 - 3}, \quad y = \frac{(1 - 2\lambda_1)\lambda_2 + 4\lambda_1(\lambda_1 - 1)}{4\lambda_1 - 3}. \tag{4.17}$$

We have at least four cases:

Case A: $s = 0$ and $\lambda_2 = 0$. We have $x = 2\lambda_1/(4\lambda_1 - 3)$ and $y = 4\lambda_1(\lambda_1 - 1)/(4\lambda_1 - 3)$. The third condition $3 - x^2 - 2y + s^2 = 0$ means that

$$3 - \frac{(2\lambda_1)^2}{(4\lambda_1 - 3)^2} - 2\left[\frac{4\lambda_1(\lambda_1 - 1)}{4\lambda_1 - 3}\right] = 0,$$

which gives a solution $\lambda_1 = 1/2, 3/2$, and $9/8$.

For $\lambda_1 = 1/2$, we have $x = -1$ and $y = 1$. This corresponds to a minimum $f(x, y) = 1$. For $\lambda_1 = 3/2$, we have $x = 1$ and $y = 1$. The objective is $f(x, y) = 3$. For $\lambda_1 = 9/8$, we have $x = 3/2$ and $y = 3/8$, which gives $f(x, y) = 189/64$. Thus, the minimum so far is at $x = -1$ and $y = 1$ with $f = 1$.

Case B: $\lambda_1 = 0$ and $\lambda_2 = 0$, we have $x = y = 0$, but they do not satisfy $x^2 + 2y \geq 3$.

Case C: $\lambda_1 = 0$ and $t = 0$. We have $x = y = -\lambda_2/3$. From $x + y + t^2 - 7 = 0$, we have $\lambda_2 = -21/2$, which gives $x = y = 7/2$ and $f(x, y) = 147/4$.

Case D: $s = 0$ and $t = 0$. No real solutions can satisfy both $x^2 + 2y = 3$ and $x + y = 7$.

Finally, based on the above cases, the optimal minimum solution is $x = -1$ and $y = 1$, which gives $f_{\min} = 1$.

As we can see from the above examples, the products such as $\lambda_1 s = 0$ and $\lambda_2 t = 0$ appear. This is essentially the complementary slackness, and we will discuss this formally in the section about KKT conditions later in this chapter.

## 4.4 Generalized Reduced Gradient Method

For a simple equality constraint, if one of the variables can be solved analytically, then a better and more accurate approach is to solve it and reduce the dimensionality of the problem. For example, the following problem

$$\text{minimize } f(x, y) = x^2 + y^2, \qquad \text{subject to} \quad y = x^2 + 1, \tag{4.18}$$

can be converted into

$$\text{minimize } f(x) = x^2 + (x^2 + 1)^2 \quad (x \in \mathbb{R}). \tag{4.19}$$

Obviously, the minimum occurs at $x = 0$ (and thus $y = 1$) with $f_{\min} = 1$. This reduces a 2D problem into a one-dimensional problem.

**Example 4.6** As another example, from

$$\text{minimize } f(x, y, z) = x^2 + 2y^2 + 3z^2, \tag{4.20}$$

subject to

$$x + y + z = 3, \quad x - y = 1, \tag{4.21}$$

we can solve $y$ and $z$, and we have

$$y = x - 1, \quad z = 3 - x - y = 3 - x - (x - 1) = 4 - 2x. \tag{4.22}$$

Thus, we can write the above optimization problem as

$$\text{minimize} f(x) = x^2 + 2(x - 1)^2 + 3(4 - 2x)^2, \tag{4.23}$$

which is unconstrained in the domain $x \in \mathbb{R}$.

Obviously, there are more than one way to reduce the above problem, and we can also reduce the problem in terms of $y$ or $z$. Once the problem is reduced, we can solve it using any good techniques such as the gradient-based methods. This is the basic idea of the reduced gradient method.

In general, an optimization problem in an $n$-dimensional space with a nonlinear objective subject to linear equality constraints in the form $\boldsymbol{A}\boldsymbol{x} = \boldsymbol{b}$, we have

$$\text{minimize} f(\boldsymbol{x}), \tag{4.24}$$

subject to

$$\boldsymbol{A}\boldsymbol{x} = \boldsymbol{b}, \tag{4.25}$$

where $\boldsymbol{A}$ is an $m \times n$ matrix for $m$ linear equality constraints. If $\boldsymbol{A}$ has such a structure that its $p$ columns are linearly independent, we can potentially decompose $\boldsymbol{A}$ into $\boldsymbol{A} = [\boldsymbol{P}, \boldsymbol{Q}]$. At the same time, the solution $\boldsymbol{x}$ is decomposed or partitioned into $\boldsymbol{x} = (\boldsymbol{x}_p, \boldsymbol{x}_q)$. In most textbooks, $\boldsymbol{x}_p$ is called the basis variables in LP if each component is nonnegative, while $\boldsymbol{x}_q$ are called non-basis variables.

With such assumptions and decomposition, we can write $\boldsymbol{A}\boldsymbol{x} = \boldsymbol{b}$ as

$$\boldsymbol{A}\boldsymbol{x} = \boldsymbol{P}\boldsymbol{x}_p + \boldsymbol{Q}\boldsymbol{x}_q = \boldsymbol{b}. \tag{4.26}$$

If $\boldsymbol{P}$ is invertible, we have

$$\boldsymbol{P}^{-1}\boldsymbol{P}\boldsymbol{x}_p + \boldsymbol{P}^{-1}\boldsymbol{Q}\boldsymbol{x}_q = \boldsymbol{P}^{-1}\boldsymbol{b} \tag{4.27}$$

or

$$\boldsymbol{x}_p = \boldsymbol{P}^{-1}\boldsymbol{b} - \boldsymbol{P}^{-1}\boldsymbol{Q}\boldsymbol{x}_q, \tag{4.28}$$

which means that we can eliminate all the $\boldsymbol{x}_p$ and thus reduce the original problem to a lower-dimensional one

$$\text{minimize} f(\boldsymbol{x}_q), \tag{4.29}$$

which is an unconstrained optimization problem. It is worth pointing out that the objective $f(\boldsymbol{x}_q)$ will in general be different from $f(\boldsymbol{x})$ with a reduced dimensionality.

**Example 4.7** Let us revisit the previous example, we have

$$\text{minimize } f(\boldsymbol{x}) = x^2 + 2y^2 + 3z^2, \tag{4.30}$$

subject to

$$\boldsymbol{A}\begin{pmatrix} x \\ y \\ z \end{pmatrix} = \boldsymbol{b}, \quad \boldsymbol{A} = \begin{pmatrix} 1 & 1 & 1 \\ 1 & -1 & 0 \end{pmatrix}, \quad \boldsymbol{b} = \begin{pmatrix} 3 \\ 1 \end{pmatrix}. \tag{4.31}$$

Since all the columns are linearly independently, there are more than one way to decompose $\boldsymbol{A}$. For example, we can write

$$\boldsymbol{A} = \begin{pmatrix} 1 & 1 & 1 \\ 1 & -1 & 0 \end{pmatrix} = [\boldsymbol{P}, \boldsymbol{Q}], \quad \boldsymbol{P} = \begin{pmatrix} 1 & 1 \\ 1 & -1 \end{pmatrix}, \quad \boldsymbol{Q} = \begin{pmatrix} 1 \\ 0 \end{pmatrix},$$

and

$$\boldsymbol{x} = [x,\ y,\ z]^{\mathrm{T}} = [\boldsymbol{x}_p,\ \boldsymbol{x}_q]^{\mathrm{T}}, \quad \boldsymbol{x}_p = [x, y]^{\mathrm{T}}, \quad \boldsymbol{x}_q = z.$$

Since

$$\boldsymbol{P}^{-1} = \frac{1}{2}\begin{pmatrix} 1 & 1 \\ 1 & -1 \end{pmatrix},$$

we have

$$\boldsymbol{x}_p = \begin{pmatrix} x \\ y \end{pmatrix} = \boldsymbol{P}^{-1}\boldsymbol{b} - \boldsymbol{P}^{-1}\boldsymbol{Q}\boldsymbol{x}_q$$
$$= \frac{1}{2}\begin{pmatrix} 1 & 1 \\ 1 & -1 \end{pmatrix}\begin{pmatrix} 3 \\ 1 \end{pmatrix} - \frac{1}{2}\begin{pmatrix} 1 & 1 \\ 1 & -1 \end{pmatrix}\begin{pmatrix} 1 \\ 0 \end{pmatrix} z = \begin{pmatrix} 2 - \frac{z}{2} \\ 1 - \frac{z}{2} \end{pmatrix},$$

which reduces the original problem to

$$\text{minimize} f(z) = (2 - \frac{z}{2})^2 + 2(1 - \frac{z}{2})^2 + 3z^2. \tag{4.32}$$

This methodology can also be extended to incorporate inequality constraints by introducing slack variables as discussed earlier. For $k$ inequalities, we have to introduce $k$ slack variables to convert all the inequalities into their corresponding equalities. Then, we can use the reduced gradient method.

However, in many applications, the equality constraints are highly nonlinear and thus not easily solvable. In this case, some linearization using Taylor series can be a good approximation in a neighborhood around a known $\boldsymbol{x}_*$. From $h(x) = 0$, we have

$$h(\boldsymbol{x}) \approx h(\boldsymbol{x}_*) + \nabla h(\boldsymbol{x}_*)^{\mathrm{T}}(\boldsymbol{x} - \boldsymbol{x}_*). \tag{4.33}$$

Since $\boldsymbol{x}_*$ is known, both quantities $h(\boldsymbol{x}_*)$ and $\nabla h(\boldsymbol{x}_*)$ are essentially constants. Thus, the above equation is linear, so we can use the method for decomposing $\boldsymbol{A}\boldsymbol{x} = \boldsymbol{b}$ as discussed earlier. This is the essence of the generalized reduced gradient (GRG) method.

It is worth poining out that the GRG has been implemented in many software packages, including the popular Excel Solver.

In general, for the discussion of optimality conditions, we need the so-called Karush–Kuhn–Tucker (KKT) conditions.

## 4.5 KKT Conditions

Let us consider the following, generic, nonlinear optimization problem

$$\underset{\boldsymbol{x}\in\mathbb{R}^n}{\text{minimize}}\ f(\boldsymbol{x}),$$

$$\text{subject to } h_i(\boldsymbol{x}) = 0\ (i = 1, \dots, M),$$

$$g_j(\boldsymbol{x}) \leq 0\ \ (j = 1, \dots, N). \tag{4.34}$$

First, we can change all the $N$ inequality $g_j(x)$ into their corresponding equalities by introducing $N$ slack variables $s_j(j = 1, 2, \dots, N)$ and we have

$$\Psi_j(\boldsymbol{x}, s_j) = g_j(\boldsymbol{x}) + s_j^2 = 0 \quad (j = 1, 2, \dots, N). \tag{4.35}$$

Now we can define a generalized Lagrangian

$$\mathcal{L} = f(\boldsymbol{x}) + \sum_{i=1}^{M} \lambda_i h_i + \sum_{j=1}^{N} \mu_j g_j, \tag{4.36}$$

and its stationary conditions lead to

$$\nabla \mathcal{L} = \nabla f(\boldsymbol{x}) + \sum_{i=1}^{M} \lambda_i \nabla h_i(\boldsymbol{x}) + \sum_{j=1}^{N} \mu_j \nabla g_j(\boldsymbol{x}) = 0, \tag{4.37}$$

where we have used the fact that $s_j(j = 1, 2, \dots, N)$ are independent of $\boldsymbol{x}$, thus $\nabla \Psi_j = \nabla g_j(\boldsymbol{x})$. Such independence also means that

$$\frac{\partial \mathcal{L}}{\partial s_j} = 2\mu_j s_j = 0 \quad (j = 1, 2, \dots, N). \tag{4.38}$$

Loosely speaking, we can say that $\mu_j s_j = 0$ is equivalent to $\mu_j s_j^2 = 0$ if we multiply both sides by $s_j$. Thus, from Eq. (4.35), we have

$$\mu_j s_j^2 = \mu_j(0 - g_j) = -\mu_j g_j, \tag{4.39}$$

which gives

$$\mu_j g_j(\boldsymbol{x}) = 0 \quad (j = 1, 2, \ldots, N). \tag{4.40}$$

This is commonly referred to as complementary slackness or complementarity.

In addition, since the minimization of $\mathcal{L}$ should be equivalent to the minimization of $f(\boldsymbol{x})$, it is required that $\mu_j \geq 0$ $(j = 1, 2, \ldots, N)$ due to $g(\boldsymbol{x}) \leq 0$. That is, to encourage the satisfaction of all inequalities, while penalizing any violation of these conditions.

In the above discussions when we use a slack variable to convert an inequality $g(\boldsymbol{x}) \leq 0$ into an equality in a higher-dimensional space, we have used the form $g(\boldsymbol{x}) + s^2 = 0$ for ease of discussion. Obviously, we can use other forms such as $g(\boldsymbol{x}) + s = 0$ and imposing $s \geq 0$ to achieve the same results. Alternatively, we can also use $g(\boldsymbol{x}) + |s| = 0$ without $s \geq 0$. This form will lead to the same result if we use

$$\frac{\mathrm{d}|s|}{\mathrm{d}s} = \frac{s}{|s|}. \tag{4.41}$$

There is a geometrical meaning for such conversion. For example, a region of $y \geq x^2$ on the $(x, y)$ plane is extended to a 3D surface $x^2 - y + s = 0$ in the $(x, y, s)$ space. Alternatively, we can view the 2D region as the projection of the 3D surface on the plane at $s = 0$. Since there are two branches of the 3D surface in the way we wrote, for uniqueness of one-to-one project, we should impose $s \geq 0$. That is, $s \geq 0$ means $g(\boldsymbol{x}) \leq 0$, while $s \leq 0$ means that $g(\boldsymbol{x}) \geq 0$.

All the above conditions can be summarized as follows. If all the functions are continuously differentiable, at a local minimum $\boldsymbol{x}_*$, there exist constants $\lambda_1, \ldots, \lambda_M$ and $\mu_1, \ldots, \mu_N$ such that the following KKT optimality conditions hold.

Stationarity conditions:

$$\nabla f(\boldsymbol{x}_*) + \sum_{i=1}^{M} \lambda_i \nabla h_i(\boldsymbol{x}_*) + \sum_{j=1}^{N} \mu_j \nabla g_j(\boldsymbol{x}_*) = 0. \tag{4.42}$$

Primal feasibility:

$$h_i(\boldsymbol{x}_*) = 0, \quad g_j(\boldsymbol{x}_*) \leq 0 \quad (i = 1, 2, \ldots, M; \; j = 1, 2, \ldots, N). \tag{4.43}$$

Complementary slackness:

$$\mu_j g_j(\boldsymbol{x}_*) = 0 \;\; (j = 1, 2, \ldots, N). \tag{4.44}$$

Dual feasibility:

$$\mu_j \geq 0 \quad (j = 1, 2, \ldots, N). \tag{4.45}$$

The last non-negativity conditions hold for all $\mu_j$, though there is no constraint on the sign of $\lambda_i$. Thus, it is obvious that the constants satisfy the following condition:

$$\sum_{j=1}^{N} \mu_j + \sum_{i=1}^{M} |\lambda_i| \geq 0. \tag{4.46}$$

One way to look at KKT conditions is that it can be considered as a generalized method of Lagrange multipliers.

The condition $\mu_j g_j(\boldsymbol{x}_*) = 0$ in Eq. (4.44) means either $\mu_j = 0$ or $g_j(\boldsymbol{x}_*) = 0$. The later case $g_j(\boldsymbol{x}_*) = 0$ for any particular $j$ means that the inequality becomes active or tight, and thus becoming an equality. For the former case $\mu_j = 0$, the inequality for a particular $j$ holds and is not tight; however, $\mu_j = 0$ means that this corresponding inequality can be ignored. Therefore, those inequalities that are not tight are ignored, while inequalities which are tight become equalities; consequently, the constrained problem with equality and inequality constraints now essentially becomes a modified constrained problem with selected equality constraints. This is the beauty of the KKT conditions. The main issue remains to identify which inequality becomes tight, and this depends on the individual optimization problem.

The KKT conditions form the basis for mathematical analysis of nonlinear optimization problems, but the numerical implementation of these conditions is not easy, and often inefficient. From the numerical point of view, the penalty method is more straightforward to implement.

## 4.6 Penalty Method

For a nonlinear optimization problem with equality and inequality constraints, a common method of incorporating constraints is the penalty method, which essentially converts a constrained optimization problem into an unconstrained one by penalizing any violation of constraints.

For the optimization problem

$$\begin{aligned} &\underset{\boldsymbol{x}\in\mathbb{R}^n}{\text{minimize}} f(\boldsymbol{x}), \quad \boldsymbol{x} = (x_1, \dots, x_n)^{\mathrm{T}} \in \mathbb{R}^n, \\ &\text{subject to } h_i(\boldsymbol{x}) = 0 \ (i = 1, \dots, M), \\ &\qquad g_j(\boldsymbol{x}) \leq 0 \ \ (j = 1, \dots, N), \end{aligned} \tag{4.47}$$

the idea is to define a penalty function so that the constrained problem is transformed into an unconstrained problem. One commonly used penalty formulation is

$$g(\boldsymbol{x}) = f(\boldsymbol{x}) + P(\boldsymbol{x}), \tag{4.48}$$

where the penalty term $P(\boldsymbol{x})$ is defined by

$$P(\boldsymbol{x}) = \sum_{j=1}^{N} \nu_j \max\{0, g_j(\boldsymbol{x})\} + \sum_{i=1}^{M} \mu_i |h_i(\boldsymbol{x})|. \tag{4.49}$$

Here, $\mu_i > 0, \nu_j > 0$ are penalty constants or penalty factors. The advantage of this method is to transform the constrained optimization problem into an unconstrained one. That is, all the constraints are incorporated into the new objective function. However, this introduces more free parameters (or hyper-parameters) whose values need to be defined so as to solve the problem appropriately.

Obviously, there are other forms of penalty functions, which may provide smoother penalty functions. For example, we can define

$$\Pi(\boldsymbol{x}, \mu_i, \nu_j) = f(\boldsymbol{x}) + \sum_{i=1}^{M} \mu_i h_i^2(\boldsymbol{x}) + \sum_{j=1}^{N} \nu_j \max\{0, g_j(\boldsymbol{x})\}^2, \tag{4.50}$$

where $\mu_i \gg 1$ and $\nu_j \geq 0$ which should be large enough, depending on the solution quality needed.

As we can see, when an equality constraint is met, its effect or contribution to $\Pi$ is zero. However, when it is violated, it is penalized heavily as it increases $\Pi$ significantly. Similarly, there is no penalty when inequality constraints are true or become tight/exact. For the ease of numerical implementation, we can also use index functions $H$ to rewrite above penalty function as

$$\Pi = f(\boldsymbol{x}) + \sum_{i=1}^{M} \mu_i H_i[h_i(\boldsymbol{x})] h_i^2(\boldsymbol{x}) + \sum_{j=1}^{N} \nu_j H_j[g_j(\boldsymbol{x})] g_j^2(\boldsymbol{x}), \tag{4.51}$$

Here, $H_i$ and $H_j$ are index functions. More specifically, $H_i[h_i(\boldsymbol{x})] = 1$ if $h_i(\boldsymbol{x}) \neq 0$, and $H_i = 0$ if $h_i(\boldsymbol{x}) = 0$. Similarly, $H_j[g_j(\boldsymbol{x})] = 0$ if $g_j(\boldsymbol{x}) \leq 0$ is true, while $H_j = 1$ if $g_j(\boldsymbol{x}) > 0$.

In principle, the numerical accuracy depends on the values of $\mu_i$ and $\nu_j$ which should be reasonably large. But, how large is large enough? As most computers have a machine precision of $\epsilon = 2^{-52} \approx 2.2 \times 10^{-16}$, $\mu_i$ and $\nu_j$ should be close to the order of $10^{15}$. Obviously, it could cause numerical problems if they are too large.

In addition, for simplicity of implementation, we can use $\mu = \mu_i$ for all $i$ and $\nu = \nu_j$ for all $j$. That is, we can use a simplified

$$\Pi(\boldsymbol{x}, \mu, \nu) = f(\boldsymbol{x}) + \mu \sum_{i=1}^{M} H_i[h_i(\boldsymbol{x})] h_i^2(\boldsymbol{x}) + \nu \sum_{j=1}^{N} H_j[g_j(\boldsymbol{x})] g_j^2(\boldsymbol{x}).$$

It is worth pointing out that the right values of penalty factors can help to make the implementation very efficient. However, what values are appropriate may be problem-specific. If the values are too small, it may lead to under penalty, while too many large values may lead to over penalty. In general, for most applications, $\mu$ and $\nu$ can be taken as $10^3$–$10^{15}$. We will use these values in our implementation.

In addition, there is no need to fix the values of penalty parameters. In fact, the variations of such hyper-parameters with time or iterations may be advantageous. Variations such as the cooling-schedule-like reduction of these parameters have shown to work in practice. These techniques with time-dependent penalty parameters belong an active research area of dynamic penalty function methods.

## Exercises

**4.1** Solve the following constrained optimization problem:

$$\text{minimize } f(x,y) = x^2 + y^2 + 2xy, \quad (x,y) \in \mathbb{R}^2,$$

subject to

$$y - x^2 + 2 = 0.$$

**4.2** Find the maximum value of

$$f(x,y) = (|x| + |y|)\exp[-x^2 - y^2],$$

using a suitable method introduced in this chapter.

**4.3** Solve the inequality constrained problem

$$\text{minimize } f(x,y) = x^2 + 5y^2 + xy,$$

subject to

$$x^2 + y^2 \leq 1.$$

Does this inequality have any effect on the minimum solution?

## Bibliography

Boyd, S.P. and Vandenberghe, L. (2004). *Convex Optimization*. Cambridge, UK: Cambridge University Press.

Broyden, C.G. (1970). The convergence of a class of double-rank minimization algorithms. *IMA Journal of Applied Mathematics* 6: 76–90.

Conn, A.R., Gould, N.I.M., and Toint, P.L. (2000). *Trust-Region Methods*. SIAM & MPS. Philadelphia: SIAM.

Karush, W. (1939). Minima of functions of several variables with inequalities as side constraints. MSc dissertation. Department of Mathematics, University of Chicago, Chicago.

Klerk, E.D., Roos, C., and Terlaky, T. (2004). *Nonlinear Optimization.* CO367 Notes. University of Waterloo.

Kuhn, H.W. and Tucker, A.W. (1951). Nonlinear programming. In: *Proceedings of the Second Berkeley Symposium*, 481–492. Berkeley: University of California Press.

Lagrange, J.L. (1811). *Méchnique Analytique.* Paris: Ve Courcier.

Lasdon, L.S., Fox, R.L., and Watner, M.W. (1973). *Nonlinear Optimization Using the Generalized Reduced Gradient Method.* Technical Report (Tech. Memo. No. 325). Cleveland: Case Western Reserve University.

Nocedal, J. and Wright, S.J. (2006). *Numerical Optimization*, 2e. New York: Springer-Verlag.

Powell, M.J.D. (1970). A new algorithm for unconstrained optimization. In: *Nonlinear Programming* (ed. J.B. Rosen, O.L. Mangasarian, and K. Ritter), 31—65. New York: Academic Press.

Shanno, D.F. (1970). Conditioning of quasi-Newton methods for function minimization. *Mathematics of Computation* 25: 647–656.

Smith, A.E. and Coit, D.W. (1997). Constraint-handling techniques – penalty functions. In: *Handbook of Evolutionary Computation* (ed. H. Baeck, D. Fogel, and Z. Michalewicz), chapter C5.2. Bristol, UK: Joint publication of Institute of Physics Publishing and Oxford University Press.

Sundaram, R.K. (1996). Inequality constraints and the theorem of Khun and Tucker. In: *A First Course in Optimization Theory*, 145–171. New York: Cambridge University Press.

Vapnyarskii, I.B. (1994). Lagrange multipliers. In: *Encyclopedia of Mathematics* (ed. M. Hazewinkel), vol. 5. Dordrecht: Kluwer Academic.

Wright, S.J. (1997). *Primal-Dual Interior-Point Methods.* Philadelphia: Society for Industrial and Applied Mathematics.

# 5

# Optimization Techniques: Approximation Methods

Optimization techniques should be sufficiently effective in solving optimization problems, yet they should also be sufficiently simple for ease of implementation. Therefore, good approximations can be used to approximate certain quantities so as to ease computation and potentially speed up the convergence of algorithms.

## 5.1 BFGS Method

The widely used BFGS method is an abbreviation of the Broydon–Fletcher–Goldfarb–Shanno method, and it is a quasi-Newton method for solving unconstrained nonlinear optimization. It is based on the basic idea of replacing the full Hessian matrix $\boldsymbol{H}$ by an approximate matrix $\boldsymbol{B}$ in terms of an iterative updating formula with rank-one matrices as its increment.

Briefly speaking, a rank-one matrix is a matrix which can be written as $\boldsymbol{r} = \boldsymbol{a}\boldsymbol{b}^{\mathrm{T}}$ where $\boldsymbol{a}$ and $\boldsymbol{b}$ are vectors, which has at most one nonzero eigenvalue and this eigenvalue can be calculated by $\boldsymbol{b}^{\mathrm{T}}\boldsymbol{a}$. However, this does not require that $\boldsymbol{A}$ is a square matrix.

**Example 5.1** For example, matrix

$$\boldsymbol{A} = \begin{pmatrix} 4 & 5 & 6 \\ 8 & 10 & 12 \\ 12 & 15 & 18 \end{pmatrix} \tag{5.1}$$

is a rank-one matrix because its second row is the twice of the first row, and the third row is the first row multiplied by 3. Therefore, there is only one independent row, which gives a rank one. In addition, the three eigenvalues are 32, 0, 0, which means that there is only one nonzero eigenvalue. Furthermore, $\boldsymbol{A}$ can be written as

*Optimization Techniques and Applications with Examples*, First Edition. Xin-She Yang.

$$\boldsymbol{A} = \begin{pmatrix} 4 & 5 & 6 \\ 8 & 10 & 12 \\ 12 & 15 & 18 \end{pmatrix} = \begin{pmatrix} 1 \\ 2 \\ 3 \end{pmatrix} \begin{pmatrix} 4 & 5 & 6 \end{pmatrix} = \boldsymbol{a}\boldsymbol{b}^{\mathrm{T}}, \tag{5.2}$$

where

$$\boldsymbol{a} = \begin{pmatrix} 1 \\ 2 \\ 3 \end{pmatrix}, \quad \boldsymbol{b} = \begin{pmatrix} 4 \\ 5 \\ 6 \end{pmatrix}. \tag{5.3}$$

The nonzero eigenvalue of $\boldsymbol{A}$ can be obtained by

$$\boldsymbol{b}^{\mathrm{T}}\boldsymbol{a} = \begin{pmatrix} 4 & 5 & 6 \end{pmatrix} \begin{pmatrix} 1 \\ 2 \\ 3 \end{pmatrix} = 4 \times 1 + 5 \times 2 + 6 \times 3 = 32. \tag{5.4}$$

To minimize a function $f(\boldsymbol{x})$ with no constraint, the search direction $\boldsymbol{s}_k$ at each iteration is determined by

$$\boldsymbol{B}_k \boldsymbol{s}_k = -\nabla f(\boldsymbol{x}_k), \tag{5.5}$$

where $\boldsymbol{B}_k$ is the approximation to the Hessian matrix at $k$th iteration. Then, a line search is performed to find the optimal stepize $\beta_k$ so that the new trial solution is determined by

$$\boldsymbol{x}_{k+1} = \boldsymbol{x}_k + \beta_k \boldsymbol{s}_k. \tag{5.6}$$

Introducing two new variables

$$\boldsymbol{u}_k = \boldsymbol{x}_{k+1} - \boldsymbol{x}_k = \beta_k \boldsymbol{s}_k, \qquad \boldsymbol{v}_k = \nabla f(\boldsymbol{x}_{k+1}) - \nabla f(\boldsymbol{x}_k), \tag{5.7}$$

we can update the new estimate as

$$\boldsymbol{B}_{k+1} = \boldsymbol{B}_k + \frac{\boldsymbol{v}_k \boldsymbol{v}_k^{\mathrm{T}}}{\boldsymbol{v}_k^{\mathrm{T}} \boldsymbol{u}_k} - \frac{(\boldsymbol{B}_k \boldsymbol{u}_k)(\boldsymbol{B}_k \boldsymbol{u}_k)^{\mathrm{T}}}{\boldsymbol{u}_k^{\mathrm{T}} \boldsymbol{B}_k \boldsymbol{u}_k}. \tag{5.8}$$

The procedure of the BFGS method is outlined in Algorithm 5.1.

---

**Algorithm 5.1** The pseudocode of the BFGS method.

---

Choose an initial guess $\boldsymbol{x}_0$ and approximate $\boldsymbol{B}_0$ (e.g. $\boldsymbol{B}_0 = \boldsymbol{I}$)
**while** (criterion)
    Calculate $\boldsymbol{s}_k$ by solving $\boldsymbol{B}_k \boldsymbol{s}_k = -\nabla f(\boldsymbol{x}_k)$
    Find an optimal step size $\beta_k$ by a line search method
    Update $\boldsymbol{x}_{k+1} = \boldsymbol{x}_k + \beta_k \boldsymbol{s}_k$
    Calculate $\boldsymbol{u}_k$, $\boldsymbol{v}_k$ and update $\boldsymbol{B}_{k+1}$ using Eqs. (5.7) and (5.8)
**end** for while
Set $k = k + 1$

---

## 5.2 Trust-Region Method

The fundamental ideas of the trust-region method have developed over many years with many seminal papers by a dozen of pioneers. A good history review of the trust-region methods can be found in the book by Conn et al. (2000). Briefly speaking, the first important work was due to Levenberg in 1944, which proposed the usage of the addition of a multiple of the identity matrix to the Hessian matrix as a damping measure to stabilize the solution procedure for nonlinear least-squares problems. Later, Marquardt in 1963 independently pointed out the link between such damping in the Hessian and the reduction of the step size in a restricted region. Slightly later in 1966, Goldfeld et al. introduced an explicit updating formula for the maximum step size. Then, in 1970, Powell proved the global convergence for the trust-region method, though it is believed that the term "trust region" was coined by Dennis in 1978, as earlier literature used various terminologies such as the region of validity, confidence region, and restricted step method.

In the trust-region algorithm, a fundamental step is to approximate the nonlinear objective function by using truncated Taylor expansions, often the quadratic form

$$\phi_k(\boldsymbol{x}) \approx f(\boldsymbol{x}_k) + \nabla f(\boldsymbol{x}_k)^{\mathrm{T}}\boldsymbol{u} + \frac{1}{2}\boldsymbol{u}^{\mathrm{T}}\boldsymbol{H}_k\boldsymbol{u}, \tag{5.9}$$

in a so-called trust region $\Omega_k$ defined by

$$\Omega_k = \{\boldsymbol{x} \in \mathbb{R}^n \Big| \|\boldsymbol{\Gamma}(\boldsymbol{x} - \boldsymbol{x}_k)\| \leq \Delta_k\}, \tag{5.10}$$

where $\Delta_k$ is the trust-region radius. Here, $\boldsymbol{H}_k$ is the local Hessian matrix. $\boldsymbol{\Gamma}$ is a diagonal scaling matrix that is related to the scalings of the optimization problem. Thus, the shape of the trust region is a hyperellipsoid, and an elliptical region in 2D centered at $\boldsymbol{x}_k$. If the parameters are equally scaled, then $\boldsymbol{\Gamma} = \boldsymbol{I}$ can be used.

The approximation to the objective function in the trust region will make it simpler to find the next trial solution $\boldsymbol{x}_{k+1}$ from the current solution $\boldsymbol{x}_k$. Then, we intend to find $\boldsymbol{x}_{k+1}$ with a sufficient decrease in the objective function. How good the approximation $\phi_k$ is to the actual objective $f(\boldsymbol{x})$ can be measured by the ratio of the achieved decrease to the predicted decrease

$$\gamma_k = \frac{f(\boldsymbol{x}_k) - f(\boldsymbol{x}_{k+1})}{\phi_k(\boldsymbol{x}_k) - \phi_k(\boldsymbol{x}_{k+1})}. \tag{5.11}$$

If this ratio is close to unity, we have a good approximation and then should move the trust region to $\boldsymbol{x}_{k+1}$.

Now the question is what radius should we use for the newly updated trust region centered at $\boldsymbol{x}_{k+1}$? Since the move is successful, and the decrease is significant, we should be bold enough to expand the trust region a little.

**Algorithm 5.2** Pseudocode of a trust-region method.

Start at an initial guess $\boldsymbol{x}_0$ and radius $\Delta_0$ of the trust region $\Omega_0$.
Initialize algorithm constants: $0 < \alpha_1 \leq \alpha_2 < 1,\ 0 < \beta_1 \leq \beta_2 < 1$.
**while** (stop criterion)
Construct an approximate model $\phi_k(\boldsymbol{x})$ for the objective $f(\boldsymbol{x}_k)$ in $\Omega_k$.
Find a trial point $\boldsymbol{x}_{k+1}$ with a sufficient model decrease inside $\Omega_k$.
Calculate the ratio $\gamma_k$ of the achieved versus predicted decrease:
$\gamma_k = \frac{f(\boldsymbol{x}_k)-f(\boldsymbol{x}_{k+1})}{\phi_k(\boldsymbol{x}_k)-\phi_k(\boldsymbol{x}_{k+1})}$.
**if** $\gamma_k \geq \alpha_1$,
Accept the move and update the trust region: $\boldsymbol{x}_k \leftarrow \boldsymbol{x}_{k+1}$;
**if** $\gamma_k \geq \alpha_2$, $\Delta_{k+1} \in [\Delta_k, \infty)$; **end if**
**if** $\gamma_k \in [\alpha_1, \alpha_2)$, $\Delta_{k+1} \in [\beta_2\Delta_k, \Delta_k]$; **end if**
**else**
Discard the move and reduce the trust-region radius $\Delta_{k+1}$;
$\Delta_{k+1} \in [\beta_1\Delta_k, \beta_2\Delta_k]$.
**end**
Update $k = k + 1$
**end**

A standard measure of such significance in decrease is to use a parameter $\alpha_1 \approx 0.01$. If $\gamma_k > \alpha_1$, the achieved decrease is noticeable, so we should accept the move (i.e. $\boldsymbol{x}_{k+1} \leftarrow \boldsymbol{x}_k$). What radius should we now use? Conventionally, we use another parameter $\alpha_2 > \alpha_1$ as an additional criterion. If $\gamma_k$ is about $O(1)$ or $\gamma_k \geq \alpha_2 \approx 0.9$, we say that decrease is significant, and we can boldly increase the trust-region radius. Typically, we choose a value $\Delta_{k+1} \in [\Delta_k, \infty)$. The actual choice may depend on the problem, though typically $\Delta_{k+1} \approx 2\Delta_k$. If the decrease is noticeable but not so significant, that is $\alpha_1 < \gamma_k \leq \alpha_2$, we should shrink the trust region so that $\Delta_{k+1} \in [\beta_2\Delta_k, \Delta_k]$ or

$$\beta_2\Delta_k < \Delta_{k+1} < \Delta_k \qquad (0 < \beta_2 < 1). \tag{5.12}$$

Obviously, if the decrease is too small or $\gamma_k < \alpha_1$, we should abandon the move as the approximation is not good enough over this larger region. We should seek a better approximation on a smaller region by reducing the trust-region radius

$$\Delta_{k+1} \in [\beta_1\Delta_k, \beta_2\Delta_k], \tag{5.13}$$

where $0 < \beta_1 \leq \beta_2 < 1$, and typically $\beta_1 = \beta_2 = 1/2$, which means half the original size is used first. The main steps are summarized in Algorithm 5.2, and the typical values of the parameters are

$$\alpha_1 = 0.01, \quad \alpha_2 = 0.9, \qquad \beta_1 = \beta_2 = \frac{1}{2}. \tag{5.14}$$

## 5.3 Sequential Quadratic Programming

### 5.3.1 Quadratic Programming

A special type of nonlinear programming is quadratic programming (QP) whose objective function is a quadratic form

$$f(\boldsymbol{x}) = \frac{1}{2}\boldsymbol{x}^{\mathrm{T}}\boldsymbol{Q}\boldsymbol{x} + \boldsymbol{b}^{\mathrm{T}}\boldsymbol{x} + \boldsymbol{c}, \tag{5.15}$$

where $\boldsymbol{b}$ and $\boldsymbol{c}$ are constant vectors. $\boldsymbol{Q}$ is a symmetric square matrix. The constraints can be incorporated using Lagrange multipliers and Karush-Kuhn-Tucker (KKT) formulation. This is a convex optimization problem if all constraints are linear, and any optimal solution found by an algorithm is also the global optimal solution.

### 5.3.2 SQP Procedure

Sequential (or successive) quadratic programming (SQP) represents one of the state-of-the-art and most popular methods for nonlinear constrained optimization. It is also one of the robust methods. For a general nonlinear optimization problem

$$\text{minimize} f(\boldsymbol{x}), \tag{5.16}$$

$$\text{subject to } h_i(\boldsymbol{x}) = 0 \qquad (i = 1, \dots, M), \tag{5.17}$$

$$g_j(\boldsymbol{x}) \leq 0 \qquad (j = 1, \dots, N). \tag{5.18}$$

The fundamental idea of SQP is to approximate the computationally extensive full Hessian matrix using a quasi-Newton updating method. Subsequently, this generates a subproblem of QP (called QP subproblem) at each iteration, and the solution to this subproblem can be used to determine the search direction and next trial solution.

Using the Taylor expansions, the above problem can be approximated, at each iteration $k$, as the following problem:

$$\text{minimize } \frac{1}{2}\boldsymbol{s}^{\mathrm{T}}\nabla^2 L(\boldsymbol{x}_k)\boldsymbol{s} + \nabla f(\boldsymbol{x}_k)^{\mathrm{T}}\boldsymbol{s} + f(\boldsymbol{x}_k), \tag{5.19}$$

$$\text{subject to } \nabla h_i(\boldsymbol{x}_k)^{\mathrm{T}}\boldsymbol{s} + h_i(\boldsymbol{x}_k) = 0 \qquad (i = 1, \dots, M), \tag{5.20}$$

$$\nabla g_j(\boldsymbol{x}_k)^T\boldsymbol{s} + g_j(\boldsymbol{x}_k) \leq 0 \qquad (j = 1, \dots, N), \tag{5.21}$$

whose Lagrange function, also called merit function, is defined by

$$
\begin{aligned}
L(\boldsymbol{x}) &= f(\boldsymbol{x}) + \sum_{i=1}^{M} \lambda_i h_i(\boldsymbol{x}) + \sum_{j=1}^{N} \mu_j g_j(\boldsymbol{x}) \\
&= f(\boldsymbol{x}) + \boldsymbol{\lambda}^{\mathrm{T}} \boldsymbol{h}(\boldsymbol{x}) + \boldsymbol{\mu}^{\mathrm{T}} \boldsymbol{g}(\boldsymbol{x}),
\end{aligned} \tag{5.22}
$$

where $\boldsymbol{\lambda} = (\lambda_1, \dots, \lambda_M)^{\mathrm{T}}$ is the vector of Lagrange multipliers, and $\boldsymbol{\mu} = (\mu_1, \dots, \mu_N)^{\mathrm{T}}$ is the vector of KKT multipliers. Here, we have used the notations

$$
\boldsymbol{h} = [h_1(\boldsymbol{x}), \dots, h_M(\boldsymbol{x})]^{\mathrm{T}}, \quad \boldsymbol{g} = [g_1(\boldsymbol{x}), \dots, g_N(\boldsymbol{x})]^{T}. \tag{5.23}
$$

To approximate the Hessian $\nabla^2 L(\boldsymbol{x}_k)$ by a positive definite symmetric matrix $\boldsymbol{H}_k$, the standard Broydon–Fletcher–Goldfarbo–Shanno (BFGS) approximation of the Hessian can be used (Algorithm 5.3), and we have

$$
\boldsymbol{H}_{k+1} = \boldsymbol{H}_k + \frac{\boldsymbol{v}_k \boldsymbol{v}_k^{\mathrm{T}}}{\boldsymbol{v}_k^{\mathrm{T}} \boldsymbol{u}_k} - \frac{\boldsymbol{H}_k \boldsymbol{u}_k \boldsymbol{u}_k^{\mathrm{T}} \boldsymbol{H}_k^{\mathrm{T}}}{\boldsymbol{u}_k^{\mathrm{T}} \boldsymbol{H}_k \boldsymbol{u}_k}, \tag{5.24}
$$

where

$$
\boldsymbol{u}_k = \boldsymbol{x}_{k+1} - \boldsymbol{x}_k, \tag{5.25}
$$

and

$$
\boldsymbol{v}_k = \nabla L(\boldsymbol{x}_{k+1}) - \nabla L(\boldsymbol{x}_k). \tag{5.26}
$$

The QP subproblem is solved to obtain the search direction

$$
\boldsymbol{x}_{k+1} = \boldsymbol{x}_k + \alpha \boldsymbol{s}_k, \tag{5.27}
$$

using a line search method.

Alternatively, the constraints can also be incorporated by using a penalty function

$$
\Phi(\boldsymbol{x}) = f(\boldsymbol{x}) + \rho \left[ \sum_{i=1}^{M} |h_i(\boldsymbol{x})| + \sum_{j=1}^{N} \max\{0, g_j(\boldsymbol{x})\} \right], \tag{5.28}
$$

where $\rho > 0$ is the penalty parameter.

---

**Algorithm 5.3** Procedure of sequential quadratic programming.

---

Choose a starting point $\boldsymbol{x}_0$ and approximation $\boldsymbol{H}_0$ to the Hessian.
**repeat** $k = 1, 2, \dots$
  Solve a QP subproblem: $\mathrm{QP}_k$ to get the search direction $\boldsymbol{s}_k$
  Given $\boldsymbol{s}_k$, find $\alpha$ so as to determine $\boldsymbol{x}_{k+1}$
  Update the approximate Hessian $\boldsymbol{H}_{k+1}$ using the BFGS scheme
  $k = k + 1$
**until** (stop criterion)

---

It is worth pointing out that any SQP method requires a good choice of $\boldsymbol{H}_k$ as the approximate Hessian of the Lagrangian $L$. Obviously, if $\boldsymbol{H}_k$ is exactly calculated as $\nabla^2 L$, SQP essentially becomes Newton's method for solving the optimality condition. A popular way to approximate the Lagrangian Hessian is to use a quasi-Newton scheme as we used the BFGS formula described earlier.

In this chapter, we have outlined several widely used algorithms without providing any examples. The main reason is that the description of such an example may be lengthy, which also depends on the actual implementation. However, there are both commercial and open-source software packages for all these algorithms. For example, the Matlab optimization toolbox implemented all these algorithms.

## 5.4 Convex Optimization

Convex optimization is a special class of nonlinear optimization that has become a central part of engineering optimization, data mining, and machine learning (mainly via regression and data fitting). The reasons are twofold: any local optimum is also the global optimum for convex optimization, and many engineering optimization problems can be reformulated as convex optimization problems. In addition, there are rigorous theoretical basis and important theorems for convex optimization. Most algorithms such as interior-point methods are not only theoretically solid, but also computationally efficient.

From earlier chapters, we know that KKT conditions play a central role in nonlinear optimization as they are optimality conditions. For a generic nonlinear optimization

$$\begin{aligned} &\underset{\boldsymbol{x}\in\mathbb{R}^n}{\text{minimize}}\ f(\boldsymbol{x}),\\ &\text{subject to } h_i(\boldsymbol{x}) = 0 \qquad (i = 1, \dots, M),\\ &\qquad\qquad g_j(\boldsymbol{x}) \le 0 \qquad (j = 1, \dots, N), \end{aligned} \tag{5.29}$$

its corresponding KKT optimality conditions can be written as

$$\nabla f(\boldsymbol{x}_*) + \sum_{i=1}^{M} \lambda_i \nabla h_i(\boldsymbol{x}_*) + \sum_{j=1}^{N} \mu_j \nabla g_j(\boldsymbol{x}_*) = 0, \tag{5.30}$$

and

$$g_j(\boldsymbol{x}_*) \le 0, \qquad \mu_j g_j(\boldsymbol{x}_*) = 0 \qquad (j = 1, 2, \dots, N), \tag{5.31}$$

with the non-negativity condition

$$\mu_j \ge 0 \qquad (j = 0, 1, \dots, N). \tag{5.32}$$

However, there is no restriction on the sign of $\lambda_i$. Using componentwise notation for simplicity and letting $\boldsymbol{\mu} = (\mu_1, \ldots, \mu_N)^{\mathrm{T}}$, the above non-negativity condition can be written as

$$\boldsymbol{\mu} \succeq 0. \tag{5.33}$$

Though various notation conventions are used in the literature, our notations here are similar to those used in Boyd and Vandenberghe's excellent book on convex optimization. We summarize these KKT conditions in a similar form so as to make things easier if you wish to pursue further studies in these areas.

Under some special requirements, the KKT conditions can guarantee a global optimality. That is, if $f(\boldsymbol{x})$ is convex and twice-differentiable, all $g_j(\boldsymbol{x})$ are convex functions and twice-differentiable, and all $h_i(\boldsymbol{x})$ are affine, then the Hessian matrix

$$\boldsymbol{H} = \nabla^2 \mathcal{L}(\boldsymbol{x}_*) = \nabla^2 f(\boldsymbol{x}_*) + \sum_i \lambda_i \nabla^2 h_i(\boldsymbol{x}_*) + \sum_j \mu_j \nabla^2 g_j(\boldsymbol{x}_*) \tag{5.34}$$

is positive definite. We then have a convex optimization problem

$$\begin{aligned} &\underset{\boldsymbol{x} \in \mathbb{R}^n}{\text{minimize}} \; f(\boldsymbol{x}), \\ &\text{subject to } \; g_j(\boldsymbol{x}) \leq 0 \qquad (j = 1, 2, \ldots, N), \\ &\boldsymbol{A}\boldsymbol{x} = \boldsymbol{b}, \end{aligned} \tag{5.35}$$

where $\boldsymbol{A}$ and $\boldsymbol{b}$ are an $M \times M$ matrix and a column vector with $M$ elements, respectively. Then the KKT conditions are both sufficient and necessary conditions. Consequently, there exists a solution $\boldsymbol{x}_*$ that satisfies the above KKT optimality conditions, and this solution $\boldsymbol{x}_*$ is in fact the global optimum. In this special convex case, the KKT conditions become

$$\boldsymbol{A}\boldsymbol{x}_* = \boldsymbol{b}, \qquad g_j(\boldsymbol{x}_*) \leq 0 \qquad (j = 1, 2, \ldots, N), \tag{5.36}$$

$$\mu_j g_j(\boldsymbol{x}_*) = 0 \qquad (j = 1, 2, \ldots, N), \tag{5.37}$$

$$\nabla f(\boldsymbol{x}_*) + \sum_{j=1}^{N} \mu_j \nabla g_j(\boldsymbol{x}_*) + \boldsymbol{A}^{\mathrm{T}} \boldsymbol{\lambda} = 0, \tag{5.38}$$

$$\boldsymbol{\mu} \succeq 0. \tag{5.39}$$

This optimization problem (5.35) or its corresponding KKT conditions can be solved efficiently by using the interior-point methods that employ Newton's method to solve a sequence of either equality constrained problems or modified versions of the above KKT conditions. The interior-point methods are a class of methods, and the convergence of a popular version, called the barrier method, has been proved mathematically. Newton's method can also solve the equality constrained problem directly, rather than transforming it into an unconstrained one.

Convex optimization has become increasingly important in engineering, data mining, machine learning, and many other disciplines. This is largely due to the significant development in polynomial time algorithms such as the interior-point methods to be introduced later. In addition, many problems in science and engineering can be reformulated as some equivalent convex optimization problems. A vast list of detailed examples can be found in more advanced literature such as Boyd and Vandenberghe's book.

So let us first briefly introduce a few examples before we proceed to discuss Newton's method for solving equality constrained problems and the main idea of the interior-point method.

As a first example, we have to solve a linear system

$$\boldsymbol{A}\boldsymbol{u} = \boldsymbol{b}, \tag{5.40}$$

where $\boldsymbol{A}$ is an $m \times n$ matrix, $\boldsymbol{b}$ is a vector of $m$ components, and $\boldsymbol{u}$ is the unknown vector. The uniqueness of the solution requires $m \geq n$. To solve this equation, we can try to minimize the residual

$$\boldsymbol{r} = \boldsymbol{A}\boldsymbol{u} - \boldsymbol{b}, \tag{5.41}$$

which will lead to the exact solution if $\boldsymbol{r} = 0$. Therefore, this is equivalent to the following minimization problem

$$\text{minimize} \quad ||\boldsymbol{A}\boldsymbol{u} - \boldsymbol{b}||_2. \tag{5.42}$$

In the special case when $\boldsymbol{A}$ is symmetric and positive semidefinite, we can rewrite it as a convex quadratic function

$$\text{minimize} \quad \frac{1}{2}\boldsymbol{u}^{\mathrm{T}}\boldsymbol{A}\boldsymbol{u} - \boldsymbol{b}^{\mathrm{T}}\boldsymbol{u}. \tag{5.43}$$

In many applications, we often have to carry out some curve-fitting using some experimental data set. A very widely used method is the method of least squares to be introduced later in this book. For a set of $m$ observations $(x_i, y_i)$ where $i = 1, 2, \ldots, m$, we often try to best-fit the data to a polynomial function

$$p(x) = \alpha_1 + \alpha_2 x + \cdots + \alpha_n x^{n-1}, \tag{5.44}$$

by minimizing the errors or residuals

$$\boldsymbol{r} = [r_1, \ldots, r_m]^{\mathrm{T}} = [p(x_1) - y_1, \ldots, p(x_n) - y_m]^{\mathrm{T}}, \tag{5.45}$$

so as to determine the coefficients

$$\boldsymbol{\alpha} = \left(\alpha_1,\ \alpha_2,\ \ldots,\ \alpha_n\right)^{\mathrm{T}}. \tag{5.46}$$

That is to minimize the square of the $L_2$-norm

$$\text{minimize}\ ||\boldsymbol{r}||_2^2 = ||\boldsymbol{A}\boldsymbol{\alpha} - \boldsymbol{b}||_2^2 = r_1^2 + \cdots + r_m^2. \tag{5.47}$$

This is equivalent to the minimization of the following convex quadratic function:

$$f(\boldsymbol{x}) = \boldsymbol{\alpha}^{\mathrm{T}}\boldsymbol{A}^{\mathrm{T}}\boldsymbol{A}\boldsymbol{\alpha} - 2\boldsymbol{b}^{\mathrm{T}}\boldsymbol{A}\boldsymbol{\alpha} + \boldsymbol{b}^{\mathrm{T}}\boldsymbol{b}. \tag{5.48}$$

The stationary condition leads to

$$\nabla f(\boldsymbol{x}) = 2A^{\mathrm{T}}\boldsymbol{A}\boldsymbol{\alpha} - 2A^{\mathrm{T}}\boldsymbol{b} = 0, \tag{5.49}$$

which gives the standard normal equation commonly used in the method of least-squares

$$\boldsymbol{A}^{\mathrm{T}}\boldsymbol{A}\boldsymbol{\alpha} = \boldsymbol{A}^{\mathrm{T}}\boldsymbol{b}, \tag{5.50}$$

whose solution can be written as

$$\boldsymbol{\alpha} = (\boldsymbol{A}^{\mathrm{T}}\boldsymbol{A})^{-1}\boldsymbol{A}^{\mathrm{T}}\boldsymbol{b}. \tag{5.51}$$

For a given data set, the goodness of best-fit typically increases as the increase in the degree of the polynomial (or the model complexity), but this often introduces strong oscillations. Therefore, there is a fine balance between the goodness of the fit and the complexity of the mathematical model, and this balance can be achieved by using penalty or regularization. The idea is to choose the simpler model if the goodness of the best-fit is essentially the same. Here, the goodness of the fit is represented by the sum of the residual squares or the $L_2$-norm. For example, Tikhonov regularization method intends to minimize

$$\text{minimize} \quad ||\boldsymbol{A}\boldsymbol{\alpha} - \boldsymbol{b}||_2^2 + \gamma||\boldsymbol{\alpha}||_2^2, \qquad \gamma \in (0, \infty), \tag{5.52}$$

where the first term of the objective is the standard least squares, while the second term intends to penalize the more complex models. Here, $\gamma$ is called the penalty or regularization parameter. This problem is equivalent to the following convex quadratic optimization:

$$\text{minimize} \quad f(\boldsymbol{x}) = \boldsymbol{\alpha}^{\mathrm{T}}(\boldsymbol{A}^{\mathrm{T}}\boldsymbol{A} + \gamma\boldsymbol{I})\boldsymbol{\alpha} - 2\boldsymbol{b}^{\mathrm{T}}\boldsymbol{A}\boldsymbol{\alpha} + \boldsymbol{b}^{\mathrm{T}}\boldsymbol{b}, \tag{5.53}$$

where $\boldsymbol{I}$ is the identity matrix. The optimality condition requires that

$$\nabla f(\boldsymbol{x}) = 2(A^{\mathrm{T}}\boldsymbol{A} + \gamma\boldsymbol{I})\boldsymbol{\alpha} - 2A^{\mathrm{T}}\boldsymbol{b} = 0, \tag{5.54}$$

which leads to

$$(\boldsymbol{A}^{\mathrm{T}}\boldsymbol{A} + \gamma\boldsymbol{I})\boldsymbol{\alpha} = \boldsymbol{A}^{\mathrm{T}}\boldsymbol{b}, \tag{5.55}$$

whose solution is

$$\boldsymbol{\alpha} = (\boldsymbol{A}^{\mathrm{T}}\boldsymbol{A} + \gamma\boldsymbol{I})^{-1}\boldsymbol{A}^{\mathrm{T}}\boldsymbol{b}. \tag{5.56}$$

It is worth pointing out that the inverse does exist due to the fact that $\boldsymbol{A}^{\mathrm{T}}\boldsymbol{A} + \gamma\boldsymbol{I}$ is full rank for all $\gamma > 0$.

As a final example, the minimax or Chebyshev optimization uses the $L_\infty$-norm and intends to minimize

$$\text{minimize} \quad \|\boldsymbol{A\alpha} - \boldsymbol{b}\|_\infty = \max\{|r_1|, \dots, |r_m|\}. \tag{5.57}$$

Here, the name is clearly from the minimization of the maximum absolute value of the residuals.

## 5.5 Equality Constrained Optimization

For an equality constrained convex optimization problem, Newton's method is very efficient. As we have seen earlier in this book, inequalities can be converted into equalities by using slack variables. Thus, for simplicity, we now discuss the case when there is no inequality constraint at all (or $N = 0$) in Eq. (5.35), and we have the following equality constrained problem:

$$\begin{aligned} &\text{minimize} \quad f(\boldsymbol{x}) \\ &\text{subject to} \quad \boldsymbol{Ax} = \boldsymbol{b}, \end{aligned} \tag{5.58}$$

where the objective function $f(\boldsymbol{x})$ is convex, continuous, and twice-differentiable. We often assume that there are fewer equality constraints than the number of design variables; that is, it is required that the $M \times n$ matrix $\boldsymbol{A}$ has a rank $M$ (i.e. rank($\boldsymbol{A}$)=$M < n$).

The KKT conditions (5.36)–(5.39) for this problem simply become

$$\boldsymbol{Ax}_* = \boldsymbol{b}, \qquad \nabla f(\boldsymbol{x}_*) + \boldsymbol{A}^{\mathrm{T}}\boldsymbol{\lambda} = 0, \tag{5.59}$$

which forms a set of $M + n$ equations for $M + n$ variables $\boldsymbol{x}_*$ ($n$ variables) and $\boldsymbol{\lambda}$ ($M$ variables). We can solve this set of KKT conditions. In this case, the pair $(\boldsymbol{x}_*, \boldsymbol{\lambda})$ is often called the primal-dual pair. By solving the above KKT conditions, we can obtain a unique set of $(\boldsymbol{x}_*, \boldsymbol{\lambda}_*)$, and $\boldsymbol{\lambda}_*$ is referred to as the optimal dual variable for this problem.

However, instead of solving the KKT conditions, it is often more desirable to solve the original problem (5.58) directly, especially for the case when the matrix $\boldsymbol{A}$ is sparse. This is because the sparsity is often destroyed by converting the equality constrained problem into an unconstrained one in terms of, for example, the Lagrange multipliers.

Newton's method often provides the superior convergence over many other algorithms; however, it uses second derivative information directly. The move or descent direction is determined by the first derivative, and descent step or Newton step $\Delta\boldsymbol{x}_k$ has to be determined by solving an approximate quadratic equation.

If $f(\boldsymbol{x})$ is not convex, we can still approximate it by a local quadratic form from a known, current solution $\boldsymbol{x}_k$ by expanding it with respect to $\Delta\boldsymbol{x}_k = \boldsymbol{x} - \boldsymbol{x}_k$, and we have

$$f(\boldsymbol{x}) = f(\boldsymbol{x}_k) + (\boldsymbol{G})^{\mathrm{T}} \Delta\boldsymbol{x}_k + \frac{1}{2}(\Delta\boldsymbol{x}_k)^{\mathrm{T}} \boldsymbol{H}\,(\Delta\boldsymbol{x}_k) + \cdots, \tag{5.60}$$

where $\boldsymbol{G} = \nabla f(\boldsymbol{x}_k)$ and $\boldsymbol{H} = \nabla^2 f(\boldsymbol{x}_k)$. The equality constraint becomes

$$\boldsymbol{A}(\boldsymbol{x}_k + \Delta\boldsymbol{x}_k) = \boldsymbol{A}\boldsymbol{x}_k + \boldsymbol{A}\Delta\boldsymbol{x}_k = \boldsymbol{b}. \tag{5.61}$$

With the approximation $\boldsymbol{A}\boldsymbol{x}_k \approx \boldsymbol{b}$, the above condition becomes

$$\boldsymbol{A}(\Delta\boldsymbol{x}_k) = 0. \tag{5.62}$$

Thus, the original optimization problem becomes

$$\text{minimize} \quad f(\boldsymbol{x}_k) + \boldsymbol{G}^{\mathrm{T}} \Delta\boldsymbol{x}_k + \frac{1}{2}(\Delta\boldsymbol{x}_k)^{\mathrm{T}} \boldsymbol{H}\,(\Delta\boldsymbol{x}_k), \tag{5.63}$$

subject to

$$\boldsymbol{A}(\Delta\boldsymbol{x}_k) = 0. \tag{5.64}$$

Its KKT conditions are

$$\boldsymbol{A}\,\Delta\boldsymbol{x}_k = 0, \quad \boldsymbol{H}(\Delta\boldsymbol{x}_k) + \boldsymbol{G} + \boldsymbol{A}^{\mathrm{T}}\boldsymbol{\lambda} = 0, \tag{5.65}$$

which can be written as a compact matrix form

$$\begin{pmatrix} \boldsymbol{H} & \boldsymbol{A}^{\mathrm{T}} \\ \boldsymbol{A} & 0 \end{pmatrix} = \begin{pmatrix} \Delta\boldsymbol{x}_k \\ \boldsymbol{\lambda} \end{pmatrix} = \begin{pmatrix} -\boldsymbol{G} \\ 0 \end{pmatrix}, \tag{5.66}$$

or more explicitly as

$$\begin{pmatrix} \nabla^2 f(\boldsymbol{x}_k) & \boldsymbol{A}^{\mathrm{T}} \\ \boldsymbol{A} & 0 \end{pmatrix} \begin{pmatrix} \Delta\boldsymbol{x}_k \\ \boldsymbol{\lambda} \end{pmatrix} = \begin{pmatrix} -\nabla f(\boldsymbol{x}_k) \\ 0 \end{pmatrix}, \tag{5.67}$$

where $\boldsymbol{\lambda}$ is an optimal dual variable associated with the approximate quadratic optimization problem.

Similar to an unconstrained minimization problem, the Newton step can be determined by

$$\Delta\boldsymbol{x}_k = -\frac{\nabla f(\boldsymbol{x}_k)}{\nabla^2 f(\boldsymbol{x}_k)} = -\frac{\boldsymbol{G}}{\boldsymbol{H}} = -[\nabla^2 f(\boldsymbol{x}_k)]^{-1} \nabla f(\boldsymbol{x}_k). \tag{5.68}$$

However, for the equality constraint, the feasible descent direction has to be modified slightly. A direction $\boldsymbol{s}$ is feasible if $\boldsymbol{A}\boldsymbol{s} = 0$. Consequently, $\boldsymbol{x} + \alpha\boldsymbol{s}$ where $\alpha > 0$ is also feasible under the constraint $\boldsymbol{A}\boldsymbol{x} = \boldsymbol{b}$, and $\alpha$ can be obtained using a line search method. Boyd and Vandenberghe have proved that a good estimate of $f(\boldsymbol{x}) - f_*$ where $f_*$ is the optimal value at $\boldsymbol{x} = \boldsymbol{x}_*$, and thus a stopping criterion, is $\theta^2(\boldsymbol{x})/2$ with $\theta(\boldsymbol{x})$ being given by

$$\theta(\boldsymbol{x}) = [\Delta\boldsymbol{x}_k^{\mathrm{T}} \boldsymbol{H} \Delta\boldsymbol{x}_k]^{1/2}. \tag{5.69}$$

Finally, we can summarize Newton's method for the equality constrained optimization as shown in Algorithm 5.4.

**Algorithm 5.4** Newton's method for the equality constrained optimization.

Initial guess $\boldsymbol{x}_0$ and accuracy $\epsilon$
**while** $(\theta^2/2 > \epsilon)$
  Calculate $\Delta\boldsymbol{x}_k$ using Eq. (5.68)
  Using a line search method to determine $\alpha$
  Update $\boldsymbol{x}_{k+1} = \boldsymbol{x}_k + \alpha\Delta\boldsymbol{x}_k$
  Estimate $\theta$ using Eq. (5.69)
**end**

Now for optimization with inequality and equality constraints, if we can somehow remove the inequality constraints by incorporating them in the objective, then the original nonlinear convex optimization problem (5.35) can be converted into a convex one with equality constraints only. Subsequently, we can use the above Newton algorithm to solve it. This conversion typically requires some sort of barrier functions.

## 5.6 Barrier Functions

In optimization, a barrier function is essentially a continuous function that becomes singular or unbounded when approaching the boundary of the feasible region. Such singular behavior, often difficult to deal with in many mathematical problems, can be used as an advantage to impose constraints by significantly penalizing any potential violation of such constraints.

In order to apply standard optimization techniques such as Newton's method for equality constrained optimization, we often intend to reformulate the constrained optimization with inequality constraints by directly writing the inequality constraints in the objective function. One way to achieve this is to use the indicator function $I_-$ which is defined by

$$I_-[u] = \begin{cases} 0 & \text{if } u \leq 0, \\ \infty & \text{if } u > 0. \end{cases} \tag{5.70}$$

Now the convex optimization problem (5.35) can be rewritten as

$$\begin{cases} \text{minimize } f(\boldsymbol{x}) + \sum_{i=1}^{N} I_-[g_i(\boldsymbol{x})] \\ \\ \text{subject to } \boldsymbol{A}\boldsymbol{x} = \boldsymbol{b}. \end{cases} \tag{5.71}$$

It seems that the problem is now solvable, but the indicator has singularity and is not differentiable. Consequently, the objective function is not differentiable, in general, and many derivative-based methods including Newton's method cannot be used for such optimization problems.

The main difficulty we have to overcome is to use a continuous function with similar property as the indicator function when approaching the boundary.

That is why barrier functions serve this purpose well. A good approximation to an indicator function should be smooth enough, and we often use the following logarithm function

$$\bar{I}_{-}(u) = -\frac{1}{t}\log(-u) \qquad (u < 0), \tag{5.72}$$

where $t > 0$ is an accuracy parameter for the approximation. Indeed, the function $\bar{I}_{-}$ is convex, non-decreasing, and differentiable. Now we can rewrite the above convex optimization as

$$\text{minimize} f(\boldsymbol{x}) + \sum_{i=1}^{N} -\frac{1}{t}\log[-g_i(\boldsymbol{x})], \tag{5.73}$$

$$\text{subject to} \quad \boldsymbol{A}\boldsymbol{x} = \boldsymbol{b}, \tag{5.74}$$

which can be solved using Newton's method. Here, the function

$$\psi(\boldsymbol{x}) = -\sum_{i=1}^{N}\log[-g_i(\boldsymbol{x})], \tag{5.75}$$

is called the logarithmic barrier function for this optimization problem, or simply the log barrier (see Figure 5.1). Its gradient can be expressed as

$$\nabla\psi(\boldsymbol{x}) = \sum_{i=1}^{N}\frac{1}{-g_i(\boldsymbol{x})}\nabla g_i(\boldsymbol{x}), \tag{5.76}$$

and its Hessian is

$$\nabla^2\psi(\boldsymbol{x}) = \sum_{i=1}^{N}\frac{1}{[g_i(\boldsymbol{x})]^2}\nabla g_i(\boldsymbol{x})\nabla g_i(\boldsymbol{x})^{\mathrm{T}} + \sum_{i=1}^{N}\frac{1}{-g_i(\boldsymbol{x})}\nabla^2 g_i(\boldsymbol{x}). \tag{5.77}$$

These are frequently used in the optimization literature. It is worth pointing out that $-g_i(\boldsymbol{x})$ is always nonnegative due to the fact that $g_i(\boldsymbol{x}) \leq 0$. In addition, since $g_i(\boldsymbol{x})$ are convex, and $-\log()$ is convex, from the composition rule of convex functions, it is straightforward to check that $\psi(\boldsymbol{x})$ is indeed convex, and also twice continuously differentiable.

It is worth pointing out that this formulation is useful in implementation. From a purely mathematical point of view, we can use $\mu = 1/t > 0$ in the above formulation. Thus, we have

$$\text{minimize} \quad f(\boldsymbol{x}) - \mu\sum_{i=1}^{N}\log[-g_i(\boldsymbol{x})], \tag{5.78}$$

subject to

$$\boldsymbol{A}\boldsymbol{x} = \boldsymbol{b}. \tag{5.79}$$

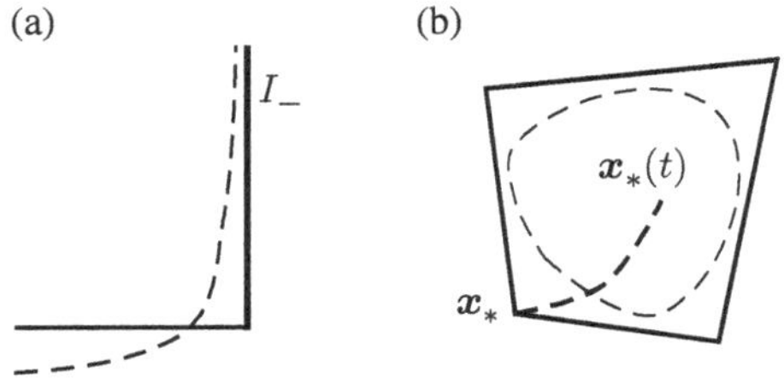


**Figure 5.1** The barrier method: (a) log barrier near a boundary and (b) central path for $n = 2$ and $N = 4$.

The optimal solution should obtained when $\mu \to 0$ (or $t \to \infty$). Now let us see a simple example.

**Example 5.2** Let us solve the minimization of $f(x, y) = x^2 + y^2$ subject to $x + y \geq 2$.

We can write the inequality $x + y \geq 2$ as a barrier function, and we have

$$\phi(x, y) = f(x, y) - \mu \log(x + y - 2) = (x^2 + y^2) - \mu \log(x + y - 2).$$

Its gradient becomes

$$\frac{\partial \phi}{\partial x} = 2x - \frac{\mu}{x + y - 2} = 0,$$

$$\frac{\partial \phi}{\partial y} = 2y - \frac{\mu}{x + y - 2} = 0.$$

Due to the symmetry, it is obvious that $x = y$ from the above equation. Thus, we have

$$2x - \frac{\mu}{2x - 2} = 0,$$

which gives

$$4x^2 - 4x - \mu = 0.$$

Its solution is

$$x_* = y_* = \frac{4 \pm 4\sqrt{1 + \mu}}{8} = \frac{1 \pm \sqrt{1 + \mu}}{2}.$$

In the limit of $\mu \to 0$, we have

$$x_* = y_* = 1 \quad \text{or} \quad x_* = y_* = 0.$$

The solution at $x_* = y_* = 1$ is the true optimal solution, while $x_* = y_* = 0$ is not a feasible solution because it does not satisfy the constraint $x + y \geq 2$. However, this point corresponds to the optimal solution without the constraint.

Using this log barrier and $t$ as a parameter, we can write the convex optimization problem (5.71) as

$$\begin{aligned} &\text{minimize } f(\boldsymbol{x}) + \tfrac{1}{t}\psi(\boldsymbol{x}) \\ &\text{subject to } \boldsymbol{A}\boldsymbol{x} = \boldsymbol{b}. \end{aligned} \tag{5.80}$$

From the above equations, we can see that $t \to \infty$, the modified objective approaches the original objective as the second term becomes smaller and smaller, as the log barrier is an approximation whose accuracy is controlled by a parameter $t$. As $t$ increases (or $\mu = 1/t$ decreases), the approximation improves; however, the Hessian varies more dramatically near the boundary, and thus makes it difficult to use Newton's method. Now the question is: how large $t$ should be enough? This is equivalent to asking: how small $\mu$ is small enough?

To get around this problem, we can use a sequence of problems (5.80) by increasing the values of $t$ gradually, and starting the optimization from the previous solution at the previous value of $t$. As $t > 0$, the multiplication of the objective by $t$ will not affect the optimality of the problem. So we can rewrite the above problem as

$$\begin{aligned} &\text{minimize } tf(\boldsymbol{x}) + \psi(\boldsymbol{x}) \\ &\text{subject to } \boldsymbol{A}\boldsymbol{x} = \boldsymbol{b}. \end{aligned} \tag{5.81}$$

For any given $t$, there should be an optimal solution $\boldsymbol{x}_*$ to this problem. Due to the convexity of the problem (if $f(\boldsymbol{x})$ is convex), we can assume that this solution is unique for a specific value of $t$. As a result, the solution $\boldsymbol{x}_*$ is a function of $t$, that is $\boldsymbol{x}_*(t)$. In essence, the search is carried out near the boundaries, inside the feasible region. Here, for convex optimization (5.35), the solution $\boldsymbol{x}_*(t)$ corresponds to a point, usually an interior point, and the set of all the points $\boldsymbol{x}_*(t)$ for $t > 0$ will trace out a path (see Figure 5.1), called central path associated with the original problem (5.35). In Figure 5.1, the central path is sketched for the case of $n = 2$ (two variables) and $N = 4$ (four inequality constraints).

The point $\boldsymbol{x}_*(t)$ is strictly feasible and satisfies the following condition:

$$\boldsymbol{A}\boldsymbol{x}_*(t) = \boldsymbol{b}, \qquad g_i[\boldsymbol{x}_*(t)] < 0, \tag{5.82}$$

for $i = 1, \dots, N$. It is worth pointing out that we cannot include the equal sign in the inequality $g_i \leq 0$ because $\psi \to \infty$ as $g_i = 0$. This means that the inequalities hold strictly, and consequently the points are interior points. Thus, the name for the interior-point methods.

Applying the optimality condition for the equality constrained problem (5.81) in the same manner as Eq. (5.59), we have

$$t\nabla f[\boldsymbol{x}_*(t)] + \nabla\psi[\boldsymbol{x}_*(t)] + \boldsymbol{A}^{\mathrm{T}}\boldsymbol{\lambda} = 0 \tag{5.83}$$

or

$$t\nabla f[\boldsymbol{x}_*(t)] + \sum_{i=1}^{N} \frac{1}{-g_i[\boldsymbol{x}_*(t)]} \nabla g_i[\boldsymbol{x}_*(t)] + \boldsymbol{A}^{\mathrm{T}}\boldsymbol{\lambda} = 0. \tag{5.84}$$

The central path has an interesting property, that is, every central point will yield a dual feasible point with a lower bound on the optimal value $f_*$. For any given tolerance or accuracy, the optimal value of $t$ can be determined by

$$t = \frac{N}{\epsilon}. \tag{5.85}$$

For an excellent discussion concerning these concepts, readers can refer to Boyd and Vanderberghe's book. Now let us introduce and summarize the barrier-based interior-point method.

## 5.7 Interior-Point Methods

The interior-point methods are a class of algorithms using the idea of central paths, and the huge success of the methods are due to the fact that the algorithm complexity is polynomial. Various names have been used. For example, when the first version of the methods was pioneered by Fiacco and McCormick in the 1960s, the term sequential unconstrained minimization techniques were used. In modern literature, we tend to use the barrier method, path-following method, or interior-point method. A significant step in the development of interior-point methods was the seminal work by Karmarkar in 1984 outlining the new polynomial-time algorithm for linear programming. The convergence of the interior-point polynomial methods was proved by Nesterov and Nemirovskii for a wider class of convex optimization problems in the 1990s. Here, we will introduce the simplest well-known barrier method.

**Algorithm 5.5** The procedure of the barrier method for convex optimization.

Start a guess in the strictly feasible region $\boldsymbol{x}_0$
Initialize $t_0 > 0$, $\beta > 1$, and $\epsilon > 0$
**while** $(N/t > \epsilon)$
  **Inner loop:**
    Start at $\boldsymbol{x}_k$
    Solve min $tf(\boldsymbol{x}_k) + \psi(\boldsymbol{x}_k)$ subject to $\boldsymbol{A}\boldsymbol{x}_k = \boldsymbol{b}$ to get $\boldsymbol{x}_*(t)$
  **end loop**
  Update $\boldsymbol{x}_k \leftarrow \boldsymbol{x}_*(t)$
  $t = \beta t$
**end**

The basic procedure of the barrier methods consists of two loops: an inner loop and an outer loop as shown in Algorithm 5.5. The method starts with a guess which is strictly inside the feasible region at $t = t_0$. For a given accuracy or tolerance $\epsilon$ and a parameter $\beta > 1$ for increasing $t$, a sequence of optimization problems are solved by increasing $t$ gradually. At each step, the main task of the inner loop tries to find the best solution $\boldsymbol{x}_*(t)$ to the problem (5.81) for a given $t$, while the outer loop updates the solutions and increases $t$ sequentially. The whole iteration process stops once the prescribed accuracy is reached.

Now the question is what value of $\beta$ should be used. If $\beta$ is too large, then the change of $t$ at each step in the outer loop is large, then the initial $\boldsymbol{x}_k$ is far from the actual optimal solution for a given $t$. It takes more steps to find the solution $\boldsymbol{x}_*(t)$. On the other hand, if $\beta$ is near 1, then the change in $t$ is small, and the initial solution $\boldsymbol{x}_k$ for each value of $t$ is good, so a smaller number of steps is needed to find $\boldsymbol{x}_*(t)$. However, as the increment in $t$ is small, to reach the same accuracy, more jumps or increments in $t$ in the outer loop are needed. Therefore, it seems that a fine balance is needed in choosing $\beta$ such that the overall number of steps are minimal. This itself is an optimization problem. In practice, any value $\beta = 3\text{–}100$ are acceptable, though $\beta = 10\text{–}20$ works well for most applications as suggested in Boyd and Vandenberghe's book.

One of the important results concerning the barrier method is the convergence estimate, and it has been proved that for given $\beta$, $\epsilon$, and $N$, the number of steps for the outer loop required is exactly

$$K = \left\lceil \frac{\log(\frac{N}{t_0\epsilon})}{\log \beta} \right\rceil, \tag{5.86}$$

where $\lceil k \rceil$ means to take the next integer greater than $k$. For example, for $N = 500$, $\epsilon = 10^{-9}$, $t_0 = 1$, and $\beta = 10$, we have

$$K = \left\lceil \frac{\log(500/10^{-9})}{\log 10} \right\rceil = 12. \tag{5.87}$$

This is indeed extremely efficient.

It is worth pointing out that the interior-point method requires the convexity of $f(\boldsymbol{x})$ and $-g_j(\boldsymbol{x})$. Otherwise, some modifications and approximations are needed. For example, we can use the local approximations of the objective function and constraints as we discussed earlier. In addition, the initial step of finding a feasible interior solution is not always straightforward.

There are many other methods concerning the interior-point methods, including geometric programming, phase I method, phase II method, and primal-dual interior-point methods. Interested readers can refer to more advanced literature at the end of this chapter.

## 5.8 Stochastic and Robust Optimization

The optimization problems we have discussed so far implicitly assume that the parameters of objective or constraints are exact and deterministic; however, in reality there is always some uncertainty. In engineering optimization, this is often the case. For example, material properties always have certain degree of inhomogeneity and stochastic components. The objective and the optimal solution may depend on some uncertainty, and they are often very sensitive to such variations and uncertainty. For example, the optimal solution at $B$ in Figure 5.2 is very sensitive to uncertainty. If there is any small change in the solution, the global minimum is no longer the global minimum as the perturbed solution may be not so good as the solution at $A$ where the solution is more stable. We say the optimal solution at $B$ is not robust. In practical applications, we are trying to find not only the best solution but also the most robust solution. Because, only robust solutions can have realistic applications. In this present case, we usually seek good-quality solutions around $A$, rather than the non-robust optimal solution at $B$.

In this case, we are dealing with stochastic optimization. In general, we can write a stochastic optimization problem as

$$\text{minimize } f(\boldsymbol{x}, \xi), \tag{5.88}$$

$$\text{subject to} \quad G(\boldsymbol{x}, \xi) \in \Omega, \tag{5.89}$$

where $G(\boldsymbol{x}, \xi)$ consists of all the constraints in the well-structure subset $\Omega$. Here $\xi$ is a random variable with a certain, probably unknown, probability distribution.

There are three ways to approach this kind of optimization problem: sensitivity analysis, stochastic programming/optimization, and robust optimization. Sensitivity analysis intends to solve this optimization problem with a fixed choice of $\xi$, and then analyze the sensitivity of the solution to the uncertainty by sensitivity techniques such as Monte Carlo methods. It is relatively simple to implement, but the solution may not be optimal or near the true range.

Stochastic programming tries to use the information of the probability density function for $\xi$ while focusing the mean of the objective. If we know the

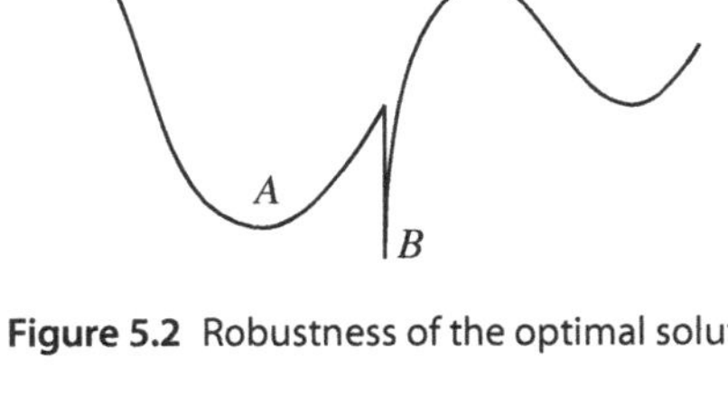


**Figure 5.2** Robustness of the optimal solution.

distribution of the random variable $\xi$ such as the Guassian or normal distribution $N(\mu, \sigma^2)$ with a mean of $\mu$ and variance of $\sigma^2$. We can then try to solve

$$\text{minimize} \quad E_\xi[f(\boldsymbol{x}, \xi)], \tag{5.90}$$

$$\text{subject to} \quad E_\xi[G(\boldsymbol{x}, \xi)] \in \Omega, \tag{5.91}$$

where $E_\xi[f]$ is the expectation of the function $f$ averaging over $\xi$. Typically, this expectation involves multidimensional integrals which are analytically intractable, and we have to use Monte Carlo integration techniques. It is worth pointing out that we have to modify the objective even more in some cases to include the uncertainty explicitly so that we have $\min E_\xi[f(\boldsymbol{x}, \xi)] + \lambda\sigma$ where $\lambda > 0$ is a parameter. The advantage of stochastic programming is that it has taken accounting the probability information into consideration; however, the implementation is often difficult, especially if there are highly nonlinear constraints.

A good alternative to tackle this problem is to use the robust optimization where it tries to study the optimization problem with uncertainty inputs and intends to seek good solutions with reasonably good performance. Even so, most problems are not tractable and some approximations are needed. However, under some strict conditions such as $\Omega$ is a well-structured convex cone $\boldsymbol{K}$, and $f(\boldsymbol{x}, .)$ is convex with respect to $\boldsymbol{x}$, this problem is largely tractable and this is the main topic of robust conic optimization. If we further assume that $G(\boldsymbol{x})$ does not depend on $\xi$ (no uncertainty in constraints), then the problem is generally solvable. In the rest of the section, we will introduce briefly the stochastic robust least squares as it is very important in engineering applications.

Now we try to solve

$$\text{minimize} \quad ||\boldsymbol{A}\boldsymbol{u} - \boldsymbol{b}||_2^2, \tag{5.92}$$

where $\boldsymbol{A}(\xi)$ depends on the random variable $\xi$ with some uncertainty such that

$$\boldsymbol{A}(\xi) = \overline{\boldsymbol{A}} + \boldsymbol{\epsilon}. \tag{5.93}$$

Here, the mean $\overline{\boldsymbol{A}}$ is fixed, and uncertainty or random matrix $\boldsymbol{\epsilon}$ has a zero mean. That is

$$E_\xi[\boldsymbol{\epsilon}] = 0. \tag{5.94}$$

The stochastic robust least-squares problem becomes the minimization of the expectation

$$\text{minimize} \quad E_\xi||\boldsymbol{A}\boldsymbol{u} - \boldsymbol{b}||_2^2. \tag{5.95}$$

From the basic definition of mean, we know that

$$\begin{aligned} E_\xi||\boldsymbol{A}\boldsymbol{u} - \boldsymbol{b}||_2^2 &= E_\xi||(\overline{\boldsymbol{A}} + \boldsymbol{\epsilon})\boldsymbol{u} - \boldsymbol{b}||_2^2 = E_\xi||(\overline{\boldsymbol{A}}\boldsymbol{u} - \boldsymbol{b}) + \boldsymbol{\epsilon}\boldsymbol{u}||_2^2 \\ &= ||\overline{\boldsymbol{A}} - \boldsymbol{b}||_2^2 + E_\xi[\boldsymbol{u}^{\mathrm{T}}\boldsymbol{\epsilon}^{\mathrm{T}}\boldsymbol{\epsilon}\boldsymbol{u}] = ||\overline{\boldsymbol{A}} - \boldsymbol{b}||_2^2 + \boldsymbol{u}^{\mathrm{T}}\boldsymbol{Q}\boldsymbol{u}^{\mathrm{T}}, \end{aligned} \tag{5.96}$$

where $Q = E_{\xi}[\epsilon^{\mathrm{T}}\epsilon]$. Using these results, we can reformulate the optimization as

$$\text{minimize} \quad ||\overline{A} - b||_2^2 + ||Q^{1/2}u||_2^2, \tag{5.97}$$

which becomes the Tikhonov regularization problem in the case of $Q = \gamma I$. That is,

$$\text{minimize} \quad ||\overline{A} - b||_2^2 + \gamma||u||_2^2, \tag{5.98}$$

which is a convex optimization problem and can be solved by the relevant methods such as the interior-point methods. We will discuss more about regularization methods later in this book.

## Exercises

**5.1** A common technique for solving some optimization problems is to convert it to an equivalent problem with a linear objective. Write the following generic convex optimization as a linear programming problem

$$\text{minimize} \quad f(\boldsymbol{x}),$$
$$\text{subject to} \quad g(\boldsymbol{x}) \leq 0, \qquad h(\boldsymbol{x}) = 0.$$

**5.2** Solve the following problem using the barrier method

$$\min f(x, y) = x^2 + (y-1)^2,$$

subject to

$$x^2 + y^2 \leq 9, \quad y \geq x^2 + 2.$$

Will the solution change if an extra constraint $x^2 + y^2 \leq 4$ is added?

**5.3** Write a simple program using two different methods to find the optimal solution of De Jong's function

$$f(\boldsymbol{x}) = \sum_{i=1}^{n} x_i^2 \qquad (-5.12 \leq x_i \leq 5.12),$$

for the cases of $n = 5$ and $n = 50$.

## Bibliography

Antoniou, A. and Lu, W.S. (2007). *Practical Optimization: Algorithms and Engineering Applications*. New York: Springer.

Boyd, S.P. and Vandenberghe, L. (2004). *Convex Optimization*. Cambridge, UK: Cambridge University Press.

Cipra, B.A. (2000). The best of the 20th century: editors name top 10 algorithms. *SIAM News* 33 (4): 1–2.

Conn, A.R., Gould, N.I.M., and Toint, P.L. (2000): *Trust-Region Methods*. SIAM & MPS. Philadelphia: SIAM.

Dennis, J.E. (1978). A brief introduction to quasi-Newton methods. In: *Numerical Analysis: Proceedings of Symposia in Applied Mathematics* (ed. G.H. Golub & J. Oliger), vol. 22, 19–52. Providence, RI: American Mathematical Society.

Fiacco, A.V. and McCormick, G.P. (1969). *Nonlinear Programming: Sequential Unconstrained Minimization Techniques*. New York: Wiley.

Fletcher, R. (1970). A new approach to variable metric algorithm. *Computer Journal* 13 (3): 317–322.

Fletcher, R. (1987). *Practical Methods of Optimization*, 2e. New York: Wiley.

Goldfarb, D. (1970). A family of variable metric updates derived by variational means. *Mathematics of Computation* 24 (109): 23–26.

Goldfeld, S.M., Quandt, R.E., and Trotter, H.F. (1966). Maximization by quadratic hill-climbing. *Econometrica* 34 (3): 541–551.

Gould, N.I.M., Hribar, M.E., and Nocedal, J. (2001). On the solution of equality constrained quadratic programming problems arising in optimization. *Journal of Convex Analysis* 23 (4): 1376–1395.

Karmarkar, N. (1984). A new polynomial-time algorithm for linear programming. *Combinatorica* 4 (4): 373–395.

Karmarkar, N. and Ramakrishna, K. G. (1991). Computational results of an interior-point algorithm for large scale linear programming. *Mathematical Programming* 52 (1–3): 555–586.

Marquardt, D. (1963). An algorithm for least-squares estimation of nonlinear parameters. *SIAM Journal on Applied Mathematics* 11 (2): 431–441.

Nesterov, Y. and Nemirovskii, A. (1994). *Interior-Point Polynomial Methods in Convex Programming*. Philadelphia: Society for Industrial and Applied Mathematics.

Nocedal, J. and Wright, S.J. (1999). *Numerical Optimization*. New York: Springer.

Pardalos, P.M. and Vavasis, S.A. (1991). Quadratic programming with one negative Eigenvalue is NP-hard. *Journal of Global Optimization* 1 (1): 15–22.

Powell, M.J.D. (1970). A new algorithm for unconstrained optimization. In: *Nonlinear Programming* (ed. J.B. Rosen, O.L. Mangasarian, and K. Ritter), 31—65. New York: Academic Press.

Shanno, D.F. and Kettler, P.C. (1970). Optimal conditioning of quasi-Newton methods. *Mathematics of Computation* 24 (111): 657–664.

Wright, S.J. (1997). *Primal-Dual Interior-Point Methods*. Philadelphia: Society for Industrial and Applied Mathematics.

Zdenek, D. (2009). *Optimal Quadratic Programming Algorithms: With Applications to Variational Inequalities*. Heidelberg: Springer.

# Part III

# Applied Optimization

# 6

# Linear Programming

Linear programming is a class of powerful mathematical programming techniques which are widely used in business planning, engineering design, airline scheduling, transportation, and many other applications.

## 6.1 Introduction

The basic idea in linear programming is to find the maximum or minimum of a linear objective under linear constraints.

For example, an Internet service provider (ISP) can provide two different services $x_1$ and $x_2$. The first service is, say, the fixed monthly rate with limited download limits and bandwidth, while the second service is the higher rate with no download limit. The profit of the first service is $\alpha x_1$ while the second is $\beta x_2$, though the profit of the second product is higher $\beta > \alpha > 0$, so the total profit is

$$P(\boldsymbol{x}) = \alpha x_1 + \beta x_2, \qquad \frac{\beta}{\alpha} > 1, \tag{6.1}$$

which is the objective function because the aim of the ISP company is to increase the profit as much as possible. Suppose the provided service is limited by the total bandwidth of the ISP company, thus at most $n_1 = 16$ (in 1 000 000 units) of the first and at most $n_2 = 10$ (in 1 000 000 units) of the second can be provided per unit time, say, each day. Therefore, we have

$$x_1 \leq n_1, \qquad x_2 \leq n_2. \tag{6.2}$$

If the management of each of the two service packages take the same staff time, so that a maximum of $n = 20$ (in 1 000 000 units) can be maintained, which means

$$x_1 + x_2 \leq n. \tag{6.3}$$

The additional constraints are that both $x_1$ and $x_2$ must be nonnegative since negative numbers are unrealistic. We now have the following constraints:

$$0 \leq x_1 \leq n_1, \qquad 0 \leq x_2 \leq n_2. \tag{6.4}$$

*Optimization Techniques and Applications with Examples*, First Edition. Xin-She Yang.

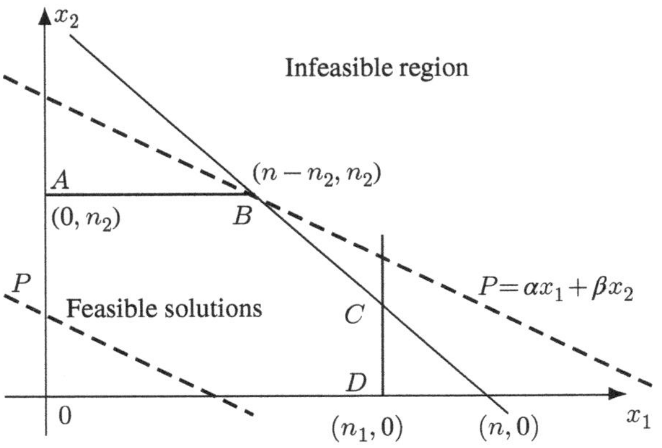


**Figure 6.1** Schematic representation of linear programming. If $\alpha = 2$, $\beta = 3$, $n_1 = 16$, $n_2 = 10$, and $n = 20$, then the optimal solution is at $B(10, 10)$.

The problem now is to find the best $x_1$ and $x_2$ so that the profit $P$ is a maximum. Mathematically, we have

$$\begin{aligned} \underset{(x_1,x_2)\in\mathbb{N}^2}{\text{maximize}} \quad & P(x_1,x_2) = \alpha x_1 + \beta x_2, \\ \text{subject to} \quad & x_1 + x_2 \le n, \\ & 0 \le x_1 \le n_1, \ 0 \le x_2 \le n_2. \end{aligned} \tag{6.5}$$

The feasible solutions to the problem (6.5) can graphically be represented as the inside region of the polygon *OABCD* as shown in Figure 6.1. As the aim is to maximize the profit $P$, thus the optimal solution is at the extreme point $B$ with $(n - n_2, n_2)$ and $P = \alpha(n - n_2) + \beta n_2$.

For example, if $\alpha = 2$, $\beta = 3$, $n_1 = 16$, $n_2 = 10$, and $n = 20$, then the optimal solution occurs at $x_1 = n - n_2 = 10$ and $x_2 = n_2 = 10$ with the total profit $P = 2 \times (20 - 10) + 3 \times 10 = 50$ thousand pounds.

Since the solution ($x_1$ and $x_2$) must be integers, an interesting thing is that the solution in this problem is independent of $\beta/\alpha$ if and only if $\beta/\alpha > 1$. However, the profit $P$ does depend on the parameters $\alpha$ and $\beta$.

In general, the number of feasible solutions is infinite if $x_1$ and $x_2$ are real numbers. Even for integers $x_1, x_2 \in \mathbb{N}$, the number of feasible solutions is quite large. Therefore, there is a need to use a systematic method to find the optimal solution. In order to find the best solution, we first plot out all the constraints as straight lines, and all the feasible solutions satisfying all the constraints form the inside region of the polygon *OABCD*. The vertices of the polygon form the set of extreme points. Then, we plot the objective function $P$ as a family of parallel lines (shown as dashed lines) so as to find the maximum value of $P$. Obviously, the highest value of $P$ corresponds to the case when the objective line goes through the extreme point $B$. Therefore, $x_1 = n - n_2$ and $x_2 = n_2$ at the point $B$

are the best solutions. The current example is relatively simple because it has only two decision variables and three constraints, which can be solved easily using a graphic approach. For more complicated problems, we need a formal approach. One of the most widely used methods is the simplex method.

## 6.2 Simplex Method

The simplex method was introduced by George Dantzig in 1947. The simplex method essentially works in the following way: for a given linear optimization problem such as the example of the ISP service we discussed earlier, it assumes that all the extreme points are known. If the extreme points are not known, the first step is to determine these extreme points or to check whether there are any feasible solutions. With known extreme points, it is easy to test whether or not an extreme point is optimal using the algebraic relationship and the objective function. If the test for optimality is not passed, then move to an adjacent extreme point to do the same test. This process stops until an optimal extreme point is found or the unbounded case occurs.

### 6.2.1 Slack Variables

Mathematically, the simplex method first transforms the constraint inequalities into equalities by using slack variables.

To convert an inequality such as

$$5x_1 + 6x_2 \le 20, \tag{6.6}$$

we can use a new variable $x_3$ or $s_1 = 20 - 5x_1 - 6x_2$ so that the original inequality becomes an equality

$$5x_1 + 6x_2 + s_1 = 20, \tag{6.7}$$

with an auxiliary non-negativeness condition

$$s_1 \ge 0. \tag{6.8}$$

Such a variable is referred to as a slack variable.

Thus, the inequalities in our example

$$x_1 + x_2 \le n, \qquad 0 \le x_1 \le n_1, \qquad 0 \le x_2 \le n_2 \tag{6.9}$$

can be written, using three slack variables $s_1, s_2, s_3$, as the following equalities:

$$x_1 + x_2 + s_1 = n, \tag{6.10}$$

$$x_1 + s_2 = n_1, \qquad x_2 + s_3 = n_2, \tag{6.11}$$

and

$$x_i \geq 0 \ (i = 1, 2), \quad s_j \geq 0 \ (j = 1, 2, 3). \tag{6.12}$$

The original problem (6.5) becomes

$$\underset{\boldsymbol{x} \in \mathbb{N}^5}{\text{maximize}} P(\boldsymbol{x}) = \alpha x_1 + \beta x_2 + 0 s_1 + 0 s_2 + 0 s_3,$$

$$\text{subject to} \begin{pmatrix} 1 & 1 & 1 & 0 & 0 \\ 1 & 0 & 0 & 1 & 0 \\ 0 & 1 & 0 & 0 & 1 \end{pmatrix} \begin{pmatrix} x_1 \\ x_2 \\ s_1 \\ s_2 \\ s_3 \end{pmatrix} = \begin{pmatrix} n \\ n_1 \\ n_2 \end{pmatrix},$$

$$x_i \geq 0 \qquad (i = 1, 2, \dots, 5), \tag{6.13}$$

which has two control variables $(x_1, x_2)$ and three slack variables $x_3 = s_1, x_4 = s_2$, and $x_5 = s_3$.

### 6.2.2 Standard Formulation

In general, a linear programming problem can be written in the following standard form:

$$\underset{\boldsymbol{x} \in \mathbb{R}^n}{\text{maximize}} f(\boldsymbol{x}) = Z = \sum_{i=1}^{n} \alpha_i x_i = \boldsymbol{\alpha}^{\mathrm{T}} \boldsymbol{x},$$

$$\text{subject to } \boldsymbol{A}\boldsymbol{x} = \boldsymbol{b}, \quad \boldsymbol{x} = (x_1, x_2, \dots, x_n)^{\mathrm{T}}, \quad x_i \geq 0 \ (i = 1, \dots, n), \tag{6.14}$$

where $\boldsymbol{A}$ is an $m \times n$ matrix, $\boldsymbol{b} = (b_1, \dots, b_m)^{\mathrm{T}}$. This problem has $n$ variables, and $m$ equalities and all $n$ variables are nonnegative. In the standard form, all constraints are expressed as equalities and all variables including slack variables are nonnegative.

A basic solution to the linear system $\boldsymbol{A}\boldsymbol{x} = \boldsymbol{b}$ of $m$ linear equations in terms of $m$ basic variables in the standard form is usually obtained by setting $n - m$ variables equal to zero, and subsequently solving the resulting $m \times n$ linear system to get a unique solution of the remaining $m$ variables. The $m$ variables (that are not bound to zero) are called the basic variables of the basic solution. The $n - m$ variables at zero are called non-basic variables. Any basic solution to this linear system is referred to as a basic feasible solution (BFS) if all its variables are nonnegative. The important property of the BFSs is that there is a unique corner point (extreme point) for each BFS, and there is at least one BFS for each corner or extreme point. These corner or extreme points are points on the intersection of two adjacent boundary lines such as $A$ and $B$ in Figure 6.1.

Two BFSs are said to be adjacent if they have $m - 1$ basic variables in common in the standard form.

Suppose $m = 500$, even the simplest integer equalities $x_i + x_j = 1$, where $i, j = 1, 2, \ldots, 500$, would give a huge number of combinations $2^{500}$. Thus, the number of BFSs will be the order of $2^{500} \approx 3 \times 10^{150}$, which is larger than the number of particles in the whole universe. This huge number of BFSs and extreme points necessitates a systematic and efficient search method. The simplex method is a powerful method to carry out such linear programming tasks.

### 6.2.3 Duality

For every LP problem (called primal), there is a dual problem that is associated with the primal. This is called duality.

For a primal LP problem

$$\text{maximize} \quad \boldsymbol{\alpha}^{\mathrm{T}} \boldsymbol{x}, \tag{6.15}$$

subject to

$$\boldsymbol{A}\boldsymbol{x} \leq \boldsymbol{b}, \quad \boldsymbol{x} \geq 0, \tag{6.16}$$

its dual problem is given by

$$\text{minimize} \quad \boldsymbol{b}^{\mathrm{T}} \boldsymbol{y}, \tag{6.17}$$

subject to

$$\boldsymbol{A}^{\mathrm{T}}\boldsymbol{y} \geq \boldsymbol{\alpha}, \quad \boldsymbol{y} \geq 0, \tag{6.18}$$

where $\boldsymbol{y} = (y_1, y_2, \ldots, y_m)^{\mathrm{T}}$ is called a dual variable vector, in contrast to the primal variable vector $\boldsymbol{x} = (x_1, \ldots, x_n)^{\mathrm{T}}$.

If we write the above problems in a more detailed form, we have

$$\text{maximize} \quad \sum_{j=1}^{n} \alpha_j x_j, \quad \text{(primal)}, \tag{6.19}$$

subject to

$$\sum_{j=1}^{n} a_{ij} x_j \leq b_i, \quad (i = 1, 2, \ldots, m), \quad x_j \geq 0 \ (j = 1, 2, \ldots, n), \tag{6.20}$$

and

$$\text{minimize} \quad \sum_{i=1}^{m} b_i y_i, \quad \text{(dual)}, \tag{6.21}$$

subject to

$$\sum_{i=1}^{m} a_{ji} y_i \geq \alpha_j, \quad (j = 1, 2, \ldots, n), \quad y_i \geq 0 \ (i = 1, 2, \ldots, m). \tag{6.22}$$

Mathematically speaking, the dual problems are closely related to their primal problems. Sometimes, it may be the case that finding a solution to a dual problem might be easier. In addition, the duality can provide an in-depth insight into the problem itself.

In general, if $\boldsymbol{x}^*$ is an optimal solution to the primal problem with an optimal prime value $p^*$ and $\boldsymbol{y}^*$ is an optimal solution to its dual with an optimal dual value $d^*$, we have

$$p^* = \boldsymbol{\alpha}^{\mathrm{T}}\boldsymbol{x}^*, \quad d^* = \boldsymbol{b}^{\mathrm{T}}\boldsymbol{y}^*, \tag{6.23}$$

and

$$p^* \leq d^*, \tag{6.24}$$

or

$$\sum_{j=1}^{n} \alpha_j x_j^* \leq \sum_{i=1}^{m} b_i y_i^*. \tag{6.25}$$

The duality gap is defined by

$$g^* = p^* - d^*. \tag{6.26}$$

In the special case that the above equality holds (i.e. $p^* = d^*$), we say the duality gap is zero. That is, the optimal value of the primal objective is the same as the optimal value of the dual problem. We will provide an example in the exercise of this chapter.

Duality has rigorous and elegant mathematical theory, and interested readers can refer to more advanced literature.

### 6.2.4 Augmented Form

The linear optimization problem (6.14) can be easily rewritten as the following standard augmented form or the canonical form

$$\begin{pmatrix} 1 & -\boldsymbol{\alpha}^{\mathrm{T}} \\ 0 & \boldsymbol{A} \end{pmatrix} \begin{pmatrix} Z \\ \boldsymbol{x} \end{pmatrix} = \begin{pmatrix} 0 \\ \boldsymbol{b} \end{pmatrix}, \tag{6.27}$$

with the objective to maximize $Z$. In this canonical form, all the constraints are expressed as equalities for all nonnegative variables. All the right-hand sides for all constraints are also nonnegative, and each constraint equation has a single basic variable. The intention of writing in this canonical form is to identify basic feasible solutions, and move from one BFS to another via a so-called pivot operation. Geometrically speaking, this means to find all the corners or extreme points first, then evaluate the objective function by going through some (ideally, a small fraction) of the extreme points so as to determine if the current BFS can be improved or not.

## 6.3 Worked Example by Simplex Method

In the framework of the canonical form, the basic steps of the simplex method are: (i) to find a BFS to start the algorithm. Sometimes, it might be difficult to start, which may either imply there is no feasible solution or that it is necessary to reformulate the problem in a slightly different way by changing the canonical form so that a BFS can be found; (ii) to see if the current BFS can be improved (even marginally) by increasing the non-basic variables from zero to nonnegative values; (iii) stop the process if the current feasible solution cannot be improved, which means that it is optimal. If the current feasible solution is not optimal, then move to an adjacent BFS. This adjacent BFS can be obtained by changing the canonical form via elementary row operations.

The pivot manipulations are based on the fact that a linear system will remain an equivalent system by multiplying a nonzero constant with a row and adding it to the other row. This procedure continues by going to the second step and repeating the evaluation of the objective function. The optimality of the problem will be reached, or we can stop the iteration if the solution becomes unbounded in the event that we can improve the objective indefinitely.

Now we come back to our ISP example, if we use $\alpha = 2$, $\beta = 3$, $n_1 = 16$, $n_2 = 10$, and $n = 20$, we then have

$$\begin{pmatrix} 1 & -2 & -3 & 0 & 0 & 0 \\ 0 & 1 & 1 & 1 & 0 & 0 \\ 0 & 1 & 0 & 0 & 1 & 0 \\ 0 & 0 & 1 & 0 & 0 & 1 \end{pmatrix} \begin{pmatrix} Z \\ x_1 \\ x_2 \\ s_1 \\ s_2 \\ s_3 \end{pmatrix} = \begin{pmatrix} 0 \\ 20 \\ 16 \\ 10 \end{pmatrix}, \tag{6.28}$$

where $x_1, x_2, s_1, \dots, s_3 \geq 0$. Now the first step is to identify a corner point or BFS by setting non-isolated variables $x_1 = 0$ and $x_2 = 0$ (thus the basic variables are $s_1, s_2, s_3$). We now have

$$s_1 = 20, \ s_2 = 16, \ s_3 = 10. \tag{6.29}$$

The objective function $Z = 0$ corresponds to the corner point $O$ in Figure 6.1. In the present canonical form, the corresponding column associated with each basic variable has only one nonzero entry (marked by a box) for each constraint equality, and all other entries in the same column are zero. A nonzero value is usually converted into 1 if it is not unity. This is shown as follows:

$$\begin{matrix} Z & x_1 & x_2 & s_1 & s_2 & s_3 \end{matrix} \tag{6.30}$$
$$\begin{pmatrix} 1 & -2 & -3 & 0 & 0 & 0 \\ 0 & 1 & 1 & \boxed{1} & 0 & 0 \\ 0 & 1 & 0 & 0 & \boxed{1} & 0 \\ 0 & 0 & 1 & 0 & 0 & \boxed{1} \end{pmatrix}.$$

When we change the set or the bases of basic variables from one set to another, we will aim to convert to a similar form using pivot row operations. There are two ways of numbering this matrix. One way is to call the first row [1 –2 –3 0 0 0] as the 0th row, so that all other rows correspond to their corresponding constraint equation. The other way is simply to use its order in the matrix, so [1 –2 –3 0 0 0] is simply the first row. We will use this standard notation.

Now the question is whether we can improve the objective by increasing one of the non-basic variables $x_1$ and $x_2$? Obviously, if we increase $x_1$ by a unit, then $Z$ will also increase by 2 units. However, if we increase $x_2$ by a unit, then $Z$ will increase by 3 units. Since our objective is to increase $Z$ as much as possible, we choose to increase $x_2$. As the requirement of the non-negativeness of all variables, we cannot increase $x_2$ without limit. So we increase $x_2$ while holding $x_1 = 0$, and we have

$$s_1 = 20 - x_2, \; s_2 = 16, \; s_3 = 10 - x_2. \tag{6.31}$$

Thus, the highest possible value of $x_2$ is $x = 10$ when $s_1 = s_3 = 0$. If $x_2$ increases further, both $s_1$ and $s_3$ will become negative; thus, it is no longer a BFS.

The next step is either to set $x_1 = 0$ and $s_1 = 0$ as non-basic variables or to set $x_1 = 0$ and $s_3 = 0$. Both cases correspond to the point $A$ in our example, so we simply choose $x_1 = 0$ and $s_3 = 0$ as non-basic variables, and the basic variables are thus $x_2$, $s_1$, and $s_2$. Now we have to do some pivot operations so that $s_3$ will be replaced by $x_2$ as a new basic variable. Each constraint equation has only a single basic variable in the new canonical form. This means that each column corresponding to each basic variable should have only a single nonzero entry (usually 1). In addition, the right-hand sides of all the constraints are nonnegative and increase the value of the objective function at the same time. In order to convert the third column for $x_2$ to the form with only a single nonzero entry 1 (all other coefficients in the column should be zero), we first multiply the fourth row by 3 and add it to the first row, and the first row becomes

$$Z - 2x_1 + 0x_2 + 0s_1 + 0s_2 + 3s_3 = 30. \tag{6.32}$$

Then, we multiply the fourth row by $-1$ and add it to the second row, and we have

$$0Z + x_1 + 0x_2 + s_1 + 0s_2 - s_3 = 10. \tag{6.33}$$

So the new canonical form becomes

$$\begin{pmatrix} 1 & -2 & 0 & 0 & 0 & 3 \\ 0 & 1 & 0 & 1 & 0 & -1 \\ 0 & 1 & 0 & 0 & 1 & 0 \\ 0 & 0 & 1 & 0 & 0 & 1 \end{pmatrix} \begin{pmatrix} Z \\ x_1 \\ x_2 \\ s_1 \\ s_2 \\ s_3 \end{pmatrix} = \begin{pmatrix} 30 \\ 10 \\ 16 \\ 10 \end{pmatrix}, \tag{6.34}$$

where the third, fourth, and fifth columns (for $x_2$, $s_1$, and $s_2$, respectively) have only one nonzero coefficient. All the values on the right-hand side are non-negative. From this canonical form, we can find the BFS by setting non-basic variables equal to zero. This is to set $x_1 = 0$ and $s_3 = 0$. We now have the BFS

$$x_2 = 10, \ s_1 = 10, \ s_2 = 16, \tag{6.35}$$

which corresponds to the corner point $A$. The objective $Z = 30$.

Now again the question is whether we can improve the objective by increasing the non-basic variables. As the objective function is

$$Z = 30 + 2x_1 - 3s_3, \tag{6.36}$$

$Z$ will increase 2 units if we increase $x_1$ by 1, but $Z$ will decrease $-3$ if we increase $s_3$. Thus, the best way to improve the objective is to increase $x_1$. The question is what the limit of $x_1$ is. To answer this question, we hold $s_3$ at 0, we have

$$s_1 = 10 - x_1, \ s_2 = 16 - x_1, \ x_2 = 10. \tag{6.37}$$

We can see if $x_1$ can increase up to $x_1 = 10$, after that $s_1$ becomes negative, and this occurs when $x_1 = 10$ and $s_1 = 0$. This also suggests that the new adjacent BFS can be obtained by choosing $s_1$ and $s_3$ as the non-basic variables. Therefore, we have to replace $s_1$ with $x_1$ so that the new basic variables are $x_1, x_2$, and $s_2$.

Using these basic variables, we have to make sure that the second column (for $x_1$) has only a single nonzero entry. Thus, we multiply the second row by 2 and add it to the first row, and the first row becomes

$$Z + 0x_1 + 0x_2 + 2s_1 + 0s_2 + s_3 = 50. \tag{6.38}$$

We then multiply the second row by $-1$ and add it to the third row, and we have

$$0Z + 0x_1 + 0x_2 - s_1 + s_2 + s_3 = 6. \tag{6.39}$$

Thus, we have the following canonical form

$$\begin{pmatrix} 1 & 0 & 0 & 2 & 0 & 1 \\ 0 & 1 & 0 & 1 & 0 & -1 \\ 0 & 0 & 0 & -1 & 1 & 1 \\ 0 & 0 & 1 & 0 & 0 & 1 \end{pmatrix} \begin{pmatrix} Z \\ x_1 \\ x_2 \\ s_1 \\ s_2 \\ s_3 \end{pmatrix} = \begin{pmatrix} 50 \\ 10 \\ 6 \\ 10 \end{pmatrix}, \tag{6.40}$$

whose BFS can be obtained by setting non-basic variables $s_1 = s_3 = 0$. We have

$$x_1 = 10, \ x_2 = 10, \ s_2 = 6, \tag{6.41}$$

which corresponds to the extreme point $B$ in Figure 6.1. The objective value is $Z = 50$ for this BFS. Let us see if we can improve the objective further. Since the objective becomes

$$Z = 50 - 2s_1 - s_3, \tag{6.42}$$

any increase of $s_1$ or $s_3$ from zero will decrease the objective value. Therefore, this BFS is optimal. Indeed, this is the same solution as that obtained earlier from the graph method. We can see that a major advantage is that we have reached the optimal solution after searching a certain number of extreme points, and there is no need to evaluate other extreme points. This is exactly why the simplex method is so efficient.

The case study we used here is relatively simple, but it is useful to show how the basic procedure works in linear programming. For more practical applications, there are well-established software packages that will do the work for you once you have properly set up the objective and constraints.

Though the simplex method works well in most applications in practice, it can still become time-consuming in some cases because it has been show that the complexity for worst case scenarios is $O(n^6B^2)$, where $n$ is the number of decision variables and $B$ is the length or number of bits in the input. When $n$ is large, often thousands and millions of decision variables, this can still be a huge number. A more efficient, polynomial-time is the interior-point algorithm developed by Narendra Karmarbar in 1984, and we will introduce it in the next section. The complexity of Karmarkar's interior-point algorithm is $O\left(n^{3.5}B^2\log(B)\log\log(B)\right)$.

## 6.4 Interior-Point Method for LP

From the standard LP formulation discussed earlier, we have

$$\text{maximize} \quad f(\boldsymbol{x}) = \boldsymbol{\alpha}^{\mathrm{T}}\boldsymbol{x} = \alpha_1 x_1 + \alpha_2 x_2 + \cdots + \alpha_n x_n, \tag{6.43}$$

subject to

$$\boldsymbol{A}\boldsymbol{x} = \boldsymbol{b}, \quad x_i \geq 0 \; (i = 1, 2, \dots, n), \tag{6.44}$$

where $\boldsymbol{x} = (x_1, x_2, \dots, x_n)^{\mathrm{T}}$ is the decision vector. For simplicity, we will assume that $\boldsymbol{A}$ is a full rank matrix. Otherwise, we can rewrite the constraints and clean up the equations so such that the matrix $\boldsymbol{A}$ becomes a full rank matrix.

We first define a Lagrangian by incorporating the equality constraints and using logarithmic barrier functions we have

$$L = \boldsymbol{\alpha}^{\mathrm{T}}\boldsymbol{x} - \mu \sum_{i=1}^{n} \ln x_i - \boldsymbol{\lambda}^{\mathrm{T}}(\boldsymbol{A}\boldsymbol{x} - \boldsymbol{b}), \tag{6.45}$$

where $\mu$ is the barrier strength parameter and $\boldsymbol{\lambda} = (\lambda_1, \lambda_2, \dots, \lambda_n)^{\mathrm{T}}$ is the set of Lagrange multipliers. This form essentially incorporates all the constraints into the objective and thus the original constrained LP becomes an unconstrained LP.

From the KKT conditions, we know that the optimality conditions are given by the stationarity condition. That is

$$\nabla_x L = \boldsymbol{\alpha} - \boldsymbol{A}^{\mathrm{T}}\boldsymbol{\lambda} - \mu \boldsymbol{X}^{-1}\boldsymbol{e} = 0, \tag{6.46}$$

where

$$\boldsymbol{X} = \mathrm{diag}(x_i) = \begin{pmatrix} x_1 & 0 & \dots & 0 \\ 0 & x_2 & \dots & 0 \\ \vdots & \vdots & \ddots & \vdots \\ 0 & 0 & \dots & x_n \end{pmatrix}, \tag{6.47}$$

and

$$\boldsymbol{e} = (1, \ 1, \ 1, \ \dots, \ 1)^{\mathrm{T}}, \tag{6.48}$$

which has $n$ entries that are all 1.

If we define a variable (often called a dual variable vector) $\boldsymbol{s}$,

$$\boldsymbol{s} = \mu \boldsymbol{X}^{-1}\boldsymbol{e}, \tag{6.49}$$

we have

$$\boldsymbol{XSe} = \mu \boldsymbol{e}, \tag{6.50}$$

where

$$\boldsymbol{S} = \mathrm{diag}(s_i) = \begin{pmatrix} s_1 & 0 & \dots & 0 \\ 0 & s_2 & \dots & 0 \\ \vdots & \vdots & \ddots & \vdots \\ 0 & 0 & \dots & s_n \end{pmatrix}. \tag{6.51}$$

This form seems much more complicated, but it is a compact form to write multiple conditions. In fact, $\boldsymbol{XSe} = 0$ is equivalent to

$$x_i s_i = 0 \quad (i = 1, 2, \dots, n), \tag{6.52}$$

which are the complementary slackness conditions in the KKT conditions. Therefore, $\boldsymbol{XSe} = \mu\boldsymbol{e}$ is a perturbed or modified form of such complementary slackness.

In addition, we have

$$\nabla_\lambda L = \boldsymbol{Ax} - \boldsymbol{b} = 0. \tag{6.53}$$

Therefore, the optimality conditions are

$$\begin{cases} \boldsymbol{Ax} & = \boldsymbol{b}, \\ \boldsymbol{A}^{\mathrm{T}}\boldsymbol{\lambda} + \boldsymbol{s} & = \boldsymbol{\alpha}, \\ \boldsymbol{XSe} & = \mu\boldsymbol{e}, \\ \boldsymbol{x} \geq 0, & \boldsymbol{s} \geq 0, \end{cases} \tag{6.54}$$

where we used $\boldsymbol{x} > 0$ and $\boldsymbol{s} \geq 0$ to mean $x_i \geq 0$ and $s_i \geq 0$ for $i = 1, 2, \dots, n$.

In the framework of the interior point method, the first condition (i.e. $\boldsymbol{Ax} = \boldsymbol{b}$) is the primal feasibility condition, while the second condition is the dual feasibility condition. The third condition is a modified complementary slackness condition perturbed by $\mu$.

In order to solve the above systems, we can use an iterative procedure, based on Newton's method. We can replace $\boldsymbol{x}$, $\boldsymbol{\lambda}$, and $\boldsymbol{s}$ by $\boldsymbol{x} + \Delta\boldsymbol{x}$, $\boldsymbol{\lambda} + \Delta\boldsymbol{\lambda}$, and $\boldsymbol{s} + \Delta\boldsymbol{s}$, respectively. We have

$$\begin{cases} \boldsymbol{A}(\boldsymbol{x} + \Delta\boldsymbol{x}) & = \boldsymbol{b}, \\ \boldsymbol{A}^{\mathrm{T}}(\boldsymbol{\lambda} + \Delta\boldsymbol{\lambda}) + (\boldsymbol{s} + \Delta\boldsymbol{s}) & = \boldsymbol{\alpha}, \\ (\boldsymbol{S} + \Delta\boldsymbol{S})(\boldsymbol{X} + \Delta\boldsymbol{X})\boldsymbol{e} & = \mu\boldsymbol{e}. \end{cases} \tag{6.55}$$

Since both $\boldsymbol{X}$ and $\boldsymbol{S}$ are diagonal matrices, thus $\Delta\boldsymbol{S} = \Delta\boldsymbol{s}$ and $\Delta\boldsymbol{X} = \Delta\boldsymbol{x}$. Re-arranging the above equations and ignoring the higher-order terms such as $\Delta\boldsymbol{x}\Delta\boldsymbol{s}$, we have

$$\begin{cases} \boldsymbol{A}\Delta\boldsymbol{x} = \boldsymbol{b} - \boldsymbol{A}\boldsymbol{x}, \\ \boldsymbol{A}^{\mathrm{T}}\Delta\boldsymbol{\lambda} + \Delta\boldsymbol{s} = \boldsymbol{\alpha} - \boldsymbol{A}^{\mathrm{T}}\boldsymbol{\lambda} - \boldsymbol{s}, \\ \boldsymbol{S}\Delta\boldsymbol{x} + \boldsymbol{X}\Delta\boldsymbol{s} = \mu\boldsymbol{e} - \boldsymbol{S}\boldsymbol{X}\boldsymbol{e}, \end{cases} \tag{6.56}$$

which can be written compactly as

$$\begin{pmatrix} \boldsymbol{A} & 0 & 0 \\ 0 & \boldsymbol{A}^{\mathrm{T}} & \boldsymbol{I}_n \\ \boldsymbol{S} & 0 & \boldsymbol{X} \end{pmatrix} \begin{pmatrix} \Delta\boldsymbol{x} \\ \Delta\boldsymbol{\lambda} \\ \Delta\boldsymbol{s} \end{pmatrix} = \begin{pmatrix} \boldsymbol{b} - \boldsymbol{A}\boldsymbol{x} \\ \boldsymbol{\alpha} - \boldsymbol{A}^{\mathrm{T}}\boldsymbol{\lambda} - \boldsymbol{s} \\ \mu\boldsymbol{e} - \boldsymbol{S}\boldsymbol{X}\boldsymbol{e} \end{pmatrix}, \tag{6.57}$$

where $\boldsymbol{I}_n$ is an identity matrix of size $n \times n$. Thus, the iteration formulas become

$$\begin{cases} \boldsymbol{x}_{t+1} = \boldsymbol{x}_t + \rho\Delta\boldsymbol{x}, \\ \boldsymbol{\lambda}_{t+1} = \boldsymbol{\lambda}_t + \rho\Delta\boldsymbol{\lambda}, \\ \boldsymbol{s}_{t+1} = \boldsymbol{s}_t + \rho\Delta\boldsymbol{s}, \end{cases} \tag{6.58}$$

where $\rho \in (0, 1)$ is a scalar damping factor. This iteration is often called primal-dual Newton's method, and the search direction is $(\Delta\boldsymbol{x}, \Delta\boldsymbol{\lambda}, \Delta\boldsymbol{s})$.

The errors or infeasibilities at each iteration can be estimated by

$$\begin{cases} \boldsymbol{E}_{\boldsymbol{x}} = \boldsymbol{b} - \boldsymbol{A}\boldsymbol{x}_{t+1}, \\ \boldsymbol{E}_{\lambda} = \boldsymbol{\alpha} - \boldsymbol{A}^{\mathrm{T}}\boldsymbol{\lambda}_{t+1} - \boldsymbol{s}_{t+1}, \\ \boldsymbol{E}_{\boldsymbol{S}} = \mu\boldsymbol{e} - \boldsymbol{X}_{t+1}\boldsymbol{S}_{t+1}\boldsymbol{e}. \end{cases} \tag{6.59}$$

In order to reduce $\mu$ gradually, in the implementation, we often use

$$\mu_{t+1} = \beta\mu_t, \quad \beta \in (0, 1). \tag{6.60}$$

Iterations stop if a predefined tolerance $\epsilon$ is met. That is,

$$||\boldsymbol{x}_{t+1} - \boldsymbol{x}_t|| \leq \epsilon. \tag{6.61}$$

**Algorithm 6.1** Primal-dual Newton's iterations for the interior-point method to solve LP.

Find a point $\boldsymbol{x}_0 > 0$ so that $\boldsymbol{A}\boldsymbol{x}_0 = \boldsymbol{b}$ is true.
Initialize $\mu_0 > 0$ (large enough) and $\beta \in (0, 1)$ at $t = 0$.
**While** (stopping criterion)
    Calculate $(\Delta\boldsymbol{x}, \Delta\boldsymbol{\lambda}, \Delta\boldsymbol{s})$ using Eq. (6.57)
    Update the solutions using Eq. (6.58)
    Calculate the infeasibilities using Eq. (6.59)
    Update the counter $\mu_{t+1} = \beta\mu_t$
**end**

Therefore, the interior-point method can be summarized as the pseudocode shown in Algorithm 6.1. The good news is that many software packages have this method implemented and well tested, which makes it easier to use for many applications.

There are a diverse range of applications of linear programming, from transport problem to scheduling. We will introduce such applications in the next chapter where we will discuss integer programming in detail.

## Exercises

**6.1** What is the link of linear programming to convex optimization?

**6.2** Solve the following LP:

$$\text{maximize } P = 3x + 4y,$$

subject to

$$8x + 7y \leq 50, \quad 2x + 3y \leq 20, \quad x + 4y \leq 25,$$

and $x, y \geq 0$. If we change $2x + 3y \leq 100$, will the result change? Why?

**6.3** Write the above problem in a compact form using matrices, and then formulate its dual problem.

## Bibliography

Dantzig, G.B. and Thapa, M.N. (1997). *Linear Programming 1: Introduction.* New York: Springer-Verlag.

Gass, S.I. (2011). *Linear Programming: Methods and Applications*, 5e. New York: Dover Publications Inc..
Gondzio, J. (2012). Interior point methods 25 years later. *European Journal of Operational Research* 218 (3): 587–601.
Heizer, J., Render, B., and Munson, C. (2016). *Operations Management: Sustainability and Supply Chain Management*, 12e. London: Pearson.
Hitchcock, F.L. (1941). The distribution of a product from several sources to numerous localities. *Journal of Mathematics and Physics* 20 (1–4): 224–230.
Karmarkar, N. (1984). A new polynomial-time algorithm for linear programming. *Combinatorica* 4 (4): 373–396.
Khachiyan, L. (1979). A polynomial time algorithm in learning programming. *Doklady Akademiia Nauk SSSR* 244: 1093–1096. English translation: *Soviet Mathematics Doklady* 20 (1): 191–194 (1979).
Lee, J. and Leyffer, S. (2011). *Mixed Integer Nonlinear Programming*. New York: Spinger.
Özlem, E. and Orlin, J.B. (2006). A dynamic programming methodology in very large scale neighbourhood search applied to the traveling salesman problem. *Discrete Optimization* 3 (1): 78–85.
Rebennack, S. (2008). Ellipsoid method. In: *Encyclopedia of Optimization*, 2e (ed. C.A. Floudas and P.M. Pardalos), 890–899. New York: Springer.
Robere, R. (2012). Interior point methods and linear programming, 1–15. Teaching Notes. University of Toronto.
Schrijver, A. (1998). *Theory of Linear and Integer Programming*. New York: Wiley.
Strang, G. (1987). Karmarkar's algorithm and its place in applied mathematics. *The Mathematical Intelligencer* 9 (2): 4–10.
Todd, M.J. (2002). The many facets of linear programming. *Mathematical Programming* 91 (3): 417–436.
Vanderbei, R.J. (2014). *Linear Programming: Foundations and Extensions*, 4e. New York: Springer.
Wright, S.J. (1997). *Primal-Dual Interior-Point Methods*. Philadelphia: Society for Industrial and Applied Mathematics.
Wright, M.H. (2005). The interior-point revolution in optimization: history, recent developments, and lasting consequences. *Bulletin of the American Mathematical Society* 42 (1): 39–56.
Yang, X.S. (2010). *Engineering Optimization: An Introduction with Metaheuristic Applications*. Hoboken, NJ: Wiley.

# 7

# Integer Programming

Integer programming (IP) is a special class of combinatorial optimization problems, which tends to be difficult to solve. Before we introduce IP in general, let us first review the fundamentals of linear programming (LP) introduced in detail in Chapter 6.

## 7.1 Integer Linear Programming

### 7.1.1 Review of LP

Suppose we try to solve a simple LP problem

$$\text{maximize } P = 2x + 3y, \tag{7.1}$$

subject to the following constraints:

$$x + y \leq 4, \tag{7.2}$$

$$x + 3y \leq 7, \tag{7.3}$$

$$x, y \geq 0. \tag{7.4}$$

This problem can be solved by either the corner method, graph method, or more systematically the simplex method introduced in Chapter 6.

Let us solve this simple LP problem using the graph method. The domain and constraints are represented in Figure 7.1 where the dashed lines correspond to a fixed value of the objective function $P$. Since $P$ increases as the solution points move to the right and upwards, the optimal solution should occur at the corner $B$.

Point $B$ is obtained by solving the two equalities

$$x + y = 4, \quad x + 3y = 7, \tag{7.5}$$

*Optimization Techniques and Applications with Examples*, First Edition. Xin-She Yang.

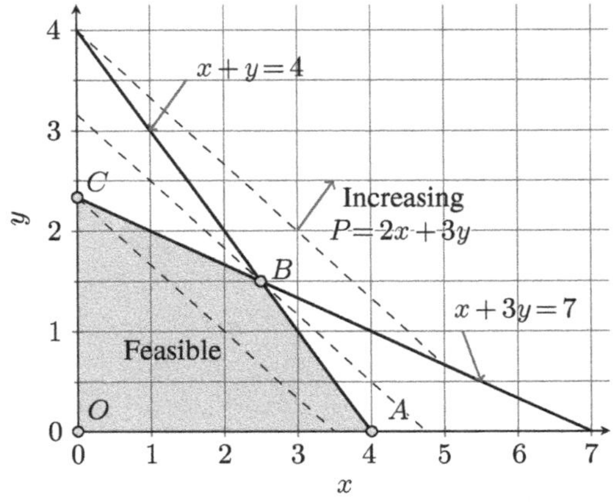


**Figure 7.1** Linear programming.

which gives

$$x = 2.5, \quad y = 1.5. \tag{7.6}$$

This corresponds to

$$P = 2x + 3y = 2 \times (2.5) + 3 \times 1.5 = 9.5, \tag{7.7}$$

which is the globally optimal solution to this simple problem.

### 7.1.2 Integer LP

The variables in LP are nonnegative real numbers, but in many real-world applications, variables can only take integer values such as the number of staff or number of products. In this case, we have to deal with IP problems.

For the simple LP problem we solved earlier, if we impose additional constraints that both variables ($x$ and $y$) are integers ($\mathbb{I}$), we have the following IP problem:

$$\text{maximize}\, P = 2x + 3y, \tag{7.8}$$

subject to

$$x + y \leq 4, \tag{7.9}$$

$$x + 3y \leq 7, \tag{7.10}$$

$$x, y \geq 0, \tag{7.11}$$

$$x, y \in \mathbb{I}. \tag{7.12}$$

On the other hand, if we assume that one variable takes only integers while the other variable can be any real number, the problem becomes a mixed-integer programming (MIP) problem. Both IP and MIP problems are NP-hard problems when the problem size is large. Therefore, solution methods can be very time-consuming. However, our focus in this section is on the pure IP problems of very small sizes.

## 7.2 LP Relaxation

How do we solve such an IP problem given earlier in Eq. (7.8)? We cannot directly use the techniques to solve LP problems as the domain for this integer program (IP) consists of discrete points marked as gray points in Figure 7.2. In this simple case, there are only 11 feasible points in the feasible domain, and it is easy to check that the global optimal solution is

$$x = 3, \quad y = 1, \qquad P = 2 \times 3 + 3 \times 1 = 9, \tag{7.13}$$

which is marked best in Figure 7.2.

But, how do we obtain the optimal solution in general? Can we relax the integer constraint by ignoring the integer requirement and solving the corresponding LP problem first, then rounding the solution to the nearest integers? This idea is basically the LP relaxation method.

If we ignore the integer requirement, the above IP problem becomes the LP problem we solved earlier. The solution to the real-variable linear program was

$$x_* = 2.5, \quad y_* = 1.5, \qquad P = 9.5. \tag{7.14}$$

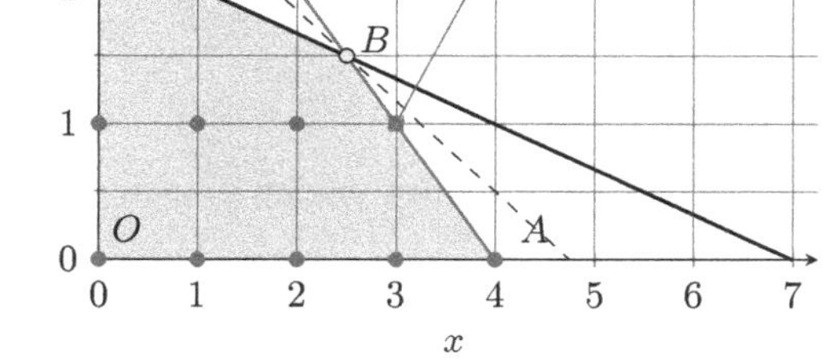

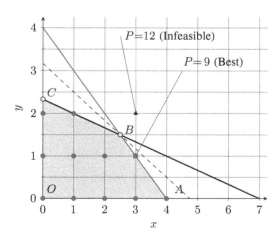

**Figure 7.2** Integer programming.

Since $x$ and $y$ have to be integers, we can now round this solution up. This is a special case, thus the above solution becomes four possibilities:

$$x = 3, \quad y = 1, \tag{7.15}$$

$$x = 3, \quad y = 2, \tag{7.16}$$

$$x = 2, \quad y = 1, \tag{7.17}$$

$$x = 2, \quad y = 2. \tag{7.18}$$

But, we have to check each case if all the constraints are satisfied. Thus, we have

$$x = 3, \quad y = 1, \ P = 9, \ \text{(feasible)}, \tag{7.19}$$

$$x = 3, \quad y = 2, P = 12, \text{(infeasible)}, \tag{7.20}$$

$$x = 2, \quad y = 1, \ P = 7, \ \text{(feasible)}, \tag{7.21}$$

$$x = 2, \quad y = 2, P = 10, \text{(infeasible)}. \tag{7.22}$$

The two infeasible solutions are outside the domain. Among the two feasible solutions, it is obvious that $x = 3$ and $y = 1$ are the global best solution with $P = 9$ (marked with a square in Figure 7.2).

It seems that the LP relaxation method can obtain the globally optimal solution as shown in the above example. However, this is a special or lucky case, but the global optimality is not necessarily achievable by such LP relaxation.

In order to show this key point, let us change the objective $P = 2x + 3y$ to a slightly different objective $Q = x + 4y$, subject to the same constraints as given earlier. Thus, we have

$$\text{maximize } Q = x + 4y, \tag{7.23}$$

subject to

$$x + y \leq 4, \tag{7.24}$$

$$x + 3y \leq 7, \tag{7.25}$$

$$x, y \geq 0, \tag{7.26}$$

$$x, y \in \mathbb{I}. \tag{7.27}$$

Since the constraints are the same, the feasible domain remains the same, and this problem is shown in Figure 7.3.

If we use the LP relaxation method by ignoring the integer constraints, we can solve the corresponding LP problem

$$\text{maximize } Q = x + 4y, \tag{7.28}$$

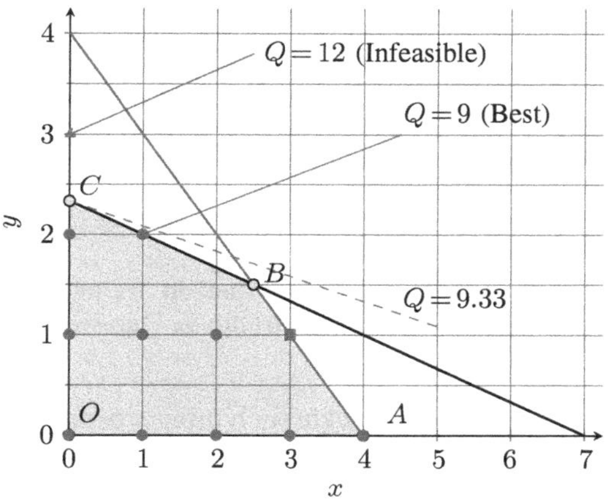


**Figure 7.3** Integer programming (with an alternative objective).

subject to

$$x + y \leq 4, \tag{7.29}$$

$$x + 3y \leq 7, \tag{7.30}$$

$$x, y \geq 0. \tag{7.31}$$

Its solution is at $C$ with

$$x_* = 0, \quad y_* = \frac{7}{3} = 2.333333, \quad P = \frac{28}{3} = 9.33333. \tag{7.32}$$

Now we round this solution to the nearest integers, we have

$$x = 0, \quad y = 2, \tag{7.33}$$

which gives

$$Q = 0 + 4 \times 2 = 8. \tag{7.34}$$

You may wonder why not round up to $x = 0$ and $y = 3$? In general, if we only round up to the nearest integers, the case $x = 0$ and $y = 3$ is usually not considered as the original constraint concerns the lower part of the boundary. Even we try to reach it by using different round-up rules such as ceiling functions, the solution for $x = 0$ and $y = 3$ is not feasible because it lies outside the feasible domain.

Now the question is: Is $Q = 8$ for $x = 0$ and $y = 2$ the best or highest we can get? By looking at the graph shown in Figure 7.3, it is clear that we can improve this solution by moving to the right at $(x = 1, y = 2)$, leading to

$$x = 1, \quad y = 2, \qquad Q = 1 + 4 \times 2 = 9, \tag{7.35}$$

which is a better solution. This solution is not achievable by the LP relaxation and the round-up method.

From the graph, we can conclude that $Q = 9$ at (1,2) is the global best solution. However, without looking at the graph, we may have to try ever possible combination to reach this conclusion. But, here there are two key issues:

1) This globally optimal solution cannot be achieved by first carrying out LP relaxation and then rounding up.
2) It is usually impractical to try every possible combination in practice because the number of combinations grows exponentially as the problem size increases.

Therefore, we have to use different approaches to tackle such integer programs in general. For this purpose, there are a few methods including the branch and bound, cut and bound, cutting planes, heuristic methods, and others. Here, we will only introduce the basic "branch and bound" method.

## 7.3 Branch and Bound

The main idea of the branch and bound idea is to divide and conquer. The branch part divides the problem into subproblems that are usually smaller problems and potentially easier to solve, while the bound part attempts to find a better solution by solving relevant subproblems and comparing the objective values (thus giving a potentially better bound). The basic procedure consists of three major steps:

1) Divide the problem into subproblems.
2) Solve each subproblem by LP relaxation.
3) Decide if a branch should be stopped.

For a given problem $P_1$, the branching division is carried out by choosing a variable with non-integer solutions (obtained by first using LP relaxation to solve $P_1$). If the optimal value for variable $x$ is $x*$ that is a non-integer, we have to divide the domain into two parts by adding an additional constraint:

1) Problem $P_2$: $x \leq \lfloor x* \rfloor$;
2) Problem $P_3$: $x \geq \lceil x* \rceil$;

then solve each of the subproblems. For example, if $x* = 2.3$, we have

$$\lfloor x* \rfloor = 2, \quad \lceil x* \rceil = 3. \tag{7.36}$$

Thus, the two new subproblems become

- Problem $P_2$: $x \leq 2$;
- Problem $P_3$: $x \geq 3$.

**Algorithm 7.1** Branch and bound method.

```
Initialization (e.g. bound, etc.);        ▷ e.g. bound = −∞ for maximization
while (still divisible) do
    Subdivide/branch into subproblems;
    Solve each subproblem by LP relaxation;
    if the LP subproblem has no solution then
        Stop;
    end if
    if the LP subproblem has a solution that is lower than the bound then
        Stop;
    end if
    if the LP subproblem has an integer solution then
        Compare and update the bound;
        Stop;
    end if
end while
```

This subdivision essentially makes the search domain smaller, which can potentially help to solve the problem quickly.

The stopping criteria for deciding a branching process are as follows:

- Stop if the new LP has no feasible solution (e.g. new branches violate any of the constraints).
- Stop if the LP problem has a lower optimal objective (for maximization problems).
- Stop if the LP problem has an integer optimal solution.

This process can also be viewed as a decision tree technique and each subproblem can be considered as a branch of the decision tree.

It is worth pointing out that for an IP, the LP relaxation will make its feasible domain slightly larger, thus the optimal solution obtained by solving the relaxed LP provides an upper bound for the IP for maximization because the solution to the IP cannot exceed the optimal solution to its corresponding LP in terms of the objective value in the same feasible domain (Algorithm 7.1).

This is an iterative procedure and, in the worse scenario, the number of iterations can be exponential in terms of its problem size. However, the optimal solutions to most IP problems can be found quite quickly in practice. We will illustrate how the method works using two examples.

**Example 7.1** Let us revisit the example problem we discussed earlier:

$$\text{Problem } P_1 : \text{maximize } Q = x + 4y, \tag{7.37}$$

subject to

$$x + y \leq 4, \tag{7.38}$$

$$x + 3y \leq 7, \tag{7.39}$$

$$x, y \geq 0, \tag{7.40}$$

$$x, y \in \mathbb{I}. \tag{7.41}$$

From the previous section, we know that the solution to the relaxed LP problem

$$\text{Relaxed LP} \quad P_1: \text{ maximize } Q = x + 4y, \tag{7.42}$$

subject to

$$x + y \leq 4, \tag{7.43}$$

$$x + 3y \leq 7, \tag{7.44}$$

$$x, y \geq 0, \tag{7.45}$$

gives the following solution:

$$x* = 0, \quad y* = \frac{7}{3} = 2.333333, \quad Q = \frac{28}{3} = 9.33333. \tag{7.46}$$

Since $x* = 0$ is an integer but $y*$ is not, so we carry out branching in terms of the $y$ variable and we have two subproblems by adding an additional constraint $y \leq 2$ or $y \geq 3$. Thus, we have two subproblems:

$$\text{Subproblem} \quad P_2: \text{ maximize } Q = x + 4y, \tag{7.47}$$

subject to

$$x + y \leq 4, \tag{7.48}$$

$$x + 3y \leq 7, \tag{7.49}$$

$$x, y \geq 0, \tag{7.50}$$

$$y \geq 3, \tag{7.51}$$

and

$$\text{Subproblem} \quad P_3: \text{ maximize } Q = x + 4y, \tag{7.52}$$

subject to

$$x + y \leq 4, \tag{7.53}$$

$$x + 3y \leq 7, \tag{7.54}$$

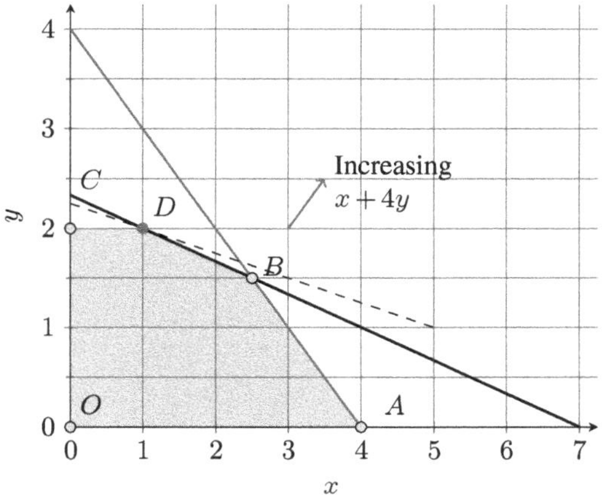


Figure 7.4 Branch and bound (Example 7.1, Subproblem $P_3$).

$$x, y \geq 0, \tag{7.55}$$

$$y \leq 2. \tag{7.56}$$

For Problem $P_2$, there is no feasible solution because $y \geq 3$ violates the constraint $x + 3y \leq 7$ for any $x \geq 0$. Thus, we can consider this problem is done.

For Problem $P_3$, the domain becomes slightly smaller and this problem is represented as the graph in Figure 7.4.

Now we use LP relaxation to solve this subproblem ($P_3$), and it is straightforward to show that its optimal solution is

$$x{*}{=}\ 1, \quad y{*}{=}\ 2, \qquad Q = 1 + 4 \times 2 = 9. \tag{7.57}$$

This corresponds to point $D$ on the graph shown in Figure 7.4. Since this is an integer solution, there is no need to subdivide any longer and we can stop this branch. As both subproblems ($P_2$ and $P_3$) have been solved, we can conclude that the optimal solution to the original IP problem has the globally optimal solution $Q = 9$ at (1,2).

It is worth pointing out that this solution $Q = 9$ is indeed smaller than the optimal solution $Q = 9.333\,33$ obtained by solving the relaxed LP problem. This problem is surprisingly simple and let us look at another example.

**Example 7.2** For the same feasible domain, optimal solutions will be different if the objective function is modified. Consequently, the branch and bound steps can also vary.

Now let us try to solve the following IP problem:

$$\text{Problem} \quad P_0: \text{ maximize } Q = 2x + 3y, \tag{7.58}$$

subject to

$$x + y \leq 4, \tag{7.59}$$

$$x + 3y \leq 7, \tag{7.60}$$

$$x, y \geq 0, \tag{7.61}$$

$$x, y \in \mathbb{I}. \tag{7.62}$$

Its corresponding relaxed LP becomes

$$\text{Relaxed LP } P_0: \text{ maximize } Q = 2x + 3y, \tag{7.63}$$

subject to

$$x + y \leq 4, \tag{7.64}$$

$$x + 3y \leq 7, \tag{7.65}$$

$$x, y \geq 0. \tag{7.66}$$

It is straightforward to show (as we did earlier) that the optimal solution is $Q = 9.5$ at

$$x{*}= 2.5, \quad y{*}= 1.5. \tag{7.67}$$

Since both $x*$ and $y*$ are not integers, we have to subdivide in terms of either $x$ or $y$. If we choose to branch by adding the additional constraint $x \leq 2$ or $x \geq 3$, we have the following two subproblems:

$$\text{Subproblem } P_1: \text{ maximize } Q = 2x + 3y, \tag{7.68}$$

subject to

$$x + y \leq 4, \tag{7.69}$$

$$x + 3y \leq 7, \tag{7.70}$$

$$x, y \geq 0, \tag{7.71}$$

$$x, y \in \mathbb{I}, \tag{7.72}$$

$$x \leq 2. \tag{7.73}$$

$$\text{Subproblem } P_2: \text{ maximize } Q = 2x + 3y, \tag{7.74}$$

subject to

$$x + y \leq 4, \tag{7.75}$$

$$x + 3y \leq 7, \tag{7.76}$$

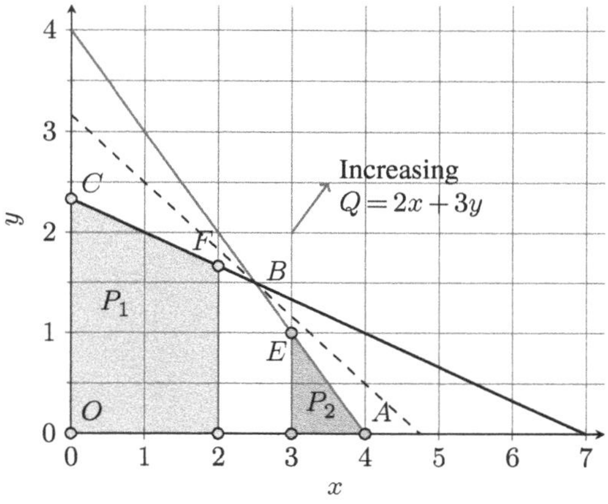


**Figure 7.5** Branch and bound (Example 7.2, subproblems).

$$x, y \geq 0, \tag{7.77}$$

$$x, y \in \mathbb{I}, \tag{7.78}$$

$$x \geq 3. \tag{7.79}$$

Both subproblems have feasible solutions and their graph representations are shown in Figure 7.5 where both subproblems are marked with gray regions.

Now we have to solve two subproblems along two different branches.

For Subproblem $P_2$, its solution by using its corresponding LP relaxation is simply

$$x* = 3, \quad y* = 1, \quad Q = 2 \times 3 + 3 \times 1 = 9. \tag{7.80}$$

Since this solution is an integer solution, there is no need to subdivide further and we can consider this branch is done. Thus, the optimal we obtain so far is $Q = 9$ at (3,1). The best bound we get so far is $Q = 9$.

For Subproblem $P_1$, its corresponding relaxed LP gives a solution

$$x*_1 = 2, \quad y*_1 = \frac{5}{3} = 1.6666, \quad Q = 9, \tag{7.81}$$

which corresponds to point $F$ in Figure 7.7. However, this solution is not an integer solution and we have to carry out further branching. By imposing additional constraint $y \geq 2$ or $y \leq 1$, we have two subproblems:

$$\text{Subproblem} \quad P_3: \text{ maximize } Q = 2x + 3y, \tag{7.82}$$

subject to

$$x + y \leq 4, \tag{7.83}$$

$$x + 3y \leq 7, \tag{7.84}$$

$$x, y \geq 0, \tag{7.85}$$

$$x, y \in \mathbb{I}, \tag{7.86}$$

$$x \leq 2, \tag{7.87}$$

$$y \geq 2, \tag{7.88}$$

$$\text{Subproblem} \quad P_4\text{: maximize } Q = 2x + 3y, \tag{7.89}$$

subject to

$$x + y \leq 4, \tag{7.90}$$

$$x + 3y \leq 7, \tag{7.91}$$

$$x, y \geq 0, \tag{7.92}$$

$$x, y \in \mathbb{I}, \tag{7.93}$$

$$x \leq 2, \tag{7.94}$$

$$y \leq 1. \tag{7.95}$$

Both subproblems have feasible solutions and their domains are marked in Figure 7.7. Now we have to solve each of the subproblems.

For Subproblem $P_3$, the relaxed LP gives the following solution:

$$x{*}_3 = 1, \quad y{*}_3 = 2, \quad Q_3 = 8, \tag{7.96}$$

which corresponds to Point $D$ in Figure 7.4. Since this is an integer solution and its objective value $Q_3 = 8$ is lower than the current best $Q = 9$, we can stop this branch. There is no need to pursue this branch any further.

For Subproblem $P_4$, it is straightforward to check that the solution to this IP problem by solving its relaxed LP is

$$x{*}_4 = 2, \quad y{*}_4 = 1, \quad Q_4 = 7, \tag{7.97}$$

which corresponds to Point $G$ in the same figure. Again this is an integer solution and the objective value is lower than $Q = 9$, we do not need to consider this branch any further.

Therefore, we can now conclude that the optimal solution to the original IP problem is

$$x{*} = 3, \quad y{*} = 1, \quad Q = 9. \tag{7.98}$$

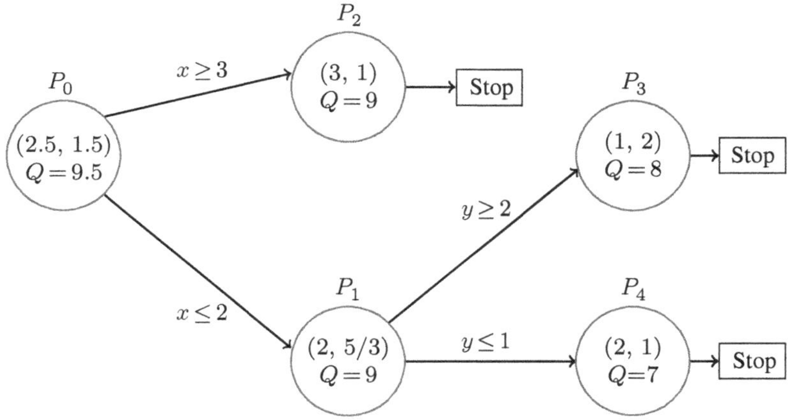


**Figure 7.6** Branch and bound tree.

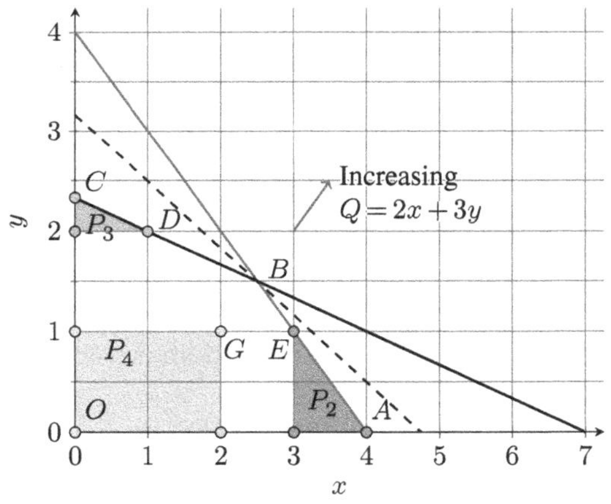


**Figure 7.7** Branch and bound (Example 7.2, continued).

This overall process can be summarized and represented as the branch and bound tree shown in Figure 7.6.

### 7.3.1 How to Branch

When we start the branch and bound process from the relaxed LP problem with the solution $x* = 2.5$ and $y* = 1.5$, we have actually carried out the branching process in terms of $x \geq 3$ and $x \leq 2$. You may wonder what happens if we carry out such branching first in terms of $y$ (instead of $x$)?

Now let us carry out such branch and bound starting with $y*= 1.5$ by adding additional constraint $y \leq 1$ or $y \geq 2$. Thus, we have two subproblems:

$$\text{Subproblem} \quad P_5: \text{ maximize } Q = 2x + 3y, \tag{7.99}$$

subject to

$$x + y \leq 4, \tag{7.100}$$

$$x + 3y \leq 7, \tag{7.101}$$

$$x, y \geq 0, \tag{7.102}$$

$$x, y \in \mathbb{I}, \tag{7.103}$$

$$y \leq 1. \tag{7.104}$$

$$\text{Subproblem} \quad P_6: \text{ maximize } Q = 2x + 3y, \tag{7.105}$$

subject to

$$x + y \leq 4, \tag{7.106}$$

$$x + 3y \leq 7, \tag{7.107}$$

$$x, y \geq 0, \tag{7.108}$$

$$x, y \in \mathbb{I}, \tag{7.109}$$

$$y \geq 2. \tag{7.110}$$

Both subproblems have feasible solutions and their feasible domains are represented in Figure 7.8.

For Subproblem $P_5$, it is easy to check that the optimal solution is

$$x*_5 = 1, \quad y*_5 = 2, \quad Q_5 = 8, \tag{7.111}$$

which is what we have obtained earlier and it corresponds to Point $D$. As this is an integer solution, we can stop this branch and record the best objective so far is $Q = 8$.

For Subproblem $P_6$, we should solve its corresponding relaxed LP

$$\text{Subproblem} \quad P_6: \text{ maximize } Q = 2x + 3y, \tag{7.112}$$

subject to

$$x + y \leq 4, \tag{7.113}$$

$$x + 3y \leq 7, \tag{7.114}$$

$$x, y \geq 0, \tag{7.115}$$

$$y \geq 2, \tag{7.116}$$

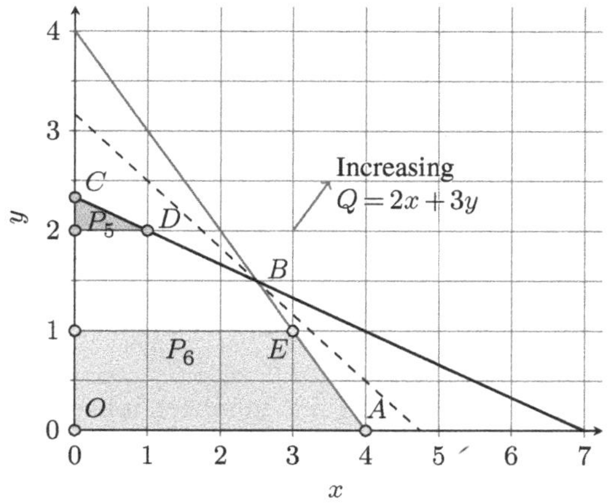


**Figure 7.8** Branch and bound (Example 7.2, alternative branching).

and its solution is

$$x*_6 = 3, \quad y*_6 = 1, \quad Q = 9, \tag{7.117}$$

which corresponds to Point $E$ in the same figure (see Figure 7.8).

Because this solution is an integer solution, we can stop this branch. Now the new best solution with a higher objective value $Q = 9$ is

$$x* = 3, \quad y* = 1, \tag{7.118}$$

which is the globally optimal solution for the original IP problem and it is the same solution we obtained earlier.

Comparing the branching steps in terms of $y$ with the branching process in terms of $x$, we only need to solve two subproblems instead of thee subproblems. Therefore, different branching decisions may lead to different branch trees and may potentially save some computational costs, so it may be worth the efforts to seek which way is better. However, there is no universal guide about how to branch at a particular point of branching trees.

## 7.4 Mixed Integer Programming

As the variables in standard LP problems are real numbers, if any of the variables is an integer variable, the LP problem becomes an MIP problem. The branch and bound method we have just discussed can be applied to solve such MIP problems without any modification. Obviously, the branch and bound should start with any of the integer variables and proceed as outlined in the previous section.

**Example 7.3** In the discussion of branch and bound methods, we have a problem

$$\text{maximize} \quad Q = x + 4y,$$

subject to

$$x + y \le 4, \; x + 3y \le 7,$$

and

$$x, y \ge 0, \quad x, y \in \mathbb{I}.$$

If we remove the integer constraint for $y$ (thus $y$ is a real number), it becomes an IP problem. Then, using the LP relaxation to solve this LP, we have

$$x_* = 0, \quad y_* = \frac{7}{3}, \quad Q_* = \frac{28}{3},$$

which is the global optimal solution for the MIP problem. In this case, the problem is solved without further branch and bound.

However, the branch and bound method may become computationally extensive when the number of integer variables is large. Other methods such as cutting planes, branch and cut, heuristic methods can be used. In general, MIP problems are usually NP-hard, thus there are no efficient methods to solve large-scale MIP problems. Recent trends tend to combine heuristic and metaheuristic methods with traditional methods. We will introduce some of these metaheuristic methods in later chapters.

## 7.5 Applications of LP, IP, and MIP

There are a diverse range of applications of LP, from transport problem to scheduling. Here, we only use five types of problems to demonstrate the ways of formulating problems in the standard forms.

### 7.5.1 Transport Problem

Transport problem concerns the optimal transport of products from $m$ suppliers (or manufacturers) to $n$ locations (such as shops, facilities, and demand centers). There is a transport cost matrix $C = [c_{ij}]$, which gives the cost for transporting one unit of product from Supplier $i$ to Demand $j$ (see Table 7.1).

Let $x_{ij}$ be the units to be transported from $S_i$ to $D_j$. Obviously, we have $x_{ij} \ge 0$ where $i = 1, 2, \ldots, m$ and $j = 1, 2, \ldots, n$.

**Table 7.1** Transport problem with $m$ suppliers and $n$ demands.

| Supply | Demand ($D_1$) | $D_2$ | ... | Demand ($D_n$) |
|---|---|---|---|---|
| $S_1$ | $c_{11}$ | $c_{12}$ | ... | $c_{1n}$ |
| $S_2$ | $c_{21}$ | $c_{22}$ | ... | $c_{2n}$ |
| $\vdots$ | $\vdots$ | $\vdots$ | $\ddots$ | $\vdots$ |
| $S_m$ | $c_{m1}$ | $c_{m2}$ | ... | $c_{mn}$ |

The objective is to minimize the overall transport cost

$$\text{minimize} f(x) = \sum_{i=1}^{m} \sum_{j=1}^{n} c_{ij} x_{ij}. \tag{7.119}$$

It makes sense that the total shipment from each supplier must be smaller or equal to each supply. That is,

$$\sum_{j=1}^{n} x_{ij} \leq S_i, \quad (i = 1, 2, \dots, m). \tag{7.120}$$

Similarly, each demand must be met; that is,

$$\sum_{i=1}^{m} x_{ij} \geq D_j, \quad (j = 1, 2, \dots, n). \tag{7.121}$$

In a special case of balanced supply and demand, the total supplies are equal to the total demands, which leads to

$$\sum_{i=1}^{m} S_i = \sum_{j=1}^{n} D_j. \tag{7.122}$$

In reality, supply and demand may not be balanced. However, it is always possible to convert it into a balanced case by using a dummy supplier (if the supply is not enough) or dummy demand (if the supply is higher than all demands). Such a dummy supplier or a demand center can be considered as a warehouse.

As any inequality can be converted into a corresponding equality by using a slack variable, we can now solely focus on the balanced supply and demand. In this case, all the inequality constraints become equality constraints. Thus, the above transport problem becomes the following standard LP transport problem:

$$\text{minimize} \sum_{i=1}^{m} \sum_{j=1}^{n} c_{ij} x_{ij}, \tag{7.123}$$

**Table 7.2** Product portfolio of a small factory.

| Product | Materials | Assembly | Transport | Profit |
|---|---|---|---|---|
| $A$ | 115 | 14 | 6 | 40 |
| $B$ | 70 | 20 | 10 | 50 |
| $C$ | 60 | 15 | 12 | 60 |
| $D$ | 35 | 5 | 2 | 20 |

subject to

$$\sum_{j=1}^{n} x_{ij} = S_i, \quad (i = 1, 2, \dots, m), \tag{7.124}$$

$$\sum_{i=1}^{m} x_{ij} = D_j, \quad (j = 1, 2, \dots, n), \tag{7.125}$$

$$\sum_{i=1}^{m} S_i = \sum_{j=1}^{n} D_j. \tag{7.126}$$

In addition, we have $x_{ij} \geq 0$ (non-negativity). This problem has $m \times n$ decision variables and $m + n + 1$ equality constraints.

From the above formulations, we can see that the transport problem is an LP problem, which can be solved by the simplex method. Many other applications such as resource allocation problems, staff assignment, and dietary problems can be formulated in a similar manner.

It is worth pointing out that the $x_{ij}$ in the above formulation can be any non-negative real numbers for many applications. However, for some applications such as staff assignment and transport problems, $x_{ij}$ must be integer values. After all, the number of staff or products cannot be 1.5 or 3.7. With an additional integer requirement, the above LP becomes an integer LP problem.

### 7.5.2 Product Portfolio

Portfolio optimization is relevant to many applications such as product mix optimization and financial portfolio optimization. Here, we will use an example to show how the portfolio of products in a small factory can be optimized.

**Example 7.4** Suppose a small factory produces a portfolio of four different products ($A, B, C, D$) with different materials costs, assembly costs, and transport costs (all in £) as shown in Table 7.2.

Obviously the resources of the factory are limited, and it can only allow maximum costs (each day) £7000 for materials, £1500 for assembly, and £600 for transport, respectively. In addition, in order to make basic profits, it must produce at least 10 units for each product.

The main aim is to design an optimal product portfolio so that the total profit is maximized on a daily basis.

Let $x_i \geq 0$ $(i = 1, 2, 3, 4)$ be the numbers (or units) of products $(A, B, C, D,$ respectively) to be produced daily. The daily profit can be calculated by

$$P(x) = 40x_1 + 50x_2 + 60x_3 + 20x_4. \tag{7.127}$$

The resource or budget for materials is limited to £7000. That is,

$$115x_1 + 70x_2 + 60x_3 + 35x_4 \leq 7000. \tag{7.128}$$

Similarly, the assembly cost is limited to £1500; that is,

$$14x_1 + 20x_2 + 15x_3 + 5x_4 \leq 1500. \tag{7.129}$$

The resource requirement for transport becomes

$$6x_1 + 10x_2 + 12x_3 + 2x_4 \leq 600. \tag{7.130}$$

We leave it as an exercise for the readers to show that the optimal portfolio is $A = 10, B = 10, C = 17,$ and $D = 118$ with total daily profit $P = 4280$. However, this is a special case that the solution is an integer vector. In general, this is not the case, and we have to impose an additional constraint that all variables must be nonnegative integers. This becomes an integer LP problem.

In general, if there are $n$ different products with $c_i$ $(i = 1, 2, \dots, n)$ profit per unit, the overall profit can be written as

$$P(x) = \sum_{i=1}^{n} c_i x_i, \tag{7.131}$$

where $x_i \geq 0$ are the units/numbers of products to be produced.

For each stage $j$ $(j = 1, 2, K)$ for producing the products (such as materials, assembly), the resource requirements mean that

$$\sum_{i=1}^{n} r_{ij} x_i \leq b_j \quad (j = 1, 2, \dots, K), \tag{7.132}$$

where $b_j$ is the budget constraint for stage $i$ and $r_{ij}$ is the unit cost of Product $i$ to be processed at stage $j$.

Therefore, the product portfolio problem can be summarized as

$$\text{maximize} \quad \sum_{i=1}^{n} c_i x_i, \tag{7.133}$$

subject to

$$\sum_{i=1}^{n} r_{ij} x_i \leq b_j \qquad (j = 1, 2, \dots, K), \tag{7.134}$$

$$x_i \geq 0 \quad (i = 1, 2, \dots, n). \tag{7.135}$$

**Table 7.3** Production scheduling.

| Season | Unit cost | Unit price |
|---|---|---|
| Spring | 12 | — |
| Summer | 15 | 23 |
| Autumn | 14 | 19 |
| Winter | — | 21 |

Other portfolio optimization can be written in a similar form, even though the constraints can be more complicated, depending on the types of problems.

### 7.5.3 Scheduling

Scheduling itself is a class of challenging problems that can be NP-hard in many applications. Such scheduling problems can also be formulated as LP problems, though the detailed formulation can be problem-dependent.

For the scheduling of simple manufacturing problems, this can be done easily, though extra care and conditions may be needed to make sure that the problem is properly formulated. Let us look at an example.

**Example 7.5** Suppose a company that manufactures a certain type of expensive products such as small aircraft. The manufacturing costs vary with season. For example, the cost of manufacturing one product is \$12 million in spring, while this product can be sold at a price of \$23 million. The details are shown in Table 7.3 where "–" means no production or sales. The company can have a capacity of producing at most 50 units per season, while their warehouse can store up to 10 units at most. Ideally, all the products should be sold by the end of the year (winter) so that nothing is left in the warehouse.

Let $x_1, x_2, x_3, x_4$ be the numbers of units produced in four seasons, respectively. It is required that $x_4 = 0$ because there is no production in the winter (otherwise, it cannot be sold by the end of the season). Let $s_i$ $(i = 1, 2, 3, 4)$ be the numbers of sales in each season. Obviously, $s_1 = 0$ because there is nothing to sell in spring. Therefore, the profit is the difference between sales income and the production costs; that is,

$$P = 23s_2 + 19s_3 + 21s_4 - (12x_1 + 15x_2 + 14x_3). \tag{7.136}$$

In order to manage the warehouse, we denote the numbers of units in the warehouse by $w_1, w_2, w_3, w_4$ for four seasons, respectively. Obviously, $0 \leq w_i \leq 10$ for $i = 1, 2, 3, 4$.

Since all products must be sold out by the end of the year, it requires that $w_4 = 0$. At the end of each season, the units in the warehouse is the number of units in the previous season plus the production in this season, subtracting the actual sales in this season. Thus, we have

$$w_1 = x_1 \quad \text{(no sales)}, \qquad w_2 = w_1 + x_2 - s_2,$$

and

$$w_3 = w_2 + x_3 - s_3, \qquad w_4 = w_3 - s_4 = 0.$$

In addition, we have

$$w_i \leq 10, \quad x_i, s_i \leq 50, \quad x_i, s_i, w_i \geq 0 \quad (i = 1, 2, 3, 4).$$

Furthermore, we have imposed that all $x_i$, $s_i$, $w_i$ are integers.

This production scheduling problem has demonstrated that even though there are only four variables for production, we have to use additional variables to model sales and warehouse. Thus, the actual formulation becomes slightly more complicated than expected. In fact, many scheduling problems can have very complicated constraints, which makes implementation quite tricky.

### 7.5.4 Knapsack Problem

Knapsack problems are a class of problem of packing items with a limited capacity, and these problems can be NP-hard. A subclass of such problem is the 0-1 knapsack problem where choices are made among $n$ items/objects so as to maximize some utility value.

Let $x_i$ $(i = 1, 2, \dots, n)$ be the decision variables for $n$ items. If $x_i = 1$, it means that $i$th item is chosen. Otherwise, we set $x_i = 0$ for not choosing that item. Each item has a utility value $p_i$ and a weight $w_i$ $(i = 1, 2, \dots, n)$. The knapsack has a fixed capacity $C > 0$.

The objective is to maximize

$$f(x) = \sum_{i=1}^{n} p_i x_i. \tag{7.137}$$

The constraint is

$$\sum_{i=1}^{n} w_i x_i \leq C. \tag{7.138}$$

Obviously, all $x_i$ are either 0 or 1.

### 7.5.5 Traveling Salesman Problem

The traveling salesman problem (TSP) concerns the tour of $n$ cities once and exactly once starting from a city and returning to the same starting city so that the total distance traveled is the minimum.

There are many different ways to formulate the TSP, and there is a vast literature on this topic. Here, we will use the binary integer LP formulation by Dantzig et al. in 1954.

Let $i = 1, 2, \ldots, n$ be $n$ cities, which can be considered as the nodes of a graph. Let $x_{ij}$ be the decision variable for connecting city $i$ to city $j$ (i.e. an edge of the graph from node $i$ to node $j$) such that $x_{ij} = 1$ means the tour starts from $i$ and ends at $j$. Otherwise, $x_{ij} = 0$ means no connection along this edge. Therefore, the cities form the set $V$ of vertices and connections form the set $E$ of the edges.

Let $d_{ij}$ be the distance between city $i$ to city $j$. Due to symmetry, we know that $d_{ij} = d_{ij}$, which means the graph is undirected. The objective is to

$$\text{minimize} \sum_{i,j \in E, i \neq j} d_{ij} x_{ij}. \tag{7.139}$$

One of the constraints is that

$$x_{ij} = 0 \quad \text{or} \quad 1, \quad (i,j) \in V. \tag{7.140}$$

Since each city $i$ should be visited once and only once, it is required that

$$\sum_{i \in V} x_{ij} = 2 \quad (\forall j \in V). \tag{7.141}$$

This means that only one edge enters a city and only one edge leaves the city, which be equivalently written as

$$\sum_{j=1}^{n} x_{ij} = 1 \quad (i \in V, i \neq j) \tag{7.142}$$

and

$$\sum_{i=1}^{n} x_{ij} = 1 \quad (j \in V, j \neq i). \tag{7.143}$$

In order to avoid unconnected subtour, for any non-empty subset $S \subset V$ of $n$ cities, it is required that

$$\sum_{i,j \in S} x_{ij} \leq |S| - 1 \quad (2 \leq |S| \leq n - 2), \tag{7.144}$$

where $|S|$ means the number of elements or the cardinality of the subset $S$. This is a binary integer LP problem, but it is NP-hard.

Many problems in practice can be cast as LP problems. Further examples include the network flow, flow shop scheduling, packaging problems, vehicle routing, coverage problem, graph coloring problem, and many others. Interested readers can refer to more advanced books listed in the references.

## Exercises

**7.1** Using branch and bound to solve the following problem:

$$\text{maximize} \quad P = 3x + 4y,$$

subject to

$$8x + 3y \leq 50, \quad 3x + 7y \leq 25, \quad x, y \in \mathbb{I}, \quad x, y \geq 0.$$

Can you use the LP relaxation to round down the solution to integers?

**7.2** Using any method to show that the solution to the problem in Eq. (7.127) is (10, 10, 17, 118).

**7.3** What is the dual problem of the previous problem? What is its optimal solution?

## Bibliography


Castillo, E., Conejo, A.J., Pedregal, P. et al. (2002). *Building and Solving Mathematical Programming Models in Engineering and Science*. New York: Wiley.

Cook, W.J., Cunningham, W.H., Pulleyblank, W.R. et al. (1997). *Combinatorial Optimization*. New York: Wiley.

Conforti, M., Cornuejols, G., and Zambelli, G. (2016). *Integer Programming*. Heidelberg: Springer.

Dantzig, G.B., Fulkerson, D.R. and Johnson, S.M. (1954). Solution of a large-scale traveling salesman problem. *Operations Research* 2: 393–410.

Dantzig, G.B. and Thapa, M.N. (1997). *Linear Programming 1: Introduction*. New York: Springer-Verlag.

Floudas, C.A. (1995). *Nonlinear and Mixed-Integer Optimization: Fundamentals and Applications*. Oxford: Oxford University Press.

Heizer, J., Render, B., and Munson, C. (2016). *Operations Management: Sustainability and Supply Chain Management*, 12e. London: Pearson.

Lee, J. and Leyffer, S. (2011). *Mixed Integer Nonlinear Programming*. New York: Spinger.

Mansini, R., Ogryczak, W., and Speranza, M.G. (2016). *Linear and Mixed Integer Programming for Portfolio Optimization*. Heidelberg: Springer.

Marchand, H., Martin, A., Weismantel, R. et al. (2002). Cutting planes in integer and mixed integer programming. *Discrete Applied Mathematics* 123 (1–3): 397–446.

Matai, R., Singh, S.P., and Mittal, M.L. (2010). Traveling salesman problem: an overview of applications, formulations, and solution approaches. In: *Traveling Salesman Problem, Theory and Applications* (ed. D. Davendra), 1–24. Croatia: InTech Europe.

Orman, A.J. and Williams, H.P. (2004). *A Survey of Different Integer Programming Formulations of the Travelling Salesman Problem*. Operational Research Working Paper. London: LSEOR, Department of Operational Research, London School of Economics and Political Science.

Pochet, Y. and Wolsey, L.A. (2009). *Production Planning by Mixed Integer Programming*. New York: Springer.

Sawik, T. (2011). *Scheduling in Supply Chains Using Mixed Integer Programming*. Hoboken, NJ: Wiley.

Schrijver, A. (1998). *Theory of Linear and Integer Programming*. New York: Wiley.

Vanderbei, R.J. (2014). *Linear Programming: Foundations and Extensions*, 4e. New York: Springer.

# 8

# Regression and Regularization

Regression can help to identify the trends in data and relationship between different quantities. Regression is one of the simplest forms of classification and supervised learning, and it is one of the most widely used data-processing techniques.

## 8.1 Sample Mean and Variance

If a sample consists of $n$ independent observations $x_1, x_2, \ldots, x_n$ on a random variable $x$ such as the noise level on a road or the price of a cup of coffee, two important and commonly used parameters are sample mean and sample variance, which can easily be estimated from the sample. The sample mean is calculated by

$$\bar{x} \equiv <x> = \frac{1}{n}(x_1 + x_2 + \cdots + x_n) = \frac{1}{n}\sum_{i=1}^{n} x_i, \tag{8.1}$$

which is essentially the arithmetic average of the values $x_i$.

The sample variance $S^2$ is defined by

$$S^2 = \frac{1}{n-1}\sum_{i=1}^{n}(x_i - \bar{x})^2. \tag{8.2}$$

Let us look at an example.

**Example 8.1** The measurements of a quantity such as the noise level on a road. The readings in dB are:

66, 73, 73, 74, 83, 70, 69, 77, 72, 75.

*Optimization Techniques and Applications with Examples*, First Edition. Xin-She Yang.

From the data, we know that $n = 10$ and the mode is 73 as 73 appears twice (all the rest only appears once). The sample mean is

$$\begin{aligned}\overline{x} &= \frac{1}{10}(x_1 + x_2 + \cdots + x_{10}) \\ &= \frac{1}{10}(66 + 73 + 73 + 74 + 83 + 70 + 69 + 77 + 72 + 75) \\ &= \frac{732}{10} = 73.2.\end{aligned}$$

The corresponding sample variance can be calculated by

$$\begin{aligned}S^2 &= \frac{1}{n-1}\sum_{i=1}^{n}(x_i - \overline{x})^2 \\ &= \frac{1}{10-1}\sum_{i=1}^{10}(x_i - 73.2)^2 = \frac{1}{9}[(66 - 73.2)^2 + (73 - 73.2)^2 \\ &\quad + \cdots + (75 - 73.2)^2] \\ &= \frac{1}{9}[(-7.2)^2 + (-0.2)^2 + \cdots + (1.8)^2] = \frac{195.6}{9} \approx 21.73.\end{aligned}$$

Thus, the standard derivation is

$$S = \sqrt{S^2} \approx \sqrt{21.73} \approx 4.662.$$

Generally speaking, if $u$ is a linear combination of $n$ independent random variables $y_1, y_2, \ldots, y_n$ and each random variable $y_i$ has an individual mean $\mu_i$ and a corresponding variance $\sigma_i^2$, we have the linear combination

$$u = \sum_{i=1}^{n} \alpha_i y_i = \alpha_1 y_1 + \alpha_2 y_2 + \cdots + \alpha_n y_n, \tag{8.3}$$

where the parameters $\alpha_i (i = 1, 2, \ldots, n)$ are the weighting coefficients. From the well-known central limit theorem, we have the mean $\mu_u$ of the linear combination

$$\mu_u = E[u] = E\left[\sum_{i=1}^{n} \alpha_i y_i\right] = \sum_{i=1}^{n} \alpha_i E[y_i] = \sum \alpha_i \mu_i. \tag{8.4}$$

Then, the variance $\sigma_u^2$ of the combination is

$$\sigma_u^2 = E[(u - \mu_u)^2] = E\left[\sum_{i=1}^{n} \alpha_i (y_i - \mu_i)^2\right], \tag{8.5}$$

which can be expanded as

$$\sigma_u^2 = \sum_{i=1}^{n} \alpha_i^2 E[(y_i - \mu_i)^2] + \sum_{i,j=1; i \neq j}^{n} \alpha_i \alpha_j E[(y_i - \mu_i)(y_j - \mu_j)], \tag{8.6}$$

where $E[(y_i - \mu_i)^2] = \sigma_i^2$. Since $y_i$ and $y_j$ are independent, we have

$$E[(y_i - \mu_i)(y_j - \mu_j)] = E[y_i - \mu_i]E[y_j - \mu_j] = 0. \tag{8.7}$$

Therefore, we get

$$\sigma_u^2 = \sum_{i=1}^{n} \alpha_i^2 \sigma_i^2. \tag{8.8}$$

The sample mean defined in Eq. (8.1) can also be viewed as a linear combination of all the $x_i$ assuming that each of which has the same mean $\mu_i = \mu$ and variance $\sigma_i^2 = \sigma^2$, and the same weighting coefficient $\alpha_i = 1/n$. Hence, the sample mean is an unbiased estimate of the sample due to the fact $\mu_{\bar{x}} = \sum_{i=1}^{n} \mu/n = \mu$. In this case, however, we have the variance

$$\sigma_{\bar{x}}^2 = \sum_{i=1}^{n} \frac{1}{n^2}\sigma^2 = \frac{\sigma^2}{n}, \tag{8.9}$$

which means the variance becomes smaller as the size $n$ of the sample increases by a factor of $1/n$.

For the sample variance $S^2$ defined earlier by

$$S^2 = \frac{1}{n-1} \sum_{i=1}^{n} (x_i - \bar{x})^2, \tag{8.10}$$

we can see that the factor is $1/(n-1)$ not $1/n$ because only $1/(n-1)$ will give the correct and unbiased estimate of the variance. From the basic introduction of probability theory in Chapter 2, we know that $E[x^2] = \mu^2 + \sigma^2$. The mean of the sample variance is

$$\mu_{S^2} = E\left[\frac{1}{n-1} \sum_{i=1}^{n} (x_i - \bar{x})^2\right] = \frac{1}{n-1} \sum_{i=1}^{n} E\left[(x_i^2 - n\bar{x}^2)\right]. \tag{8.11}$$

Using $E[\bar{x}^2] = \mu^2 + \sigma^2/n$, we get

$$\mu_{S^2} = \frac{1}{n-1} \sum_{i=1}^{n} \{E[x_i^2] - nE[\bar{x}^2]\}$$
$$= \frac{1}{n-1}\left[n(\mu^2 + \sigma^2) - n\left(\mu^2 + \frac{\sigma^2}{n}\right)\right] = \sigma^2. \tag{8.12}$$

Obviously, if we used the factor $1/n$ instead of $1/(n-1)$, we would get $\mu_{S^2} = (n-1)/n\sigma^2 < \sigma^2$, which would underestimate the sample variance. The other way to think about the factor $1/(n-1)$ is that we need at least one value to estimate the mean, and we need at least two values to estimate the variance. Thus, for $n$ observations, only $n-1$ different values of variance can be obtained to estimate the total sample variance.

## 8.2 Regression Analysis

Regression is a class of methods that are mostly based on the method of least squares and the maximum likelihood theory.

### 8.2.1 Maximum Likelihood

For a sample of $n$ values $y_1, y_2, \dots, y_n$ of a random variable $Y$ whose probability density function $p(y)$ depends on a set of $k$ parameters $\beta_1, \dots, \beta_k$, the joint probability is the product of all the probabilities, that is,

$$\Phi(\beta_1, \dots, \beta_k) = \prod_{i=1}^{n} p(y_i, \beta_1, \dots, \beta_k)$$
$$= p(y_1, \beta_1, \dots, \beta_k) \cdot p(y_2, \beta_1, \dots, \beta_k) \cdots p(y_n, \beta_1, \dots, \beta_k), \tag{8.13}$$

where $\Pi$ means the product of all its components. For example, $\Pi_{n=1}^{3} a_i = a_1 \times a_2 \times a_3$. The essence of the maximum likelihood is to maximize $\Phi$ by choosing the parameters $\beta_j$. As the sample can be considered as given values, the maximum likelihood requires the following stationary conditions:

$$\frac{\partial \Phi}{\partial \beta_j} = 0 \quad (j = 1, 2, \dots, k), \tag{8.14}$$

whose solutions for $\beta_j$ are the maximum likelihood estimates.

Regression is a special case of the method of least squares. Many other problems can be reformulated in this framework.

### 8.2.2 Regression

For experiments and observations, we usually plot one variable such as pressure or price $y$ against another variable $x$ such as time or spatial coordinates. We try to present the data in such a way that we can see some trend in the data.

For a set of $n$ data points $(x_i, y_i)$, the usual practice is to try to draw a straight line $y = a + bx$ so that it represents the major trend. Such a line is often called the regression line or the best fit line as shown in Figure 8.1.

The method of linear least squares is to try to determine the two parameters, $a$ (intercept) and $b$ (slope), for the regression line from $n$ data points, assuming that $x_i$ are known more precisely, and the values of $y_i$ obey a normal distribution around the potentially best fit line with a variance $\sigma^2$. So we have the joint probability with each being normally distributed

$$P = \prod_{i=1}^{n} p(y_i) = A \exp\left\{ -\frac{1}{2\sigma^2} \sum_{i=1}^{n} [y_i - f(x_i)]^2 \right\}, \tag{8.15}$$

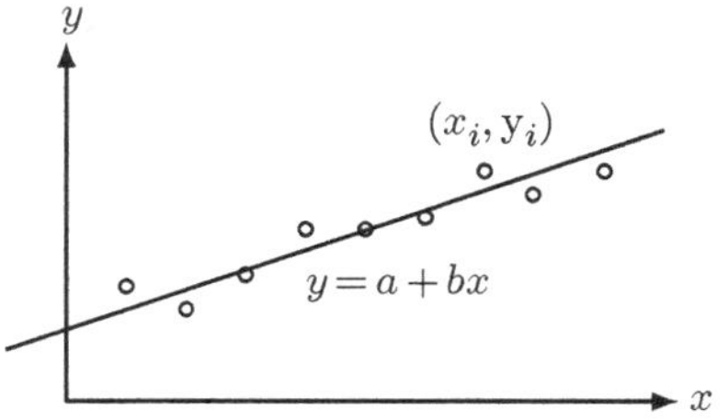


**Figure 8.1** Least square and the best fit line.

where $A$ is a constant, and $f(x)$ is the function for the regression [$f(x) = a + bx$ for the linear regression]. It is worth pointing out that the exponent

$$\psi = \frac{\sum_{i=1}^{n}[y_i - f(x_i)]^2}{\sigma^2}$$

is essentially a weighted sum of residuals or errors.

The maximization of $P$ is equivalent to the minimization of $\psi$. In order to minimize $\psi$ as a function of $a$ and $b$ via the model of $f(x) = a + bx$, its derivatives should be zero. That is

$$\frac{\partial \psi}{\partial a} = -2 \sum_{i=1}^{n} [y - (a + bx_i)] = 0 \tag{8.16}$$

and

$$\frac{\partial \psi}{\partial b} = -2 \sum_{i=1}^{n} x_i [y_i - (a + bx_i)] = 0, \tag{8.17}$$

where we have used $\sigma^2 \neq 0$ and thus omitted this factor.

By expanding these equations, we have

$$na + b \sum_{i=1}^{n} x_i = \sum_{i=1}^{n} y_i \tag{8.18}$$

and

$$a \sum_{i=1}^{n} x_i + b \sum_{i=1}^{n} x_i^2 = \sum_{i=1}^{n} x_i y_i \tag{8.19}$$

which is a system of linear equations for $a$ and $b$, and it is straightforward to obtain the solutions as

$$a = \frac{1}{n} \left[ \sum_{i=1}^{n} y_i - b \sum_{i=1}^{n} x_i \right] = \bar{y} - b\bar{x}, \tag{8.20}$$

$$b = \frac{n \sum_{i=1}^{n} x_i y_i - (\sum_{i=1}^{n} x_i)(\sum_{i=1}^{n} y_i)}{n \sum_{i=1}^{n} x_i^2 - (\sum_{i=1}^{n} x_i)^2}, \tag{8.21}$$

where

$$\bar{x} = \frac{1}{n}\sum_{i=1}^{n} x_i, \qquad \bar{y} = \frac{1}{n}\sum_{i=1}^{n} y_i. \tag{8.22}$$

If we use the following notations:

$$K_x = \sum_{i=1}^{n} x_i, \qquad K_y = \sum_{i=1}^{n} y_i, \tag{8.23}$$

and

$$K_{xx} = \sum_{i=1}^{n} x_i^2, \qquad K_{yy} = \sum_{i=1}^{n} y_i^2, \qquad K_{xy} = \sum_{i=1}^{n} x_i y_i, \tag{8.24}$$

then the above equations for $a$ and $b$ become

$$a = \frac{K_{xx}K_y - K_x K_{xy}}{nK_{xx} - (K_x)^2} \tag{8.25}$$

and

$$b = \frac{nK_{xy} - K_x K_y}{nK_{xx} - (K_x)^2}. \tag{8.26}$$

The residual error is defined by

$$\epsilon_i = y_i - (a + bx_i), \tag{8.27}$$

whose sample mean is given by

$$\mu_\epsilon = \frac{1}{n}\sum_{i=1}^{n} \epsilon_i = \frac{1}{n} y_i - a - b\frac{1}{n}\sum_{i=1}^{n} x_i = \bar{y} - a - b\bar{x} = 0. \tag{8.28}$$

The sample variance $S^2$ is

$$S^2 = \frac{1}{n-2}\sum_{i=1}^{n} [y_i - (a + bx_i)]^2 = \frac{1}{n-2}\text{RSS}, \tag{8.29}$$

where the RSS stands for the residual sum of squares given by

$$\text{RSS} = \sum_{i=1}^{n} [y_i - f(x_i)]^2 = \sum_{i=1}^{n} [y_i - (a + bx_i)]^2. \tag{8.30}$$

Here, the factor $1/(n-2)$ comes from the fact that two constraints are needed for the best fit, and therefore the residuals have $n-2$ degrees of freedom.

The correlation coefficient $r_{x,y}$ is a very useful parameter for finding any potential relationship between two sets of data $x_i$ and $y_i$ for two random variables $x$ and $y$, respectively. If $x$ has a mean $\bar{x}$ and a sample variance $S_x^2$, and $y$ has a mean $\bar{y}$ and a sample variance $S_y^2$, we have

$$\mathrm{var}(x) = S_x^2 = \frac{\sum_{i=1}^{n}(x_i - \bar{x})^2}{n-1}, \quad \mathrm{var}(y) = S_y^2 = \frac{\sum_{i=1}^{n}(y_i - \bar{y})^2}{n-1}. \tag{8.31}$$

The correlation coefficient is defined by

$$r_{x,y} = \frac{\mathrm{cov}(x,y)}{S_x S_y} = \frac{E[xy] - \bar{x}\,\bar{y}}{S_x S_y}, \tag{8.32}$$

where

$$\mathrm{cov}(x,y) = E[(x-\bar{x})(y-\bar{y})] = E[xy] - \bar{x}\,\bar{y} \tag{8.33}$$

is the covariance, which can be calculated explicitly by

$$\mathrm{cov}(x,y) = \frac{\sum_{i=1}^{n}(x_i - \bar{x})(y_i - \bar{y})}{n-1}. \tag{8.34}$$

It is worth pointing out that $S_x$ and $S_y$ must be sample variances; otherwise, the result is incorrect. In addition, we can also write

$$S_x = \mathrm{cov}(x,x), \quad S_y = \mathrm{cov}(y,y). \tag{8.35}$$

It is obvious that $\mathrm{cov}(x,y) = \mathrm{cov}(y,x)$. Thus, the covariance matrix

$$\boldsymbol{C}_s = \begin{pmatrix} \mathrm{cov}(x,x) & \mathrm{cov}(x,y) \\ \mathrm{cov}(y,x) & \mathrm{cov}(y,y) \end{pmatrix} = \begin{pmatrix} S_x & \mathrm{cov}(x,y) \\ \mathrm{cov}(x,y) & S_y \end{pmatrix} \tag{8.36}$$

is also symmetric.

If the two variables are independent or $\mathrm{cov}(x,y) = 0$, there is no correlation between them ($r_{x,y} = 0$). If $r_{x,y}^2 = 1$, then there is a linear relationship between these two variables. $r_{x,y} = 1$ is an increasing linear relationship where the increase of one variable will lead to the increase of another. On the other hand, $r_{x,y} = -1$ is a decreasing relationship when one increases while the other decreases. In general, we have $-1 \le r_{x,y} \le 1$.

For a set of $n$ data points $(x_i, y_i)$, the correlation coefficient can be calculated by

$$r_{x,y} = \frac{n\sum_{i=1}^{n} x_i y_i - \sum_{i=1}^{n} x_i \sum_{i=1}^{n} y_i}{\sqrt{\left[n\sum x_i^2 - (\sum_{i=1}^{n} x_i)^2\right]\left[n\sum_{i=1}^{n} y_i^2 - (\sum_{i=1}^{n} y_i)^2\right]}}$$

or

$$r_{x,y} = \frac{nK_{xy} - K_x K_y}{\sqrt{(nK_{xx} - K_x^2)(nK_{yy} - K_y^2)}}, \tag{8.37}$$

where $K_{yy} = \sum_{i=1}^{n} y_i^2$.

**Table 8.1** Two quantities with measured data.

| Input ($H$) | Output ($Y$) | Input ($H$) | Output ($Y$) |
|---|---|---|---|
| 90 | 270 | 300 | 910 |
| 110 | 330 | 350 | 1080 |
| 140 | 410 | 400 | 1270 |
| 170 | 520 | 450 | 1450 |
| 190 | 560 | 490 | 1590 |
| 225 | 670 | 550 | 1810 |
| 250 | 750 | 650 | 2180 |

Now let us look at an example that are measurements of two quantities ($H$ and $Y$). The data for a set of random samples are given in Table 8.1. Is there any relationship between these two quantities?

Now let us try to do a linear regression in the following form:

$$Y = a + bH.$$

**Example 8.2** From the data in Table 8.1 with $n = 14$, we can calculate

$$K_H = \sum_{i=1}^{14} H_i = 90 + 110 + \cdots + 650 = 4365,$$

$$K_Y = \sum_{i=1}^{14} Y_i = 270 + 330 + \cdots + 2180 = 13\,800,$$

$$K_{HY} = \sum_{i=1}^{14} H_i Y_i = 90 * 270 + \cdots + 650 * 2180 = 5\,654\,150,$$

$$K_{HH} = \sum_{i=1}^{14} H_i^2 = 90^2 + \cdots + 650^2 = 1\,758\,025,$$

and

$$K_{YY} = \sum_{i=1}^{14} Y_i^2 = 270^2 + \cdots + 2180^2 = 18\,211\,800.$$

Thus, we get

$$a = \frac{K_{HH}K_Y - K_H K_{HY}}{nK_{HH} - K_H^2}$$
$$= \frac{1\,758\,025 \times 13\,800 - 4\,365 \times 5\,654\,150}{14 \times 1\,758\,025 - 4365^2} \approx -75.48$$

and

$$b = \frac{nK_{HY} - K_H K_Y}{nK_{HH} - K_H^2}$$
$$= \frac{14 \times 5\,654\,150 - 4365 \times 13\,800}{14 \times 1\,758\,025 - 4365^2} \approx 3.404.$$

So the regression line becomes

$$Y = -75.48 + 3.404H.$$

Therefore, their correlation coefficient $r$ is given by

$$r = \frac{nK_{HY} - K_H K_Y}{\sqrt{(nK_{HH} - K_H^2)(nK_{YY} - K_Y^2)}}$$
$$= \frac{14 \times 5\,654\,150 - 4365 \times 13\,800}{\sqrt{(14 \times 1\,758\,025 - 4365^2)(14 \times 18\,211\,800 - 13\,800^2)}} \approx 0.999\,03.$$

This is a relatively strong correlation indeed.

The above formulations are based on the fact that the curve-fitting function $y = f(x) = a + bx$ is linear in terms of the independent variable $x$ and the parameters ($a$ and $b$). Here, the key linearity is about parameters, but not about the basis function $x$. Thus, the above technique can still be applicable to both $f(x) = a + bx + cx^2$ and $g(x) = a + b\sin(x)$ functions with some minor adjustments to be discussed later in this chapter. However, if we have a function in the form:

$$y = \ln(a + bx),$$

then the above technique cannot be applied directly, and some linearization approximations should be used.

### 8.2.3 Linearization

Sometimes, some obviously nonlinear functions can be transformed into linear forms so as to carry out linear regression, instead of more complicated nonlinear regression. However, there is no general formula for such linearization and thus it is often necessary to deal with each case individually. This can be illustrated by some examples.

**Example 8.3** For example, the following nonlinear function:

$$f(x) = \alpha e^{-\beta x} \tag{8.38}$$

can be transformed into a linear form by taking logarithms of both sides. We have

$$\ln f(x) = \ln(\alpha) - \beta x, \tag{8.39}$$

which is equivalent to $y = a + bx$ if we let $y = \ln f(x)$, $a = \ln(\alpha)$, and $b = -\beta$.
In addition, the following function:

$$f(x) = \alpha e^{-\beta x + \gamma} = A e^{-\beta x},$$

where $A = \alpha e^{\gamma}$ is essentially the same as the above function.
Similarly, function

$$f(x) = \alpha x^{\beta} \tag{8.40}$$

can also be transformed into

$$\ln[f(x)] = \ln(\alpha) + \beta \ln(x), \tag{8.41}$$

which is a linear regression $y = a + b\zeta$ between $y = \ln[f(x)]$ and $\zeta = \ln(x)$, where $a = \ln(\alpha)$ and $b = \beta$.

Furthermore, function

$$f(x) = \alpha \beta^{x} \tag{8.42}$$

can also be converted into the standard linear form

$$\ln f(x) = \ln \alpha + x \ln \beta, \tag{8.43}$$

by letting $y = \ln[f(x)]$, $a = \ln \alpha$, and $b = \ln \beta$.
It is worth pointing out that the data points involving zeros should be taken out due to the potential singularity of the logarithm. Fortunately, these points rarely occur in the regression for the functions in the above form.

**Example 8.4** If a set of data can fit to a nonlinear function

$$y = ax \exp(\frac{-x}{b}),$$

in the range of $(0, \infty)$, it is then possible to convert it to a linear regression.
As $x = 0$ is just a single point, we can leave this out. For $x \neq 0$, we can divide both sides by $x$, we have

$$\frac{y}{x} = a \exp(\frac{-x}{b}).$$

Taking the logarithm of both sides, we have

$$\ln \frac{y}{x} = \ln a - \frac{1}{b} x,$$

which is a linear regression of $y/x$ versus $x$.

In general, linearization is possible only for a small class of nonlinear functions. For nonlinear functions, we have to use either approximation or full nonlinear least squares to be introduced later in this chapter.

### 8.2.4 Generalized Linear Regression

The most widely used linear regression is the so-called generalized least square as a linear combination of basis functions. Fitting to a polynomial of degree $p$,

$$y(x) = \alpha_0 + \alpha_1 x + \alpha_2 x^2 + \cdots + \alpha_p x^p, \tag{8.44}$$

is probably the most widely used. This is equivalent to the regression to the linear combination of the basis functions 1, $x$, $x$, ..., and $x^p$. However, there is no particular reason why we have to use these basis functions. In fact, the basis functions can be any arbitrary known functions such as $\sin(x)$, $\cos(x)$, and even $\exp(x)$, and the main requirement is that they can be explicitly expressed as basis functions. In this sense, the generalized least square can be written as

$$y(x) = \sum_{j=0}^{p} \alpha_j f_j(x), \tag{8.45}$$

where the basis functions $f_j$ are known functions of $x$ without any unknown or undetermined parameters.

Now the sum of least squares is defined as

$$\psi = \sum_{i=1}^{n} \frac{[y_i - \sum_{j=0}^{p} \alpha_j f_j(x_i)]^2}{\sigma_i^2}, \tag{8.46}$$

where $\sigma_i (i = 1, 2, \ldots, n)$ are the standard deviations of the $i$th data point at $(x_i, y_i)$. There are $n$ data points in total. In order to determine the coefficients uniquely, it requires that

$$n \geq p + 1. \tag{8.47}$$

In the case of unknown standard deviations $\sigma_i$, a common practice is to set all the values $\sigma_i$ as the same constant $\sigma_i = \sigma = 1$.

Let $\boldsymbol{D} = [D_{ij}]$ be the design matrix which is given by

$$D_{ij} = \frac{f_j(x_i)}{\sigma_i}. \tag{8.48}$$

The minimum of $\psi$ is determined by

$$\frac{\partial \psi}{\partial \alpha_j} = 0 \qquad (j = 0, 1, \ldots, p). \tag{8.49}$$

That is,

$$\sum_{i=1}^{n} \frac{f_k(x_i)}{\sigma_i^2} \left[ y_i - \sum_{j=0}^{p} \alpha_j f_j(x_i) \right] = 0 \qquad (k = 0, \dots, p). \tag{8.50}$$

Rearranging the terms and interchanging the order of summations, we have

$$\sum_{j=0}^{p} \sum_{i=1}^{n} \frac{\alpha_j f_j(x_i) f_k(x_i)}{\sigma_i^2} = \sum_{i=1}^{n} \frac{y_i f_k(x_i)}{\sigma_i^2}, \tag{8.51}$$

which can be written compactly as the following matrix equation:

$$\sum_{j=0}^{p} A_{kj} \alpha_j = b_k \tag{8.52}$$

or

$$\boldsymbol{A\alpha} = \boldsymbol{b}, \tag{8.53}$$

where

$$\boldsymbol{A} = \boldsymbol{D}^T \cdot \boldsymbol{D}$$

is a $(p+1) \times (p+1)$ matrix. That is,

$$A_{kj} = \sum_{i=1}^{n} \frac{f_k(x_i) f_j(x_i)}{\sigma_i^2}. \tag{8.54}$$

Here $b_k$ is a column vector given by

$$b_k = \sum_{i=1}^{n} \frac{y_i f_k(x_i)}{\sigma_i^2}, \tag{8.55}$$

where $(k = 0, \dots, p)$. Equation (8.52) is a linear system of the so-called normal equations which can be solved using the standard methods for solving linear systems. The solution of the coefficients is $\boldsymbol{\alpha} = \boldsymbol{A}^{-1}\boldsymbol{b}$ or

$$\alpha_k = \sum_{j=0}^{p} [A]_{kj}^{-1} b_j \qquad (k = 0, \dots, p), \tag{8.56}$$

where $\boldsymbol{A}^{-1} = [A]_{ij}^{-1}$.

A special case of the generalized linear least squares is the so-called polynomial least squares when the basis functions are simple power functions $f_i(x) = x^i$, $(i = 0, 1, \dots, p)$. That is,

$$f_i(x) = 1,\ x,\ x^2, \dots,\ x^p. \tag{8.57}$$

For simplicity, we assume that $\sigma_i = \sigma = 1$. The matrix equation (8.52) simply becomes

$$\begin{pmatrix} \sum_{i=1}^n 1 & \sum_{i=1}^n x_i & \dots & \sum_{i=1}^n x_i^p \\ \sum_{i=1}^n x_i & \sum_{i=1}^n x_i^2 & \dots & \sum_{i=1}^n x_i^{p+1} \\ \vdots & & \ddots & \\ \sum_{i=1}^n x_i^p & \sum_{i=1}^n x_i^{p+1} & \dots & \sum_{i=1}^n x_i^{2p} \end{pmatrix} \begin{pmatrix} \alpha_0 \\ \alpha_1 \\ \vdots \\ \alpha_p \end{pmatrix} = \begin{pmatrix} \sum_{i=1}^n y_i \\ \sum_{i=1}^n x_i y_i \\ \vdots \\ \sum_{i=1}^n x_i^p y_i \end{pmatrix}.$$

In the simplest case when $p = 1$, it becomes the standard linear regression

$$y = \alpha_0 + \alpha_1 x = a + bx.$$

Now we have

$$\begin{pmatrix} n & \sum_{i=1}^n x_i \\ \sum_{i=1}^n x_i & \sum_{i=1}^n x_i^2 \end{pmatrix} \begin{pmatrix} \alpha_0 \\ \alpha_1 \end{pmatrix} = \begin{pmatrix} \sum_{i=1}^n y_i \\ \sum_{i=1}^n x_i y_i \end{pmatrix}. \tag{8.58}$$

Its solution is

$$\begin{pmatrix} \alpha_0 \\ \alpha_1 \end{pmatrix} = \frac{1}{\Delta} \begin{pmatrix} \sum_{i=1}^n x_i^2 & -\sum_{i=1}^n x_i \\ -\sum_{i=1}^n x_i & n \end{pmatrix} \begin{pmatrix} \sum_{i=1}^n y_i \\ \sum_{i=1}^n x_i y_i \end{pmatrix}$$

$$= \frac{1}{\Delta} \begin{pmatrix} (\sum_{i=1}^n x_i^2)(\sum_{i=1}^n y_i) - (\sum_{i=1}^n x_i)(\sum_{i=1}^n x_i y_i) \\ n \sum_{i=1}^n x_i y_i - (\sum_{i=1}^n x_i)(\sum_{i=1}^n y_i) \end{pmatrix}, \tag{8.59}$$

where

$$\Delta = n \sum_{i=1}^n x_i^2 - \left( \sum_{i=1}^n x_i \right)^2. \tag{8.60}$$

These are exactly the same coefficients as those in Eq. (8.26).

**Example 8.5** We now use a quadratic function to best fit the following data (as shown in Figure 8.2):

$$x: \ -0.98, \ 1.00, \ 2.02, \ 3.03, \ 4.00$$

$$y: \ 2.44, \ -1.51, \ -0.47, \ 2.54, \ 7.52$$

For the formula $y = \alpha_0 + \alpha_1 x + \alpha_2 x^2$, we have

$$\begin{pmatrix} n & \sum_{i=1}^n x_i & \sum_{i=1}^n x_i^2 \\ \sum_{i=1}^n x_i & \sum_{i=1}^n x_i^2 & \sum_{i=1}^n x_i^3 \\ \sum_{i=1}^n x_i^2 & \sum_{i=1}^n x_i^3 & \sum_{i=1}^n x_i^4 \end{pmatrix} \begin{pmatrix} \alpha_0 \\ \alpha_1 \\ \alpha_2 \end{pmatrix} = \begin{pmatrix} \sum_{i=1}^n y_i \\ \sum_{i=1}^n x_i y_i \\ \sum_{i=1}^n x_i^2 y_i \end{pmatrix}.$$

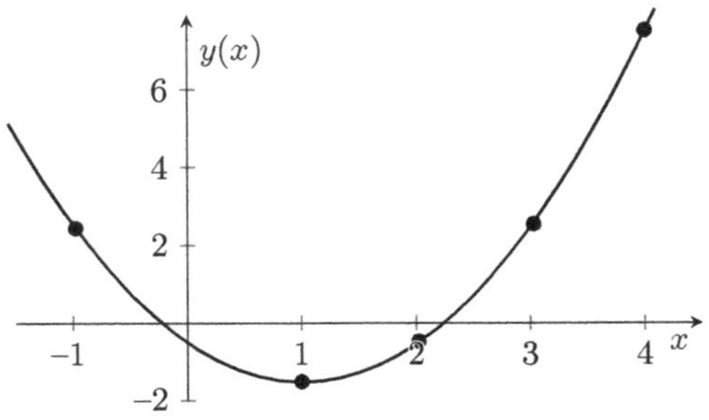


**Figure 8.2** Best fit curve for $y(x) = x^2 - 2x - 1/2$ with 2.5% noise.

Using the data set, we have $n = 5$, $\sum_{i=1}^{n} x_i = 9.07$, and $\sum_{i=1}^{n} y_i = 10.52$. Other quantities can be calculated in a similar way. Therefore, we have

$$\begin{pmatrix} 5.0000 & 9.0700 & 31.2217 \\ 9.0700 & 31.2217 & 100.119 \\ 31.2217 & 100.119 & 358.861 \end{pmatrix} \begin{pmatrix} \alpha_0 \\ \alpha_0 \\ \alpha_2 \end{pmatrix} = \begin{pmatrix} 10.52 \\ 32.9256 \\ 142.5551 \end{pmatrix}.$$

By direct inversion, we have

$$\begin{pmatrix} \alpha_0 \\ \alpha_1 \\ \alpha_2 \end{pmatrix} = \begin{pmatrix} -0.5055 \\ -2.0262 \\ 1.0065 \end{pmatrix}.$$

Finally, the best fit equation is

$$y(x) = -0.5055 - 2.0262x + 1.0065x^2,$$

which is quite close to the formula $y = x^2 - 2x - 1/2$ used to generate the original data with a random component of about 2.5%. The total RSS is RSS = 0.0045.

The fit seems to be highly accurate.

### 8.2.5 Goodness of Fit

In the above example, if we choose $p = 2$ (a quadratic polynomial), the above curve-fitting seems to work very well. But, how do we know which order of polynomials to use in the first place? In fact, the degree $p$ is a hyper-parameter for this curve fitting problem, and we have to use some additional information to find the right value for this parameter.

Suppose, we start with $p = 1$ (a straight line) and carry out the regression in the similar way as we did before. We should get a best-fit line

$$f_1(x) = 0.9373x + 0.4038 \tag{8.61}$$

with the RSS = 36.3485.

Table 8.2 Goodness of fit in terms of RSS.

| Order | $p = 1$ | $p = 2$ | $p = 3$ | $p = 4$ |
|---|---|---|---|---|
| RSS | 36.3485 | 0.0045 | 0.0080 | $6.9 \times 10^{-30}$ |

If we use $p = 3$, we have

$$f_3(x) = 0.0080x^3 + 0.9694x^2 - 2.0131x - 0.4580, \tag{8.62}$$

with the RSS = 0.002, which is the smallest RSS for $p = 1, 2, 3$.

If we used RSS as the goodness of fit, then it seems $p = 3$ gives a better fit than $p = 2$, even though the coefficient of the highest order $x^3$ is 0.0080. Now we proceed this way, what happens if we use $p = 4$?

If we use $p = 4$, we have

$$f_4(x) = 0.0101x^4 - 0.0610x^3 + 1.0778x^2 - 1.9571x - 0.5798, \tag{8.63}$$

with even a smaller RSS $= 6.9 \times 10^{-30}$. Let us summarize the above results in Table 8.2.

However, we cannot continue this way because we do not have enough data to produce well-posed coefficients if $p$ is higher than $n$. In general, as $p$ increases, the RSS at the data points can usually decreases but the oscillations between data points can increase dramatically.

In addition, the higher-order models may introduce unrealistic model parameters that are not supported by the data. This is the well-known over-fitting phenomenon, which should be avoided. We will discuss some approaches such as information criteria and regularization to deal with over-fitting in later sections.

## 8.3 Nonlinear Least Squares

As functions in most mathematical models are nonlinear, we need the nonlinear least squares in general. For given $n$ data points $(x_i, y_i)(i = 1, 2, \dots, n)$, we can fit a model $f(x_i, \boldsymbol{a})$ where $\boldsymbol{a} = (a_0, a_1, \dots, a_m)^{\mathrm{T}}$ is a vector of $m$ parameters. In the simplest linear case, we have $f(x_i, \boldsymbol{a}) = a_0 + a_1 x$. For the one-variable logistic regression to be discussed later, we have $f(x_i, \boldsymbol{a}) = 1/(1 + e^{a_0 + a_1 x})$.

In general, we have the nonlinear least squares to minimize the $L_2$-norm of the residuals $R_i = y_i - f(x_i, \boldsymbol{a})$. That is, to minimize the fitting error:

$$\text{minimize } E(\boldsymbol{a}) = \sum_{i=1}^{n} R_i^2(\boldsymbol{a}) = \sum_{i=1}^{n} [y_i - f(x_i, \boldsymbol{a})]^2 = ||R_i(\boldsymbol{a})||_2^2. \tag{8.64}$$

If we treat this as an optimization problem so as to find the best $\boldsymbol{a}$, we can use any optimization techniques such as Newton's method to solve this optimization problem. However, we can use the properties of this problem to get the solution more efficiently.

### 8.3.1 Gauss–Newton Algorithm

Let $\boldsymbol{J}$ denote the Jacobian matrix in the form

$$\boldsymbol{J} = [J_{ij}] = \frac{\partial R_i}{\partial a_j} \quad (i = 1, 2, \ldots, n; \quad j = 0, 1, 2, \ldots, m), \tag{8.65}$$

which is an $n \times (m+1)$ matrix. Then the gradient of the objective (error) function is the differentiation of $E$,

$$\frac{\partial E}{\partial a_j} = 2 \sum_{i=1}^{n} \frac{\partial R_i}{\partial a_j} R_i \quad (j = 0, 1, 2, \ldots, m), \tag{8.66}$$

which can be written in the vector form as

$$\nabla E(\boldsymbol{a}) = 2\boldsymbol{J}^{\mathrm{T}}\boldsymbol{R}, \tag{8.67}$$

where the residual vector $\boldsymbol{R}$ is given by

$$\boldsymbol{R}(\boldsymbol{a}) = \big(R_1(\boldsymbol{a}),\ R_2(\boldsymbol{a}),\ \ldots,\ R_n(\boldsymbol{a})\big)^{\mathrm{T}}. \tag{8.68}$$

Here, $\nabla E$ is a vector with $m+1$ components. Similarly, the Hessian matrix $\boldsymbol{H}$ of the objective $E$ can be written as

$$\boldsymbol{H} = \nabla^2 E = 2 \sum_{i=1}^{n} [\nabla R_i \nabla R_i^{\mathrm{T}} + R_i \nabla^2 R_i] = 2\boldsymbol{J}^{\mathrm{T}}\boldsymbol{J} + 2 \sum_{i=1}^{n} R_i \nabla^2 R_i, \tag{8.69}$$

which gives an $(m+1) \times (m+1)$ matrix. As it is expected that $R_i$ will get smaller as the goodness of fit increases, we can essentially approximate the Hessian matrix by ignoring all the higher-order terms, so we have

$$\boldsymbol{H} \approx 2\boldsymbol{J}^{\mathrm{T}}\boldsymbol{J}. \tag{8.70}$$

Now we can first solve the above nonlinear least squares (8.64) by Newton's method and we have

$$\begin{aligned} \boldsymbol{a}_{t+1} &= \boldsymbol{a}_t - \frac{\nabla E}{\nabla^2 E} \\ &= \boldsymbol{a}_t - \frac{2\boldsymbol{J}^{\mathrm{T}}\boldsymbol{R}}{2\boldsymbol{J}^{\mathrm{T}}\boldsymbol{J}} = \boldsymbol{a}_t - (\boldsymbol{J}^{\mathrm{T}}\boldsymbol{J})^{-1}\boldsymbol{J}^{\mathrm{T}}\boldsymbol{R}(\boldsymbol{a}_t). \end{aligned} \tag{8.71}$$

The initial vector $\boldsymbol{a}_0$ should be a good educated guess, though $\boldsymbol{a}_0 = (1, 1, \ldots, 1)^{\mathrm{T}}$ may work well in most cases.

Here, we have used the approximation of $\boldsymbol{H}$ by $2\boldsymbol{J}^{\mathrm{T}}\boldsymbol{J}$. In this case, this iterative method is often referred to the Gauss–Newton method, which usually has a good convergence rate. However, it is required that $\boldsymbol{J}$ should be a full rank so that the inverse $(\boldsymbol{J}^T\boldsymbol{J})^{-1}$ exists.

It is worth pointing out that in the context of line search $\boldsymbol{a}_{t+1} = \boldsymbol{a}_t + \alpha_t \boldsymbol{s}_t$ where $\boldsymbol{s}_t$ is the step size and $0 < \alpha_t \leq 1$ is the scaling parameter or learning rate, this Gauss–Newton method is equivalent to a line search with a step size

$$\boldsymbol{J}^T\boldsymbol{J}\boldsymbol{s}_t = -\boldsymbol{J}^T\boldsymbol{R}(\boldsymbol{a}_t). \tag{8.72}$$

Though the iteration does not necessarily lead to the reduction of $E$ in every iteration, it is better to choose $\alpha_t$ such that

$$E(\boldsymbol{a}_t + \alpha_t\boldsymbol{s}_t) < E(\boldsymbol{a}_t), \tag{8.73}$$

which will lead to the reduction of least square errors.

**Example 8.6** As an example, let us use the following data

$$x: 0.10,\ 0.50,\ 1.0,\ 1.5,\ 2.0,\ 2.5$$
$$y: 0.10,\ 0.28,\ 0.40,\ 0.40,\ 0.37,\ 0.32$$

to fit a model

$$y = \frac{x}{a + bx^2}, \tag{8.74}$$

where $a$ and $b$ are the coefficients to be determined by the data. The objective is to minimize the sum of the residual squares

$$S = \sum_{i=1}^{6} R_i^2 = \sum_{i=1}^{6} \left[1 - \frac{x}{a + bx^2}\right]^2, \tag{8.75}$$

where

$$R_i = y_i - \frac{x_i}{a + bx_i^2}.$$

Since

$$\frac{\partial R_i}{\partial a} = \frac{x_i}{(a + bx_i^2)^2}, \quad \frac{\partial R_i}{\partial b} = \frac{x_i^3}{(a + bx_i^2)^2}.$$

If we use the initial guess $a = 1$ and $b = 1$, then the initial residuals are

$$\boldsymbol{R} = \begin{pmatrix} 0.0010 & -0.1200 & -0.1000 & -0.0615 & -0.0300 & -0.0248 \end{pmatrix}^T.$$

The initial Jacobian matrix is

$$\boldsymbol{J} = \begin{pmatrix} 0.0980 & 0.0010 \\ 0.3200 & 0.0800 \\ 0.2500 & 0.2500 \\ 0.1420 & 0.3195 \\ 0.0800 & 0.3200 \\ 0.0476 & 0.2973 \end{pmatrix}.$$

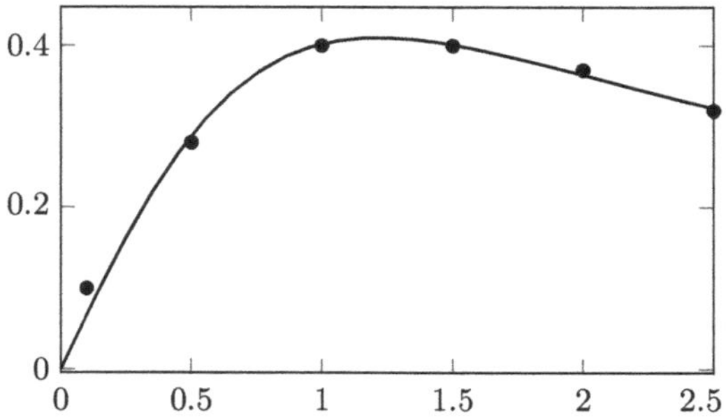


**Figure 8.3** An example of nonlinear least squares.

Thus, the first iteration using the Gauss–Newton algorithm gives

$$\begin{pmatrix} a \\ b \end{pmatrix}_1 = \begin{pmatrix} 1 \\ 1 \end{pmatrix} - (J^{\mathrm{T}}J)^{-1}JR = \begin{pmatrix} 1.3449 \\ 1.0317 \end{pmatrix}.$$

Then, by updating the new Jacobian and residuals, we have

$$a = 1.4742, \quad b = 1.0059,$$

after the second iteration. Similarly, we have

$$a = 1.4852, \quad b = 1.0022 \quad \text{(third iteration)}$$

and

$$a = 1.4854, \quad b = 1.0021 \quad \text{(fourth iteration)}.$$

In fact, this converges quickly and the parameters almost remain the same values even after 10 iterations. The data points and the best fit model are shown in Figure 8.3.

In general, the Guass–Newton method can work very well for a wide range of nonlinear curve fitting problems, even for large-scale problems. However, when the elements of the Jacobian are small (close to zeros), the matrix $J^{\mathrm{T}}J$ may become singular, thus pseudo-inverse may become ill-posed. In addition, when approaching the optimality, the gradient becomes close to zero, and the convergence becomes very slow. A possible remedy is to use the Levenberg–Marquardt algorithm.

### 8.3.2 Levenberg–Marquardt Algorithm

In essence, the Levenberg–Marquardt algorithm is more robust by using a damping term in the approximation of the Hessian, that is,

$$H \approx 2[J^{\mathrm{T}}J + \mu I], \tag{8.76}$$

where $\mu > 0$ is the damping coefficient, also called Marquardt parameter, and $\boldsymbol{I}$ is the identity matrix of size as the same size as $\boldsymbol{H}$. Thus, the iteration formula becomes

$$\boldsymbol{a}_{t+1} = \boldsymbol{a}_t - \frac{\boldsymbol{J}^{\mathrm{T}}\boldsymbol{R}(\boldsymbol{a}_t)}{\boldsymbol{J}^{\mathrm{T}}\boldsymbol{J} + \mu\boldsymbol{I}} = \boldsymbol{a}_t - (\boldsymbol{J}^{\mathrm{T}}\boldsymbol{J} + \mu\boldsymbol{I})^{-1}\boldsymbol{J}^{\mathrm{T}}\boldsymbol{R}(\boldsymbol{a}_t), \tag{8.77}$$

which is equivalent to the step size $\boldsymbol{s}_t$ given by

$$(\boldsymbol{J}^{\mathrm{T}}\boldsymbol{J} + \mu\boldsymbol{I})\boldsymbol{s}_t = -\boldsymbol{J}^{\mathrm{T}}\boldsymbol{R}(\boldsymbol{a}_t). \tag{8.78}$$

Mathematically speaking, a large $\mu$ will effectively reduce the step size (in comparison with those in the Gauss–Newton algorithm) and damps the moves so that the descent is in the right direction with the right amount. If the reduction in $E$ is sufficient, we can either keep this value of $\mu$ or reduce it. However, if the reduction $E$ is not sufficient, we can increase $\mu$. Thus, $\mu$ should vary as iteration continues and there are various schemes for varying this hyper-parameter. It is obvious that this method reduces to the standard Gauss–Newton algorithm if $\mu = 0$. Any nonzero $\mu$ essentially ensures that matrix $\boldsymbol{J}^{\mathrm{T}}\boldsymbol{J}$ is full rank, and thus the iterations can be more robust.

An alternative view of the Levenberg–Marquardt algorithm is the approximation of $\boldsymbol{H}$ in a trust region, which is equivalent to the following minimization problem:

$$\text{minimize } ||\boldsymbol{J}^{\mathrm{T}}\boldsymbol{s}_t + \boldsymbol{R}||_2^2, \quad |\boldsymbol{s}_t| \leq \Delta_t, \tag{8.79}$$

where $\Delta_k$ is the radius of the trust region. For details, readers can refer to more advanced literature.

### 8.3.3 Weighted Least Squares

In many cases, the measurement errors in data may be different or the assumption of equal variances in data may not be true. In this case, we need to weight the residuals differently, which leads to the so-called weighted least squares

$$\text{minimize } \sum_{i=1}^{n} w_i R_i^2 = \sum_{i=1}^{n} \frac{R_i^2}{\sigma_i^2} = \left|\left|\frac{R_i}{\sigma_i}\right|\right|_2^2, \tag{8.80}$$

where

$$R_i = y_i - f(x_i, \boldsymbol{a}). \tag{8.81}$$

Here, $w_i = 1/\sigma_i^2$ and $\sigma_i^2$ is the variance associated with data point $(x_i, y_i)$.

By defining a weight matrix $\boldsymbol{W}$ as

$$\boldsymbol{W} = \mathrm{diag}(w_i) = \begin{pmatrix} w_1 & 0 & \dots & 0 \\ 0 & w_2 & \dots & 0 \\ \vdots & \vdots & \ddots & \vdots \\ 0 & 0 & \dots & w_n \end{pmatrix}, \tag{8.82}$$

and following the similar derivations as above, Eq. (8.71) will become

$$\boldsymbol{a}_{t+1} = \boldsymbol{a}_t - (\boldsymbol{J}^{\mathrm{T}}\boldsymbol{W}\boldsymbol{J})^{-1}(\boldsymbol{J}^{\mathrm{T}}\boldsymbol{W}\boldsymbol{R}), \tag{8.83}$$

which is equivalent to approximating the Hessian by $\boldsymbol{H} = 2\boldsymbol{J}^2\boldsymbol{W}\boldsymbol{J}$ and the gradient by $\nabla E = 2\boldsymbol{J}^{\mathrm{T}}\boldsymbol{W}\boldsymbol{R}$.

There are other regression methods. For example, for classification purposes, the logistic regression in the form

$$y(\boldsymbol{x}) = \frac{1}{1 + \mathrm{e}^{-(a+bx)}} \tag{8.84}$$

is often used because its outputs can be interpreted as binary (0 or 1).

Another powerful regression method is the principal component analysis (PCA), which is a multivariate regression method. We will introduce these methods later in this chapter. Now let us first discuss over-fitting and information criteria.

## 8.4 Over-fitting and Information Criteria

As we can see in the above sections, curve-fitting and regression are optimization in the least squares sense because the objective is to minimize the fitting errors from the target values. In principle, these errors at known data points can become sufficiently small if higher-order polynomials are used; however, oscillations between data points become more severe. This leads to the so-called over-fitting, which subsequently gives over-complicated models. Ideally, the fitting should be guided by Occam's razor, which states that, if there are many competing models to explain or fit the same data, the one with the fewest assumptions or parameters should be selected. The issue of over-fitting is relevant to many applications such as curve-fitting, regression in general, and training of neural networks.

It is worth pointing out that the degree of polynomials to be used for curve fitting is a hyper-parameter, which needs extra information or rule to determine. Though sophisticated methods in model selection may help, some simple criteria such as the Bayesian information criterion (BIC) or Akaike information criterion (AIC) can be used to select such hyper-parameters to avoid over-fitting.

For a statistical model such as regression with $k$ parameters, the AIC is defined by

$$\mathrm{AIC} = 2k - 2\ln L, \tag{8.85}$$

where $L$ is the maximum value of the likelihood function. For $n$ data points with errors being independent, identical, normally distributions, we have

$$\mathrm{AIC} = 2k + n\ln\left(\frac{\mathrm{RSS}}{n}\right), \tag{8.86}$$

where RSS is the residual sum of squares. That is,

$$\mathrm{RSS} = \sum_{i=1}^{n} [y_i - \hat{y}_i(x_i)]^2, \tag{8.87}$$

where $y_i (i = 1, 2, \dots, n)$ are the true values, while $\hat{y}_i(x_i)(i = 1, 2, \dots, n)$ are the predicted value by the model. In principle, the minimization of AIC will give the best $k$.

However, this AIC may become inaccurate when the sample size is small, especially when $n/k < 40$, we have to use a corrected AIC, called $\mathrm{AIC_c}$, which is given by

$$\begin{aligned}\mathrm{AIC_c} &= \mathrm{AIC} + \frac{2k(k+1)}{n-k-1} \\ &= 2k + n \ln\left(\frac{\mathrm{RSS}}{n}\right) + \frac{2k(k+1)}{n-k-1}.\end{aligned} \tag{8.88}$$

In essence, the AIC is equivalent to the principle of maximum entropy.

Another information criterion is the BIC, which can be written as

$$\mathrm{BIC} = k \ln n - 2 \ln L. \tag{8.89}$$

With the same assumptions of errors obeying Gaussian distributions, we have

$$\mathrm{BIC} = k \ln n + n \ln\left(\frac{\mathrm{RSS}}{n}\right). \tag{8.90}$$

Both AIC and BIC are useful criteria, but it is difficult to say which one is better, which may also depend on the types of problems.

**Example 8.7** Now let revisit an earlier example (Example 8.5) in Section 8.2.5 using the AIC. We know $n = 5$ as there are five data points. Using the AIC criterion (8.86), we have (for $p = 1$ and $k = 2$)

$$\mathrm{AIC} = 2 \times 2 + 5 \ln(\frac{36.3485}{5}) = 13.64. \tag{8.91}$$

Similarly, we have

$$\mathrm{AIC} = -29.07, \quad -24.19, \tag{8.92}$$

for $p = 2, 3$, respectively. Among these three values, $p = 2$ has the lowest AIC. Since the value starts to increase for $p = 3$, we can conclude that the best degree of fit is $p = 2$ with $k = 3$ parameters. The results of AIC values are summarized in Table 8.3.

Though $p = 4$ can have a fourth-order polynomial fit, the leading coefficient 0.01 (see Section 8.2.5) is too small, compared with other coefficients. This case should not be considered as it is an indication of over-fitting. Ideally, a properly scaled and properly fit polynomial should have coefficients of $O(1)$. This point becomes clearer when we discuss regularization in the next section.

**Table 8.3** AIC as the goodness of fit.

| Order | $p = 1$ | $p = 2$ | $p = 3$ |
|---|---|---|---|
| $k = p + 1$ | 2 | 3 | 4 |
| RSS | 36.3485 | 0.0045 | 0.0080 |
| AICc | 13.64 | −29.07 | −24.19 |

## 8.5 Regularization and Lasso Method

Regularization is another approach to deal with over-fitting. If we use a general model

$$y = f(\boldsymbol{Z}) + \epsilon, \tag{8.93}$$

where the errors $\epsilon$ obey a zero-mean normal distribution with a variance $\sigma^2$, that is $\epsilon \sim N(0, \sigma^2)$.

For multivariate cases with $p$ components, we have

$$\boldsymbol{Z} = (Z_1, Z_2, \dots, Z_p)^{\mathrm{T}}. \tag{8.94}$$

In case of linear regression, we have

$$y = \beta_0 + \beta_1 Z_1 + \beta_2 Z_2 + \cdots + \beta_p Z_p = \boldsymbol{Z}^{\mathrm{T}} \boldsymbol{\beta} + \beta_0, \tag{8.95}$$

where $\beta_0$ is the bias, while $\boldsymbol{\beta} = (\beta_1, \ \beta_2, \dots, \beta_p)^{\mathrm{T}}$ is the coefficient vector.

The standard method of least squares is to minimize the RSS. That is,

$$\text{minimize } \sum_{i=1}^{n} (y_i - \boldsymbol{Z}_i^{\mathrm{T}} \boldsymbol{\beta} - \beta_0)^2. \tag{8.96}$$

The Ridge regression uses a penalized RSS in the form

$$\text{minimize } \sum_{i=1}^{n} (y_i - \boldsymbol{Z}_i^{\mathrm{T}} \boldsymbol{\beta} - \beta_0)^2 + \lambda \sum_{j=1}^{p} \beta_j^2, \tag{8.97}$$

where $\lambda$ is the penalty coefficient that controls the amount of regularization. The above formula can be written compactly as

$$\text{minimize } ||y - \beta_0 - \boldsymbol{Z}^{\mathrm{T}} \boldsymbol{\beta}||_2^2 + \lambda ||\boldsymbol{\beta}||_2^2, \tag{8.98}$$

where $||.||_2$ is the $L_2$-norm and this regularization term is based on the Tikhonov regularization on the parameter/coefficient vector. Here, the bias $\beta_0$ is not part of the penalty or regularization term. One of the reason is that we can always preprocess the data $y_i$ (for example, by subtracting their mean value) so that the bias $\beta_0$ becomes zero.

In case of $\lambda = 0$, there is no penalty, which degenerates to the standard least squares. In fact, $\lambda$ is a hyper-parameter, which needs to be tuned.

The Lasso method uses an $L_1$-norm in the regularization term

$$\text{minimize} \ ||y - \beta_0 - \boldsymbol{Z}^{\mathrm{T}}\boldsymbol{\beta}||_2^2 + \lambda||\boldsymbol{\beta}||_1, \tag{8.99}$$

which is equivalent to the following minimization problem:

$$\text{minimize} \ \sum_{i=1}^{n}(y_i - \beta_0 - \boldsymbol{Z}_i^{\mathrm{T}}\boldsymbol{\beta})^2, \tag{8.100}$$

subject to

$$||\boldsymbol{\beta}||_1 = |\beta_1| + |\beta_2| + \cdots + |\beta_p| \leq t, \tag{8.101}$$

where $t > 0$ is a predefined hyper-parameter. Here, $\beta_0$ is a bias, which is not be penalized in the Lasso formulation.

A hybrid method is the elastic net regularization or regression, which combines the Ridge and Lasso methods into a hybrid as

$$\text{minimize} \ ||y - \beta_0 - \boldsymbol{Z}^{\mathrm{T}}\boldsymbol{\beta}||_2^2 + \lambda_1||\boldsymbol{\beta}||_1 + \lambda_2||\boldsymbol{b}||_2^2, \tag{8.102}$$

where both the $L_1$-norm and $L_2$-norm are used with two regularization hyper-parameters $\lambda_1$ and $\lambda_2$.

## 8.6 Logistic Regression

In the above analysis, all the dependent variable values $y_i$ are continuous. In some applications such as biometrics and classifications, the dependent variable is just discrete or a simply binary categorical variable, taking two values 1 (yes) and 0 (no). In this case, a more appropriate regression tool is the logistic regression, developed by David Cox in 1958.

Before we introduce the formulation of logistic regression, let us define two functions: logistic function $S$ and logit function. A logistic function, also called Sigmoid function, is defined as

$$S(x) = \frac{1}{1 + \mathrm{e}^{-x}} = \frac{\mathrm{e}^x}{1 + \mathrm{e}^x}, \quad x \in \mathbb{R}, \tag{8.103}$$

which can be written as

$$S(x) = \frac{1}{2}\left[1 + \tanh\left(\frac{x}{2}\right)\right], \quad \tanh(x) = \frac{\mathrm{e}^x - \mathrm{e}^{-x}}{\mathrm{e}^x + \mathrm{e}^{-x}}. \tag{8.104}$$

It is easy to see that $S \to +1$ when $x \to +\infty$, while $S \to 0$ when $x \to -\infty$. Thus, the range of $S$ is $S \in (0, 1)$.

This function has an interesting property for differentiation. From the differentiation rules, we have

$$S'(x) = \left[\frac{1}{1+e^{-x}}\right]' = \frac{-1}{(1+e^{-x})^2}(-e^{-x}) = \frac{(1+e^{-x})-1}{(1+e^{-x})^2}$$
$$= \frac{1}{(1+e^{-x})} - \frac{1}{(1+e^{-x})^2} = \frac{1}{1+e^{-x}}\left[1 - \frac{1}{(1+e^{-x})}\right]$$
$$= S(x)[1 - S(x)], \tag{8.105}$$

which means that its first derivative can be obtained by multiplication. This property can be very useful for finding the weights of artificial neural networks and machine learning to be introduced in Chapter 9.

To get the inverse of the logistic function, we can rewrite Eq. (8.103) as

$$S(1 + e^{-x}) = S + Se^{-x} = 1, \tag{8.106}$$

which gives

$$e^{-x} = \frac{1-S}{S} \tag{8.107}$$

or

$$e^{x} = \frac{S}{1-S}. \tag{8.108}$$

Taking the natural logarithm, we have

$$x = \ln\left(\frac{S}{1-S}\right), \tag{8.109}$$

which is the well-known logit function in probability and statistics. In fact, the logit function can be defined as

$$\text{logit}(P) = \log\left(\frac{P}{1-P}\right) = \ln\left(\frac{P}{1-P}\right), \tag{8.110}$$

which is valid for $0 < P < 1$.

The simple logistic regression with one independent variable $x$ and a binary dependent variable $y$ with data points $(x_i, y_i)(i = 1, 2, \dots, n)$ tries to fit a model of logistic probability

$$P = \frac{1}{1 + e^{a+bx}}, \tag{8.111}$$

which can be written by using the logit function as

$$\ln\left(\frac{P}{1-P}\right) = a + bx, \tag{8.112}$$

which becomes a linear model in terms of the logit of probability $P$. In fact, the odds can be calculated from probability by

$$O_d(\text{odd}) = \frac{P}{1-P} \tag{8.113}$$

or

$$P = \frac{O_d}{1 + O_d}, \tag{8.114}$$

which means that the logistic regression can be considered as a linear model of log(odds) to $x$.

One naive way to solve the regression model (8.111) is to convert it to nonlinear least squares, and we have

$$\text{minimize} \sum_{i=1}^{n} \left[ y_i - \frac{1}{1 + e^{a+bx_i}} \right]^2, \tag{8.115}$$

so as to find the optimal $a$ and $b$. This is equivalent to fitting the logistic model to the data directly so as to minimize the overall fitting errors. This can give a solution to the parameters, but this is not the true logistic regression.

However, a more rigorous mathematical model exists for the binary outcomes $y_i$ and the objective is to maximize the log-likelihood of the model with the right parameters to explain the data. Thus, for a given data set $(x_i, y_i)$ with binary values of $y_i \in \{0, 1\}$, the proper binary logistic regression is to maximize the log-likelihood function. That is,

$$\text{maximize} \quad \log(L) = \sum_{i=1}^{n} \left[ y_i \ln P_i + (1 - y_i) \ln(1 - P_i) \right], \tag{8.116}$$

where

$$P_i = \frac{1}{1 + e^{a+bx_i}} \quad (i = 1, 2, \ldots, n). \tag{8.117}$$

This is based on the theory of the maximum likelihood probability. Since $y_i = 1$ (yes or true) or 0 (no or false), the random variable $Y$ for generating $y_i$ should obey a Bernoulli distribution for probability $P_i$. That is,

$$B_P(Y = y_i) = P_i^{y_i} (1 - P_i)^{1-y_i}, \tag{8.118}$$

so the joint probability of all data gives a likelihood function

$$L = \prod_{i=1}^{n} P(x_i)^{y_i} \left[ 1 - P(x_i) \right]^{1-y_i}, \tag{8.119}$$

whose logarithm is given above in Eq. (8.116). The maximization of $L$ is equivalent to the maximization of $\log(L)$. Therefore, the binary logistic regression is to fit the data such that the log-likelihood is maximized.

In principle, we can solve the above optimization problem (8.116) by Newton's method or any other optimization techniques. Let us use an example to explain the procedure in detail.

**Example 8.8** To fit a binary logistic regression using

$$\begin{array}{lllllll} x: & 0.1, & 0.5, & 1.0, & 1.5, & 2.0, & 2.5, \\ y: & 0, & 0, & 1, & 1, & 1, & 0, \end{array}$$

we can use the following form:

$$P_i = \frac{1}{1+\exp(a+bx_i)} \quad (i = 1, 2, \dots, 6), \tag{8.120}$$

starting with an initial value $a = 1$ and $b = 1$.

Then, we can calculate $P_i$ with $a = 1$ and $b = 1$, and we have

$$P_i = \begin{pmatrix} 0.2497 & 0.1824 & 0.1192 & 0.0759 & 0.0474 & 0.0293 \end{pmatrix}.$$

The log-likelihood for each datapoint can be calculated by

$$L_i = y_i \ln P_i + (1 - y_i) \ln(1 - P_i),$$

and we have

$$L_i = \begin{pmatrix} -0.2873 & -0.2014 & -2.1269 & -2.5789 & -3.0486 & -0.0298 \end{pmatrix},$$

with the log-likelihood objective as

$$\sum_{i=1}^{6} L_i = -8.2729.$$

If we try to modify the values of $a$ and $b$ by Newton's method, then after about 20 iterations, we should have

$$a = 0.8982, \quad b = -0.7099, \quad L_{\max} = -3.9162.$$

This means that the logistic regression model is

$$P = \frac{1}{1+\exp(0.8982 - 0.7099x)}.$$

The above logistic regression has only one independent variable. In case of multiple independent variables $\tilde{x}_1, \tilde{x}_2, \dots, \tilde{x}_m$, we can extend the above model as

$$y = \frac{1}{1 + e^{a_0 + a_1\tilde{x}_1 + a_2\tilde{x}_2 + \cdots + a_m\tilde{x}_m}}. \tag{8.121}$$

In order to write them compactly, let us define

$$\tilde{\boldsymbol{x}} = [1,\ \tilde{x}_1,\ \tilde{x}_2,\ \dots, \tilde{x}_m]^{\mathrm{T}} \tag{8.122}$$

and

$$\boldsymbol{a} = [a_0,\ a_1,\ a_2,\ \dots, a_m]^{\mathrm{T}}, \tag{8.123}$$

where we have used 1 as a variable so as to eliminate to write $a_0$ everywhere in the formulas. Thus, the logistic model becomes

$$P(\boldsymbol{a}) = \frac{1}{1 + \exp(\boldsymbol{a}^T \tilde{\boldsymbol{x}})}, \tag{8.124}$$

which is equivalent to

$$\text{logit}(P) = \ln \frac{P}{1-P} = \boldsymbol{a}^T \tilde{\boldsymbol{x}}. \tag{8.125}$$

For all the data points $\tilde{\boldsymbol{x}}_i = [1, \tilde{x}_1^{(i)}, \dots, \tilde{x}_m^{(i)}]$ with $y_i \in \{0, 1\}$ $(i = 1, 2, \dots, n)$, we have

$$\text{maximize} \quad \log(L) = \sum_{i=1}^{n} \left[ y_i \ln P_i + (1 - y_i) \ln(1 - P_i) \right], \tag{8.126}$$

where $P_i = 1/[+ \exp(\boldsymbol{a}^T \tilde{\boldsymbol{x}}_i)]$. The solution procedure is the same as before and can be obtained by any optimization algorithm.

Obviously, the binary logistic regression can be extended to the case with multiple categories, that is, $y_i$ can take $K \geq 2$ different values. In this case, we have to deal with the so-called multinomial logistic regression, and interested readers can refer to more advanced literature for details.

Now let us conclude this chapter by introducing the PCA.

## 8.7 Principal Component Analysis

For many quantities such as $X_1, X_2, \dots, X_p$ and $y$, it is desirable to represent the model as a hyperplane given by

$$y = \beta_0 + \beta_1 X_1 + \beta_2 X_2 + \cdots + \beta_p X_p = \boldsymbol{X}^T \boldsymbol{\beta} + \beta_0. \tag{8.127}$$

In the simplest case, we have

$$y = \beta_0 + \beta_1 x. \tag{8.128}$$

Let us now start with the simplest case first, we will then extend the idea and procedure to multiple variables.

For $n$ data points $(x_i, y_i)$, we can use the notations

$$\boldsymbol{x} = (x_1, x_2, \dots, x_n)^T, \tag{8.129}$$

$$\boldsymbol{y} = (y_1, y_2, \dots, y_n)^T, \tag{8.130}$$

so that we have

$$\boldsymbol{y} = \beta_0 + \beta_0 \boldsymbol{x}. \tag{8.131}$$

Now we can adjust the data so that their means are zero by subtracting their means $\bar{x}$ and $\bar{y}$, respectively. We have

$$\tilde{\boldsymbol{x}} = \boldsymbol{x} - \bar{x}, \quad \tilde{\boldsymbol{y}} = \boldsymbol{y} - \bar{y}, \tag{8.132}$$

and we have

$$\tilde{\boldsymbol{y}} + \bar{y} = \beta_0 + \beta_1(\tilde{\boldsymbol{x}} + \bar{x}), \tag{8.133}$$

which gives

$$\tilde{\boldsymbol{y}} = \beta_1 \tilde{\boldsymbol{x}}, \tag{8.134}$$

where we have used $\bar{y} = \beta_0 + \beta_1 \bar{x}$. This is essentially equivalent to removing the bias $\beta_0$ (or zero bias) because the adjusted data have zero means. In this case, the covariance can be calculated in terms of a dot product

$$\mathrm{cov}(\tilde{\boldsymbol{x}}, \tilde{\boldsymbol{y}}) = \frac{1}{n-1} \tilde{\boldsymbol{x}}^{\mathrm{T}} \tilde{\boldsymbol{y}}. \tag{8.135}$$

In fact, the covariance matrix can be written as

$$\boldsymbol{C} = \frac{1}{n-1} \begin{pmatrix} \tilde{\boldsymbol{x}}^{\mathrm{T}} \tilde{\boldsymbol{x}} & \tilde{\boldsymbol{x}}^{\mathrm{T}} \tilde{\boldsymbol{y}} \\ \tilde{\boldsymbol{y}}^{\mathrm{T}} \tilde{\boldsymbol{x}} & \tilde{\boldsymbol{y}}^{\mathrm{T}} \tilde{\boldsymbol{y}} \end{pmatrix}. \tag{8.136}$$

Now we extend this to the case of $p$ variables. If the mean of each variable $X_i$ is $\overline{X}_i (i = 1, 2, \ldots, p)$, we now use $\tilde{\boldsymbol{X}}$ to denote the adjusted data with zero means (i.e. $\boldsymbol{X}_i - \overline{X}_i$) in the following form:

$$\tilde{\boldsymbol{X}} = \begin{pmatrix} \tilde{\boldsymbol{X}}_1 & \tilde{\boldsymbol{X}}_2 & \ldots & \tilde{\boldsymbol{X}}_p \end{pmatrix}, \quad \tilde{\boldsymbol{X}}_i = \boldsymbol{X}_i - \overline{X}_i. \tag{8.137}$$

Then, $\tilde{\boldsymbol{X}}$ is an $n \times p$ matrix, and the covariance matrix can be obtained by

$$\boldsymbol{C} = \frac{1}{n-1} \tilde{\boldsymbol{X}}^{\mathrm{T}} \tilde{\boldsymbol{X}}, \tag{8.138}$$

which is a $p \times p$ symmetric matrix. If the data values are all real numbers, then $\boldsymbol{C}$ is a real, symmetric matrix whose eigenvalues are all real, and the eigenvectors of two distinct eigenvalues are orthogonal to each other.

As the covariance matrix $\boldsymbol{C}$ has a size of $p \times p$, it should in general have $p$ eigenvalues $(\lambda_1, \lambda_2, \ldots, \lambda_p)$. Their corresponding (column) eigenvectors $\boldsymbol{u}_1, \boldsymbol{u}_2$, ..., $\boldsymbol{u}_p$ should span an orthogonal matrix of size $p \times p$ by using $p$ eigenvectors (column vectors)

$$\boldsymbol{Q} = \begin{pmatrix} \boldsymbol{u}_1 & \boldsymbol{u}_2 & \ldots & \boldsymbol{u}_p \end{pmatrix}, \tag{8.139}$$

which has the properties $\boldsymbol{Q}^{\mathrm{T}} = \boldsymbol{Q}$ and $\boldsymbol{Q}^{-1} = \boldsymbol{Q}^{\mathrm{T}}$.

The component associated with the principal direction (of the eigenvector) of the largest eigenvalue $\lambda *= \max\{\lambda_i\}$ is the first main component. This corresponds to the rotation of base vectors so as to align the main component to this principal direction. The transformed data can be obtained by

$$X_Q = \tilde{X}Q. \tag{8.140}$$

It is worth pointing out that we have used column vectors here, and slightly different forms of formulas may be used if the row vectors are used, which is the case in some literature.

The PCA essentially intends to transform a set of data points (for $p$ variables that may be correlated) into a set of $k$ principal components which are a set of transformed points of linearly uncorrelated variables. It is essentially to identify the first (major) component with the most variations, then identify the next component with second-most variations. Other components can be identified in a similar manner. Thus, the main information comes from the covariance matrix $\boldsymbol{C}$ and the eigenvalues (as well as the eigenvectors) of the covariance matrix.

It is worth pointing out that since variances measure the variations, thus it is invariant with respect to a shift. That is $\text{var}(X + a) = \text{var}(X)$ where $a \in \mathbb{R}$. However, it is scaled as $\text{var}(aX) = a^2\text{var}(X)$.

The main aim is to reduce redundancy in representing data. Ideally, $k$ should be much less than $p$, but the first $k$ main components should be still sufficient to represent the original data. Therefore, PCA can be considered as a technique for dimensionality reduction.

The choice of the hyper-parameter $k$ can be tricky. If $k$ is too small, too much information is lost. If $k$ is close to $n$, almost all components are kept. So the choice of $k$ can largely depend on the quality of the representations and information needed. A heuristic approach is based on the percentage of variance (via the eigenvalues $\lambda_i$) that can be retained in the $k$ components. If we wish to keep $\mu$ as a percentage, we can use

$$\text{minimize} \quad k, \tag{8.141}$$

subject to

$$\sum_{i=1}^{k} \lambda_i \geq \mu \sum_{i=1}^{p} \lambda_i \quad (k \geq 1). \tag{8.142}$$

The typical value of $\mu$ can be 0.9 or higher.

Once $k < p$ principal components have been chosen, data can be reconstructed easily. It is worth pointing out that some scaling and whitening are needed so as to make PCA work. In addition, PCA is a linear framework, it cannot work for nonlinear systems. We will not delve into details of these topics, but interested readers can refer to more advanced literature.

Let us use a simple example to demonstrate how PCA works.

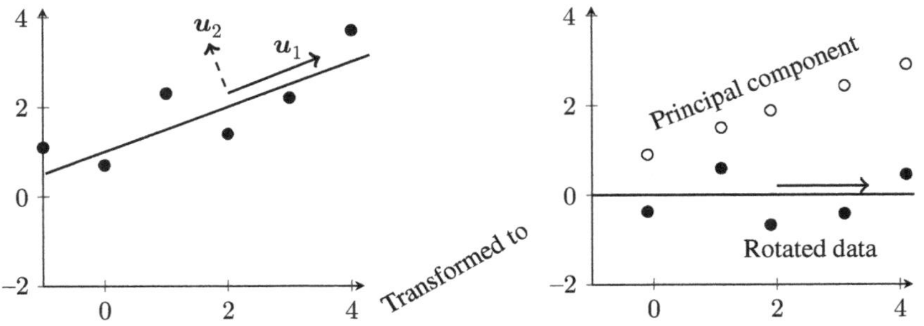


**Figure 8.4** Some random data with a trend.

**Example 8.9** For the following data (see Figure 8.4):

$$\begin{cases} x: & -1.0 \ +0.0 \ +1.0 \ +2.0 \ +3.0 \ +4.0 \\ y: & +1.1 \ +0.7 \ +2.3 \ +1.4 \ +2.2 \ +3.7 \end{cases}$$

First, we adjust the data by subtracting their means $\bar{x}$ and $\bar{y}$, respectively, we have

$$\bar{x} = \frac{1}{6}\sum_{i=1}^{6} x_i = 1.5, \quad \bar{y} = \frac{1}{6}\sum_{i=1}^{6} y_i = 1.9.$$

We have $X = x - \bar{x}$ and $Y = y - \bar{y}$

$$\begin{cases} X: & -2.5 \ -1.5 \ -0.5 \ +0.5 \ +1.5 \ +2.5 \\ Y: & -0.8 \ -1.2 \ +0.4 \ -0.5 \ +0.3 \ +1.8 \end{cases}$$

that have zero means. We can then calculate the covariance matrix

$$\boldsymbol{C} = \begin{pmatrix} 3.500 & 1.660 \\ 1.660 & 1.164 \end{pmatrix}.$$

Since cov$(x, y)$ is positive [so is cov$(X, Y)$=cov$(x, y)$], it is expected that $y$ increases with $x$ or vice versa.

The two eigenvalues of $\boldsymbol{C}$ are

$$\lambda_1 = 4.36, \quad \lambda_2 = 0.302.$$

Their corresponding eigenvectors are

$$\boldsymbol{u}_1 = \begin{pmatrix} 0.887 \\ 0.461 \end{pmatrix}, \quad \boldsymbol{u}_2 = \begin{pmatrix} -0.461 \\ 0.887 \end{pmatrix},$$

and these two vectors are orthogonal. That is, $\boldsymbol{u}_1^T \boldsymbol{u}_2 = \boldsymbol{u}_1 \cdot \boldsymbol{u}_2 = 0$. They span an orthogonal matrix

$$\boldsymbol{Q} = \begin{pmatrix} 0.887 & -0.461 \\ 0461 & 0.887 \end{pmatrix}.$$

Using the adjusted data

$$\tilde{X} = \begin{pmatrix} -2.5 & -0.8 \\ -1.5 & -1.2 \\ -0.5 & +0.4 \\ +0.5 & -0.5 \\ +1.5 & +0.3 \\ +2.5 & +1.8 \end{pmatrix},$$

we have the rotated data

$$\tilde{X}_Q = \tilde{X}\Omega,$$

which gives [0.44, −0.37, 0.59, −0.67, −0.42, 0.45] that are shown in Figure 8.4.

Since $\lambda_1$ is the larger eigenvalue, its corresponding eigenvector indicates that $x$ direction is the main component.

Almost all major software packages in statistics, data mining, and machine learning have implemented the algorithms we introduced here in this chapter.

## Exercises

**8.1** For a nonlinear function $y = \exp[ax^2 + b\sin x]$ where $a, b \in \mathbb{R}$, find a way to do linear regression if given data $(x_i, y_i)$ for $i = 1, 2, \ldots, n$.

**8.2** For the following data:

$$\begin{cases} x\colon & -0.9, & +1.0, & +2.1, & +2.9, & +4.0 \\ y\colon & +2.5, & -1.5, & -0.5, & +2.5, & +7.5 \end{cases}$$

use a linear polynomial of degree $p$ when $p = 1, 2, 3$ to fit the above data. Using AIC or BIC criterion to show that $p = 2$ is the best choice.

**8.3** For the data in the previous exercise, try to calculate the covariance, eigenvalues, and eigenvectors. Then, discuss which is the main component. Readers can refer to a useful website.[1]

**8.4** Carry out a standard binary logistic regression using the following data:

$$\begin{cases} x\colon & 0.5 & 2.2 & 3.7 & 4.9 & 6.6 & 8.8 & 10.0, \\ y\colon & 0 & 0 & 0 & 1 & 1 & 1 & 0. \end{cases}$$

1 http://ufldl.stanford.edu/wiki/index.php.

## Bibliography


Akaike, H. (1974). A new look at the statistical model identification. *IEEE Transactions on Automatic Control* 19 (6): 716–723.

Bishop, C. (2007). *Pattern Recognition and Machine Learning*. New York: Springer.

Box, M.J., Davies, D., and Swann, W.H. (1969). *Non-Linear Optimization Techniques*. Princeton, NJ: Oliver and Boyd for Imperial Chemical Industries Ltd.

Cox, D.R. (1958). The regression analysis of binary sequences. *Journal of the Royal Statistical Society, Series B (Methodological)* 20 (2): 215–242.

Croeze, A., Pittan, L., and Reynolds, W. (2012). *Solving Nonlinear Least-Squares Problems with the Gauss-Newton and Levenberg-Marquardt Methods*. LSU Technical Report. Baton Rouge: Louisiana State University.

Draper, N.R. and Smith, H. (1998). *Applied Regression Analysis*, 3e. New York: Wiley.

Freedman, D.A. (2009). *Statistical Models: Theory and Practice*. Cambridge, UK: Cambridge University Press.

Hosmer, D.W. and Lemeshow, S. (2000). *Applied Logistic Regression*, 2e. Berlin: Springer.

Jackson, J.E. (1991). *A User's Guide to Principal Components*. New York: Wiley.

Jolliffe, I.T. (2002). *Principal Component Analysis*, 2e, New York: Springer.

Levenberg, K. (1944). A method for the solution of certain non-linear problems in least squares. *Quarterly of Applied Mathematics* 2 (2), 164–168.

Marquardt, D. (1963). An algorithm for least-squares estimation of nonlinear parameters. *SIAM Journal on Applied Mathematics* 11 (2): 431–441.

Mayer-Schönberger, V. and Cukier, K. (2013). *Big Data: A Revolution That Will Transform How We Live, Work, and Think*. London: John Murray Publishers.

Nocedal, J. and Wright, S.J. (2006). *Numerical Optimization*, 2e. New York: Springer.

Pearson, K. (1901). On lines and planes of closest fit to systems to points in space. *Philosophical Magazine* 2 (11): 559–572.

Schwarz, G.E. (1978). Estimating the dimension of a model. *Annals of Statistics* 6 (2): 461–464.

Shlens, J. (2014). *A Tutorial on Principal Component Analysis*, Version 3, 1–12. https://arxiv.org/abs/1404.1100 (accessed 20 January 2018).

Smith, L.I. (2002). *A Tutorial on Principal Components Analysis*, Technical Report OUCS-2002-12. New Zealand: University of Otago.

Stigler, S.M. (1981). Gauss and the invention of least squares. *Annals of Statistics* 9 (3): 465–474.

Strutz, T. (2016). *Data Fitting and Uncertainty: A Practical Introduction to Weighted Least Squares and Beyond*, 2e. Heidelberg: Springer.

Tibshirani, R. (1996). Regression shrinkage and selection via the lasso. *Journal of the Royal Statistical Society, Series B (Methodological)* 58 (1): 267–288.
Tittering, D.M. and Cox, D.R. (2001). *Biometrika: One Hundred Years*. Oxford: Oxford University Press.
Wolberg, J. (2005). *Data Analysis Using the Method of Least Squares: Extracting the Most Information from Experiments*. Berlin: Springer.
Zaki, M.J. and Meira Jr., W. (2014). *Data Mining and Analysis: Fundamental Concepts and Algorithms*. Cambridge, UK: Cambridge University Press.
Zou, H. and Hastie, T. (2005). Regularization and variable selection via the elastic net. *Journal of the Royal Statistical Society, Series B* 67 (Part 2) 301–320.

# 9

# Machine Learning Algorithms

Machine learning algorithms are a class of sophisticated optimization algorithms, including both supervised learning and unsupervised learning algorithms. In general, there are a diverse range of algorithms in this category, and they are classification and clustering algorithms, regression, decision trees, artificial neural networks (ANNs), support vector machines (SVMs), Bayesian networks, Boltzmann machines, natural language processing, deep belief networks, evolutionary algorithms such as genetic algorithms, and metaheuristics such as particle swarm optimization and the firefly algorithm.

Due to the diversity of algorithms, there is a vast literature in machine learning and artificial intelligence. In Chapter 8, we have covered regression and principal component analysis. Here, in this chapter, we will briefly introduce three types of algorithms: data mining techniques, ANNs, and SVMs.

## 9.1 Data Mining

The evolution of the Internet and social media has resulted in the huge increase of data in terms of both volumes and complexity. In fact, "big data" has become a buzzword nowadays, and the so-called big data science is becoming an important area. Data mining has expanded beyond the traditional data modeling techniques such as statistical models and regression methods. Data mining now also includes clustering and classifications, feature selection and feature extraction, and machine learning techniques such as decision tree methods, hidden Markov models, ANNs, and SVMs. To introduce these methods systematically can take a whole book, and it is not possible to cover even a good fraction of these methods in a book chapter. Therefore, we will focus on some of the most widely used methods.

Clustering and classification methods are rather rich with a wide spectrum of methods. We introduce the basic $k$ means method for clustering and SVMs machines for classification. ANNs are a class of methods with different variations and variants, and ANN can have many applications in a diverse range

*Optimization Techniques and Applications with Examples*, First Edition. Xin-She Yang.

of areas, including clustering, classification, machine learning, computational intelligence, feature extraction and selection, and others. In the rest of this chapter, we will briefly introduce the essence of these methods.

### 9.1.1 Hierarchy Clustering

For a given set of $n$ observations, the aim is to divide them into some clusters (say, $k$ different clusters) so as to minimize certain clustering measures or objectives. There are many key issues here. Firstly, we usually do not know how many clusters the data may intrinsically have. Secondly, the data sets can be massive ($n \gg 1$, for example, $n = O(10^9)$ or even $n = O(10^{18})$). Thirdly, the data may not be clean enough (often with useless information and/or noisy data). Finally, the data can be incomplete, and thus may lack sufficient information needed for correct clustering. Obviously, there are other issues, too, such as time factors, unstructured data, and distance metrics.

Hierarchy clustering usually works well for small data sets. It starts with every point in its own cluster (that is, $k = n$ for $n$ data points), followed by a simple iterative procedure:

1) Each point belongs to its own cluster $k = n$.
2) While (stopping criterion),
3) Choose two nearest clusters to merge into one cluster;
4) Update $k \leftarrow k - 1$;
5) Repeat until the metric measure goes up;
6) End.

This iterative procedure can lead to one big cluster $k = 1$ in the end. But it does result in a complex decision tree, which provides an informative summary and some insight into the structures and relationship within the data. However, if a distance metric such as the Euclidean metric is defined properly, the metric will start to decrease at the initial stage when two clusters are merged. In the final stage, this metric usually starts to increase, which is an indication to stop and the number of clusters can be the true number of clusters. However, this is not straightforward in practice, and there may not exist any unique $k$ value at all.

In fact, $k$ is a hyper-parameter, which needs some tuning and parametric studies in practice.

In the case when Euclidean distance measures are used, the distance between centroids is defined as the cluster distance. The complexity of this algorithm is $O(n^3)$ where $n$ is the number of points. For $n = 10^9$, this can lead to $O(10^{27})$ floating-point operations, which is quite computationally expensive.

### 9.1.2 k-Means Clustering

The main aim of the $k$-means clustering method is to divide a set of $n$ observations into $k$ different clusters in such a way that each point belongs to the nearest cluster with the shortest distance to its corresponding cluster mean or centroid.

Suppose we have $n$ observation points $\boldsymbol{x}_1, \boldsymbol{x}_2, \dots, \boldsymbol{x}_n$ in a $d$-dimensional vector space, our aim is to partition these observations into $k$ clusters $(S_1, S_2, \dots, S_k)$ with centroid means $(\boldsymbol{\xi}_1, \boldsymbol{\xi}_2, \dots, \boldsymbol{\xi}_k)$ so that the cluster-wise sum of squares, also called within-cluster sum of squares, can be minimized. That is,

$$\text{minimize} \sum_{j=1, \boldsymbol{x}_i \in S_i}^{k} ||\boldsymbol{x}_i - \boldsymbol{\xi}_j||^2, \tag{9.1}$$

where $1 < k \leq n$ and typically $k \ll n$.

This $k$-means method for dividing $n$ points into $k$ clusters can be summarized schematically as follows:

1) Choose randomly $k$ points as the initial centroids of the $k$ clusters.
2) For each remaining point $i$,
3) Assign $i$ to the cluster with the closest centroid;
4) Update the centroid of that cluster (containing $i$);
5) End.

There are some key issues concerning this method. The choice of $k$ points as the initial centroids is not efficient. In the worst case, the $k$ points selected randomly can belong to the same cluster. One possible remedy is to choose $k$ points with the largest distances from each other. This is often carried out by starting from a random point, and then try to find the second point that is as far as possible to the first point, and then try to find the third point that is as far as possible from the previous two points. This continues until the first $k$ points are initialized. This method is an improvement over the previous random selection method, but there is still no guarantee that the choice of these initial points will lead to the best clustering solutions. Therefore, some sort of random restart and multiple runs are needed.

On the other hand, the algorithm complexity of this clustering method is typically $O(n^{kd+1}\log(n))$ where $d$ is the dimension of the data. Even for $n = 10^6$, $k = 3$, and $d = 2$, this becomes $O(10^{43})$, which is extremely computationally expensive. However, it is worth pointing out that such complexity is just theoretical, and these methods can sometime work surprisingly well in practice (at least for small data sets). In the worst cases, such algorithm complexity can become NP-hard.

For a given data set, it is difficult to know what $k$ should be used because $k$ is a hyper-parameter. Ideally, the initial choice of $k$ should be sufficiently close to the actual number of intrinsic classes in the data, and then varying around this initial guess. But in practice, this may need either some experience or other methods to get sense of data. In addition, the clustering distances can also be used to check if $k$ is a proper choice in many cases.

There are many other methods such as $k$ nearest neighbor (kNN) method, fuzzy $k$-means method, and others. Interested readers can refer to more advance literature about data mining.

### 9.1.3 Distance Metric

It is worth pointing out that distance metrics are also very important. Even with the most efficient methods, if the metric measure is not defined properly, the results may be incorrect or meaningless. Most clustering methods use the Euclidean distance and Jaccard similarity, though other distances such as the edit distance and Hamming distance are also widely used.

Briefly speaking, the Euclidean distance $d(x, y)$ between two data points $x$ and $y$ is the $L_p$-norm given by

$$d(x, y) = \left( \sum_i |x_i - y_i|^p \right)^{1/p}. \tag{9.2}$$

In most cases, $p = 2$ is used. Jaccard's similarity index of two sets $U$ and $V$ is defined as

$$J(U, V) = \frac{|U \cap V|}{|U \cup V|}, \tag{9.3}$$

which leads to $0 \leq J(U, V) \leq 1$, and the Jaccard distance is defined as

$$d_J(U, V) = 1 - J(U, V). \tag{9.4}$$

The edit distance between two strings $U$ and $V$ is the smallest number of insertions and deletions of single characters that will convert $U$ to $V$. For example, $U$ = "abcde" and $V$ = "ackdeg," the edit distance is $d(U, V) = 3$. By deleting $b$, inserting $k$ after $c$, and then inserting $g$ after $e$, string $U$ can be converted to $V$.

On the other hand, the Hamming distance is the number of components in which two vectors/strings differ. For example, the Hamming distance between 10101 and 11110 is 3. Obviously, other distance metrics are also being used in the literature.

## 9.2 Data Mining for Big Data

The big data science has become increasingly important nowadays, driven by the internet, social media, and the internet of things. Many applications are now dynamically data-driven. Comparing with traditional databases and data analytics, big data have some key characteristics and thus the techniques required to cope with such big data are also more sophisticated.

### 9.2.1 Characteristics of Big Data

The main characteristics of big data can be summarized as 5 Vs and they are: volume, velocity, variety, value, and veracity.

- Volume: the volume of data has increased dramatically in recent years, driven by the Internet, multimedia, and social media.
- Velocity: the rate of data accumulation is also increased dramatically. For example, it is estimated that there are about 20 trillion GB data added each year.
- Variety: the variety of data is also diverse and data can be structured, unstructured from different types, sources, and media. For example, digital astronomy can have large data sets of images and sky survey images, but they are mainly structured data. In comparison, big data from social media, digital economy, and internet of things can collect a huge range of different data types.

In addition, the dimensionality of big data can be high, depending many different factors, features or variables, either explicit or hidden.

The collection, storage, and processing of such big data sets pose many challenges in terms of both hardware and software. However, whatever these challenges may be, we still have to try to extract some values from such big data.

- Value: The main aim of analysis and processing is to extract some useful features so as to gain insight into the data and to make predictions. Ideally, the ultimate aim is to understand the data so as to potentially predict future events.
- Veracity: Obviously, no matter how big the data may be, they can be incomplete with noise and uncertainty, subject to dynamic changes. Even so, the quality of the data and the subsequent analysis will have different accuracy and values.

All these characteristics, in addition to high-dimensionality, make the data analysis a very challenging task, if not impossible.

### 9.2.2 Statistical Nature of Big Data

Almost all data mining techniques that use some statistical properties traditionally assume that data are independent in the statistical sense. However, in reality, most data are not statistically independent, and thus statistical foundations for such methods may not be valid. Currently, there is no rigorous theory for methods that are based on statistically dependent data. Even so, researchers just assume whatever is most appropriate for the data so as to be able to use certain techniques to process and analyze the big data.

Thus, traditional methods are still used with some modifications to deal with big data, including regression, decision tree, hidden Markov models, neural networks, SVMs, and others. However, from the statistical point of view, we should be aware of Bonferroni's principle when dealing with big data. If you try to calculate some expected number of occurrences of certain patterns and if this number is significantly larger than the number of real instances that you hope to find, then you can conclude that almost anything you find is false (which means it is an artifact in a statistical sense, rather than actual evidence). In order words, if you look too hard for interesting patterns that your data may support, you are bound to find false patterns. Thus, care should be taken when interpreting data.

### 9.2.3 Mining Big Data

For big data, apart from other challenges such as storage and retrieval, one of the challenges is that the data to be processed is much larger than the main physical memory of your computer, thus it is not possible to load all the data into a computer's main memory to process. Some kind of sampling and segment-by-segment processing may be needed, before applying any algorithms.

Recent developments show that new methods may be more suitable for large data sets. For example, the Bradley–Fayyad–Reina (BFR) algorithm and Clustering Using REpresentative (CURE) algorithm have shown good results. For more details about these two algorithms, readers can refer to the book by Leskovec et al. (2014), Bradley et al. (1998), and Guha et al. (1998).

In essence, the BFR algorithm is an extension of $k$-means to high-dimensional data. It assumes that data are distributed around centroids according to Gaussian distributions in Euclidean spaces. The essence of BFR algorithm can be outlined as follows:

1) Choose a small subset (by resampling) of the big data using either $k$-means or hierarchy methods to find the initial $k$ clusters.
2) Take each chunk data (a subset) from the big data set, do the following:
   - Assign the data points and summarize the clusters (see details below), then discard the data.
   - Compress and merge points that are close to one another (forming compressed sets).
   - Retain the data points that are not assigned or not close to one another (forming retained sets).

For any given $N$ points (in a subset), calculate the $\text{SUM}_i$ and $\text{SUMSQ}_i$ where $\text{SUM}_i$ is the vector sum of data in the $i$th dimension, and $\text{SUMSQ}_i$ is the sum of the squares of all the data points in the $i$th dimension.

Then, the centroid can be updated at $\text{SUM}_i/N$ in the $i$th dimension, and its variance in the $i$th dimension can be estimated as $\text{SUMSQ}_i$-$(\text{SUM}_i/N)^2$. The advantage of such dimension-by-dimension calculations is that both $\text{SUM}_i$ and $\text{SUMSQ}_i$ become simple sums when combining two clusters.

When deciding if a new point $\boldsymbol{x}_i$ is close enough to one of the $k$ clusters, we can use the following rules: (i) add point $\boldsymbol{x}$ to a cluster if it has the centroid closest to $\boldsymbol{x}$, or (ii) assign $\boldsymbol{x}_i$ to a cluster with the least Mahalanobis distance.

The Mahalanobis distance between $\boldsymbol{x} = (x_1, x_2, \ldots, x_n)$ to a cluster centroid $c = (c_1, c_2, \ldots, c_n)$ is

$$d_m = \sqrt{\sum_{i=1}^{n} \left( \frac{x_i - c_i}{\sigma_i} \right)^2}, \tag{9.5}$$

where $\sigma_i$ is the standard deviation of the cluster in the $i$th dimension. Therefore, this distance is a variance-based, scaled distance.

For a subset of data points (a chunk from the big data), the detailed calculations are as follows:

- For data points that are close to the centroid of a cluster, add these points to that cluster. Then, update the centroid and other metrics such as $\text{SUM}_i$ and $\text{SUMSQ}_i$.
- For points that are not close to any centroid, cluster them, together with the retained sets. Then, merge any mini-clusters when appropriate.
- For points that are assigned to a cluster (including any mini-cluster), update the centroid and other metrics (then discard such points).
- In the final stage (after going through all the subset of the data or loading the last chunk of the data), post-process the retained sets and compressed sets by either assigning each point to the cluster of the nearest centroid, or discarding them as outliers.

Though the BFR algorithm is efficient, it is mainly for data that are symmetric around clusters, thus it cannot deal with S-shapes or rings effectively. For such complicated data sets, we can use another powerful algorithm, called the CURE algorithm, which is a point-assignment algorithm in Euclidean spaces without any assumptions about the shape of clusters (unlikely the normal distribution assumption in the BFR algorithm). As a result, this algorithm can deal with data clustering of odd shapes such as bends, S-shapes, and rings.

The main steps of the CURE algorithm can be summarized as follows:

1) Choose a small sample (a small subset) of the data so as to be clustered in the main memory using any good methods such as $k$-means and hierarchy methods, which gives the initial clusters.
2) Select a small set of points from each cluster to be representative points (points should be as far from one another as possible).
3) Move each of the representative points a fixed fraction (typically 20%) of the distance between its location to the centroid of its cluster.
4) Merge two clusters if they have a pair of representation points (one from each cluster) that are sufficiently close.

5) Repeat the above steps and merge until no more close cluster to merge.
6) Carry out point assignment for all the rest of points in the big data.

Though this algorithm can be sufficiently efficient in practice, there is no guarantee for the global optimality. For large data sets, it is impractical to reach the true global optimality. Any sub-optimal or sufficient good solutions can become acceptable in practical applications.

It is worth pointing out that the above clustering calculations have been based on the Euclidean distances. Obviously, non-Euclidean distances such as Jaccard similarty, Edit distance, and Hamming distance can be used, depending on the types of problems.

Recent trends tend to combine traditional algorithms with optimization algorithms that are based on swarm intelligence. The basic idea is to use optimization techniques to optimize the centroids and then use clustering methods such as $k$-means to carry out clustering. Recent studies suggest such hybrid methods can produce very promising results. For example, the firefly algorithm can be used to do clustering and classifications with superior performance. We will introduce some of these swarm intelligence algorithms in the final part of this book.

For any methods to be efficient and useful, large matrices should be avoided and there is no need to try every possible combination. The methods used to solve a large-scale problem should be efficient enough to produce good results in a practically acceptable time scale. However, in general, there is no guarantee that the global optimality can be found.

There are other methods such as GRGPF and BIRCH algorithms for clustering big data sets. In addition, algorithms and implementations can be parallelized to speed up these algorithms. Readers can refer to more advanced literature on these topics.

## 9.3 Artificial Neural Networks

Many applications use ANNs, which is especially true in machine learning, artificial intelligence, and pattern recognition. The fundamental idea of ANNs is to learn from data and make predictions. For a given set of input data, a neural network maps the input data into some outputs. The relationships between the inputs and the outputs are quite complicated, and it is usually impossible to express such relationships in any exact or analytical forms. By comparing the outputs with the true outputs, the network system can adjust its weights so as to better match its outputs. If there is a sufficient number of data, the network can be "trained" well, and thus the trained network can even make predictions for new data, which is the essence of ANNs.

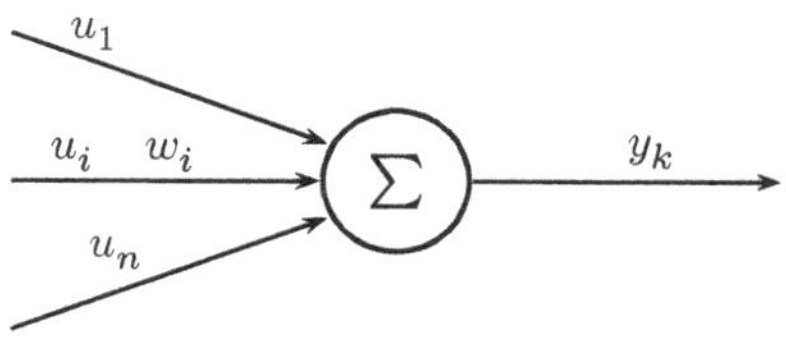


**Figure 9.1** A simple neuron.

### 9.3.1 Neuron Model

The basic mathematical model of an artificial neuron was first proposed by W. McCulloch and W. Pitts in 1943, and this fundamental model is referred to as the McCulloch–Pitts model. Other models and neural networks are based on it.

An artificial neuron with $n$ inputs or impulses and an output $y_k$ will be activated if the signal strength reaches a certain threshold $\theta$. Each input has a corresponding weight $w_i$ (see Figure 9.1). The output of this neuron is given by

$$y_l = \Phi(\xi), \quad \xi = \sum_{i=1}^{n} w_i u_i, \tag{9.6}$$

where the weighted sum $\xi$ is the total signal strength, and $\Phi$ is the so-called activation function, which can be taken as a step function. That is, we have

$$\Phi(\xi) = \begin{cases} 1 & \text{if } \xi \geq \theta, \\ 0 & \text{if } \xi < \theta. \end{cases} \tag{9.7}$$

We can see that the output is only activated to a nonzero value if the overall signal strength is greater than the threshold $\theta$.

The step activation function is binary and has discontinuity; sometimes, it is easier to use a nonlinear, smooth function for activation, called a Sigmoid function

$$S(\xi) = \frac{1}{1 + e^{-\xi}}, \tag{9.8}$$

which approaches 1 as $\xi \to \infty$, and becomes 0 as $\xi \to -\infty$. An interesting property of this function is

$$S'(\xi) = S(\xi)[1 - S(\xi)]. \tag{9.9}$$

Another useful neuron activation model is the rectified linear model, also called rectified linear unit (ReLU), which can be written as

$$y(\xi) = \xi^{+1} = \max\{0, \xi\} = \begin{cases} \xi, & \text{if } \xi > 0, \\ 0, & \text{otherwise.} \end{cases} \tag{9.10}$$

This function can be approximated by a smoother function

$$y(\xi) = \log(1 + e^{\xi}), \tag{9.11}$$

whose derivative is $y'(\xi) = e^{\xi}/(1 + e^{\xi}) = S(\xi)$.

In fact, there are many other neuron models to approximate and smoothen the activation functions, including the stochastic rectified linear model, hyperbolic tangent, exponential linear model, and others. For example, the hyperbolic tangent can be written as

$$\tanh(\xi) = \frac{e^{\xi} - e^{-\xi}}{e^{\xi} + e^{-\xi}} = \frac{1 - e^{-2\xi}}{1 + e^{-2\xi}}, \tag{9.12}$$

which approaches to $+1$ as $\xi \to \infty$ and $-1$ as $\xi \to -\infty$.

### 9.3.2 Neural Networks

A single neuron can only perform a simple task – on or off. Complex functions can be designed and performed using a network of interconnecting neurons or perceptrons. The structure of a network can be complicated, and one of the most widely used is to arrange them in a layered structure, with an input layer, an output layer, and one or more hidden layer (see Figure 9.2). The connection strength between two neurons is represented by its corresponding weight. Some ANNs can perform complex tasks and simulate complex mathematical models, even if there is no explicit functional form mathematically. Neural networks have developed over the last few decades and have been applied in almost all areas of science and engineering.

The construction of a neural network involves the estimation of the suitable weights of a network system with some training/known data sets. The task of training is to find the suitable weights $\omega_{ij}$ so that the neural networks not only can best-fit the known data, but also can predict outputs for new inputs. A good ANN should be able to minimize both errors simultaneously – the fitting/learning errors and the prediction errors.

The errors can be defined as the difference between the calculated (or predicated) output $v_k$ and real output $y_k$ for all $N$ output neurons in the least-square sense

$$E = \frac{1}{2} \sum_{k=1}^{N} (v_k - y_k)^2. \tag{9.13}$$

Here, the output $v_k$ is a function of inputs/activations and weights. This can also be written as

$$E = \frac{1}{2} ||\boldsymbol{v} - \boldsymbol{y}||_2^2, \tag{9.14}$$

where $\boldsymbol{v}$ is the output vector and $\boldsymbol{y}$ is the real or desired output vector. Here, the error function $E$ is often called the loss function of the ANN. The main aim of training a neural network is to minimize the loss or training error and also to maximize the prediction accuracy.

In order to minimize this error, we can use the standard minimization techniques to find the solutions of the weights.

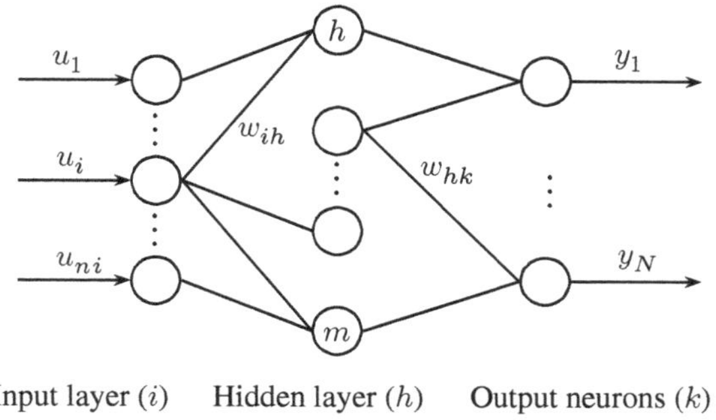


**Figure 9.2** Schematic representation of a simple feedforward neural network with $n_i$ inputs, $m$ hidden nodes, and $N$ outputs.

A simple and yet efficient technique is the steepest descent method. For any initial random weights, the weight increment for $w_{hk}$ is

$$\Delta w_{hk} = -\eta \frac{\partial E}{\partial w_{hk}} = -\eta \frac{\partial E}{\partial v_k} \frac{\partial v_k}{\partial w_{hk}}, \tag{9.15}$$

which follows the basic chain rule of differentiation. Here, $0 < \eta \leq 1$ is the learning rate. In a special case, we can use $\eta = 1$ for discussions.

From

$$S_k = \sum_{h=1}^{m} w_{hk} v_h \qquad (k = 1, 2, \ldots, N) \tag{9.16}$$

and

$$v_k = f(S_k) = \frac{1}{1 + e^{-S_k}}, \tag{9.17}$$

we have

$$f' = f(1 - f), \tag{9.18}$$

$$\frac{\partial v_k}{\partial w_{hk}} = \frac{\partial v_k}{\partial S_k} \frac{\partial S_k}{\partial w_{hk}} = v_k(1 - v_k) v_h, \tag{9.19}$$

and

$$\frac{\partial E}{\partial v_k} = (v_k - y_k). \tag{9.20}$$

Therefore, we have

$$\Delta w_{hk} = -\eta \delta_k v_h, \tag{9.21}$$

where

$$\delta_k = v_k(1 - v_k)(v_k - y_k). \tag{9.22}$$

In general, based on Eq. (9.15), the increment for any weight $w_{ij}$ related to errors is given by

$$\Delta w_{ij} = -\eta \frac{\partial E}{\partial w_{ij}}, \tag{9.23}$$

which can be written compactly as an iterative formula

$$\boldsymbol{w}^{t+1} = \boldsymbol{w}^{t} - \eta \nabla E(\boldsymbol{w}^{t}), \tag{9.24}$$

where $\nabla E$ is the gradient vector. However, for large-scale problems with many outputs, the computation of the gradient vectors can be very expensive, thus some iterative or propagation formulae are preferred.

It is worth pointing out that the topology of a neural network is also important as the ways of arranging neurons will influence the algorithm used. Different topologies will have different connections and thus different weights. Here, we have used a common feedforward structure with neurons on the previous layer (on the left) that can affect the neurons on the next layer (on the right), but the neurons on the right cannot affect the neurons on the left. Thus, the inputs are fed forward to the outputs, and this structure allows the efficient implementation of the back propagation to be introduced in the next section.

For a given structure of the neural network, there are many ways of calculating weights by supervised learning. One of the simplest and widely used methods is to use the back propagation algorithm for training neural networks, often called back propagation neural networks (BPNNs).

### 9.3.3 Back Propagation Algorithm

The basic idea of a BPNN is to start from the output layer and propagate backward so as to estimate and update the weights (see Algorithm 9.1).

From any initial random weighting matrices $w_{ih}$ (for connecting the input nodes to the hidden layer) and $w_{hk}$ (for connecting the hidden layer to the output nodes), we can calculate the outputs of the hidden layer $v_h$ as

$$v_h = \frac{1}{1 + \exp[-\sum_{i=1}^{n_i} w_{ih} u_i]} \qquad (h = 1, 2, \dots, m), \tag{9.25}$$

and the outputs for the output nodes

$$v_k = \frac{1}{1 + \exp[-\sum_{h=1}^{m} w_{hk} v_h]} \qquad (k = 1, 2, \dots, N). \tag{9.26}$$

The errors for the output nodes are given by

$$\delta_k = v_k(1 - v_k)(v_k - y_k), \qquad (k = 1, 2, \dots, N), \tag{9.27}$$

**Algorithm 9.1** Pseudocode of back propagation neural networks.

Initialize weight matrices $w_{ih}$ and $w_{hk}$ randomly
**for** all training data points
  **while** (residual errors are not small)
    Calculate the output for the hidden layer $v_h$ using Eq. (9.25)
    Calculate the output for the output layer $v_k$ using Eq. (9.26)
    Compute errors $\delta_k$ and $\delta_h$ using Eqs. (9.27) and (9.28)
    Update weights $w_{ih}$ and $w_{hk}$ via Eqs. (9.29) and (9.30)
  **end while**
**end for**

where $y_k$ $(k = 1, 2, \dots, N)$ are the data (real outputs) for the inputs $u_i$ $(i = 1, 2, \dots, n_i)$. Similarly, the errors for the hidden nodes can be written as

$$\delta_h = v_h(1 - v_h) \sum_{k=1}^{N} w_{hk}\delta_k \qquad (h = 1, 2, \dots, m). \tag{9.28}$$

The updating formulae for weights at iteration $t$ are

$$w_{hk}^{t+1} = w_{hk}^{t} - \eta\delta_k v_h, \tag{9.29}$$

and

$$w_{ih}^{t+1} = w_{ih}^{t} - \eta\delta_h u_i, \tag{9.30}$$

where $0 < \eta \leq 1$ is the learning rate.

Here we can see that the weight increments are

$$\Delta w_{ih} = -\eta\delta_h u_i, \tag{9.31}$$

with similar updating formulas for $w_{hk}$.

Again, we can write the above in the following general formula:

$$\boldsymbol{w}^{t+1} = \boldsymbol{w}^{t} - \eta\nabla E(\boldsymbol{w}^{t}), \quad E = \frac{1}{2}\sum_{k=1}^{N}\left[v_k(\boldsymbol{w}) - y_k\right]^2, \tag{9.32}$$

which applies iteratively to each layer propagating backward for multilayered networks.

It is worth pointing out that the error model we used is an $L_2$-norm (i.e. $||\boldsymbol{v} - \boldsymbol{y}||_2^2$) in terms of differences of the outputs and targets. There are other error models for neural networks. For more details, readers can refer to more advanced literature.

In a very special case when the target outputs are the same as inputs (that is to use a neural network to learn and fit the inputs to themselves), this becomes an auto-encoder network. In this case, Hinton and Salakhutdinov in 2006 used a multilayer neural network with a small central layer to reconstruct high-dimensional input vectors. The weights were adjusted and fine-tuned by gradient descent with a pre-training precedure to obtain better initial weights. They showed that their deep auto-encoder network can learn low-dimensional representations or codes that work much better than principal component analysis, which can effectively reduce the dimensionality of data.

### 9.3.4 Loss Functions in ANN

The loss function we have discussed so far is mainly the residual errors in terms of an $L_2$-norm as given in Eq. (9.14). There are many other forms of the loss functions in the literature, which may be good alternatives for certain tasks and certain types of ANNs.

If $\boldsymbol{y} = (y_1, y_2, \dots, y_N)^{\mathrm{T}}$ is a vector of the predicted values and $\bar{\boldsymbol{y}} = (\bar{y}_1, \bar{y}_2, \dots, \bar{y}_N)$ is the vector of the true values, the $L_2$ loss function is usually defined as

$$E = ||\boldsymbol{y} - \bar{\boldsymbol{y}}||_2 = ||\bar{\boldsymbol{y}} - \boldsymbol{y}||_2 = \sum_{i=1}^{N} (\bar{y}_i - y_i)^2, \tag{9.33}$$

which is essentially the same as Eq. (9.14), except for a convenient factor of $1/2$. This can also be converted to the mean square error form as

$$E_m = \frac{1}{N} \sum_{i=1}^{N} (\bar{y}_i - y_i)^2. \tag{9.34}$$

In comparison with the smooth $L_2$-norm loss, the $L_1$-norm-based loss function is given by

$$E_1 = ||\bar{\boldsymbol{y}} - \boldsymbol{y}||_1 = \sum_{i=1}^{N} |\bar{y}_i - y_i|, \tag{9.35}$$

which can also be interpreted as the mean absolute error (by multiplying a factor of $1/N$):

$$\frac{1}{N} \sum_{i=1}^{N} |\bar{y}_i - y_i|. \tag{9.36}$$

For binary classification tasks, cross entropy $E_c$ can be used as a loss function

$$E_c = -\frac{1}{N} \sum_{i=1}^{N} \left[ \bar{y}_i \log y_i + (1 - \bar{y}_i) \log(1 - y_i) \right], \tag{9.37}$$

which is a measure of the differences between the true values and predicted values in the probabilistic sense. Thus, a larger cross entropy means a larger difference.

The Kullback–Leibler (KL) divergence loss function can be defined by

$$E_{\mathrm{KL}} = \frac{1}{N}\sum_{i=1}^{N} \bar{y}_i \log(\bar{y}_i) - \frac{1}{N}\sum_{i=1}^{N} \bar{y}_i \log y_i, \tag{9.38}$$

which is the combination of the entropy (the first sum) and the cross entropy (the second term).

For other applications such the classification using the SVMs (to be introduced in the next section), the loss function can be defined as a hinge loss

$$E_h = \sum_{i=1}^{N} \max\{0, 1 - \bar{y}_i \cdot y_i\}, \tag{9.39}$$

which can also be defined with a scaling factor 1/2 as

$$E_h = \sum_{i=1}^{N} \max\{0, \frac{1}{2} - \bar{y}_i \cdot y_i\}. \tag{9.40}$$

Another related hinge loss function is the so-called squared hinge loss function

$$E_{h^2} = \sum_{i=1}^{N} \left(\max\{0, 1 - \bar{y}_i \cdot y_i\}\right)^2. \tag{9.41}$$

There are over a dozen other loss functions, including the Tanimoto loss, Chebyshev loss, and Cauchy–Schwarz divergence. For a detailed review of loss functions, readers can refer to more advanced literature such as Janocha and Czarnecki (2017) or Changhau (2017).

### 9.3.5 Stochastic Gradient Descent

As we have seen above, neural networks, especially multilayered networks, require the calculation of the gradient vectors iteratively. This can become very expensive if the numbers of weights and outputs are huge, which is true for deep learning. Therefore, some approximation or reduction in gradient calculation can speed up the learning process. Stochastic gradient descent (SGD) is one of such methods.

In the standard gradient descent, we need to calculate $n$ values of the $n$ components of the gradient vector $\nabla E$, and $n$ varies with layers. For example, $n = N$ for the output layer, while $n = m$ for the hidden layers. Instead of using the full gradient, the SGD uses a single observation or data example $\nabla E_i$ $(i = 1, 2, \dots, n)$ to approximate $\nabla E$. That is,

$$\boldsymbol{w}^{t+1} = \boldsymbol{w} - \eta \nabla E_i, \tag{9.42}$$

which can reduce the number of calculations by a factor of $n$. The choice of $i$ can be the current data set. As this gradient instance is online and random, it is thus called stochastic gradient or online gradient.

However, this extreme reduction may lead to inaccurate estimates of the true gradient. A better estimate can be obtained by a subset of stochastic gradient averaged. Alternatively, some extra term can be introduced

$$w^{t+1} = w - \eta \nabla E_i + \tau \Delta w, \tag{9.43}$$

where $0 \leq \tau \leq 1$ is the parameter controlling the inertia or momentum. This method is thus called the momentum method.

### 9.3.6 Restricted Boltzmann Machine

A very useful tool for deep learning applications is the restricted Boltzmann machine (RBM), which is a two-layer (or two-group) Boltzmann machine with $m$ visible units $v_i$ $(i = 1, 2, \dots, m)$ and $n$ hidden units $h_j$ $(j = 1, 2, \dots, n)$ where both $v_i$ and $h_j$ are binary states. The visible unit $i$ has a bias $\alpha_i$, and the hidden unit $j$ has a bias $\beta_j$, while the weight connecting them is denoted by $w_{ij}$. The restriction is that their neuron units connecting visible and hidden units form a biparte graph (see Figure 9.3), while no connection with the same group (visible or hidden) is allowed.

Using the notations and configuration given by Hinton (2010), a RBM can be represented by a pair $(\boldsymbol{v}, \boldsymbol{h})$ where $\boldsymbol{v} = (v_1, v_2, \dots, v_m)^T$ and $\boldsymbol{h} = (h_1, h_2, \dots, h_n)^T$. The energy of the system can be calculated by

$$E(\boldsymbol{v}, \boldsymbol{h}) = -\sum_{i \in \text{visible}} \alpha_i v_i - \sum_{j \in \text{hidden}} \beta_j h_j - \sum_{i,j} w_{ij} v_i h_j. \tag{9.44}$$

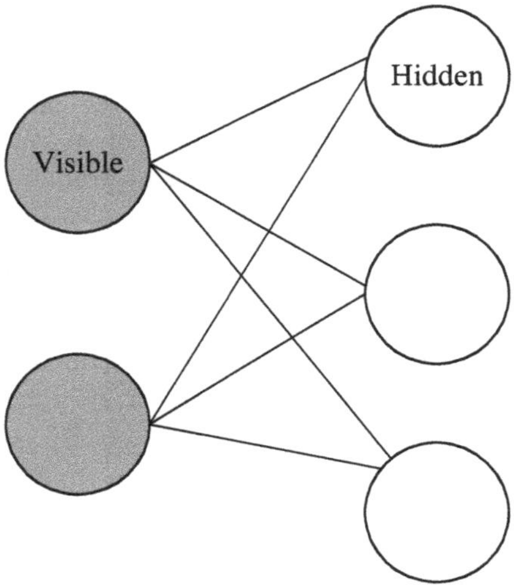


**Figure 9.3** Restricted Boltzmann machine with visible and hidden units.

The probability of a network associated with every possible pair of a visible vector $\boldsymbol{v}$ and a hidden vector $\boldsymbol{h}$ is assumed to obey the Boltzmann distribution

$$p(\boldsymbol{v}, \boldsymbol{h}) = \frac{1}{Z} e^{-E(\boldsymbol{v},\boldsymbol{h})}, \tag{9.45}$$

where $Z$ is a normalization constant, also called partition function, which is essentially the summation over all possible configurations. That is,

$$Z = \sum_{\boldsymbol{v},\boldsymbol{h}} e^{-E(\boldsymbol{v},\boldsymbol{h})}. \tag{9.46}$$

The marginal probability of a network associated with $\boldsymbol{v}$ can be calculated by sum over all possible hidden vectors in Eq. (9.45), so we have

$$p(\boldsymbol{v}) = \frac{1}{Z} \sum_{\boldsymbol{h}} e^{-E(\boldsymbol{v},\boldsymbol{h})}. \tag{9.47}$$

The essential idea of using RBM for training over a set of data (such as images) is to adjust the weights and bias values so that a training image can maximize its associated network probability (thus minimizing its corresponding energy). For a training set, the maximization of the joint probability $p(\boldsymbol{v})$ is equivalent to the maximization of the expected log probability $\log p(\boldsymbol{v})$. Since

$$\frac{\partial \log p(\boldsymbol{v})}{\partial w_{ij}} = <v_i h_j>_{\text{data}} - <v_i h_j>_{\text{RBM model}}, \tag{9.48}$$

we can calculate the adjustments in weights by using the stochastic gradient method

$$\Delta w_{ij} = \eta \left( <v_i h_j>_{\text{data}} - <v_i h_j>_{\text{RBM model}} \right), \tag{9.49}$$

where $<v_i h_j>$ means the expectation over the associated distributions. It is worth pointing out that the stochastic gradient ascent (in contrast to the SGD) is used.

The individual activation probabilities for visible and hidden units are Sigmoid function $S(x) = 1/(1 + e^{-x})$. That is,

$$p(v_i = 1|\boldsymbol{h}) = S\left( \alpha_i + \sum_j w_{ij} h_j \right) \tag{9.50}$$

and

$$p(h_j = 1|\boldsymbol{v}) = S\left( \beta_j + \sum_i w_{ij} v_i \right). \tag{9.51}$$

The RBMs form the essential part of deep belief networks with stacked RBM layers.

There are many good software packages for ANNs, and there are dozens of good books fully dedicated to theory and implementations. Therefore, we will not provide any code here.

## 9.4 Support Vector Machines

Support vector machines are a class of powerful tools which become increasingly popular in classifications, data mining, pattern recognition, artificial intelligence, and optimization.

In many applications, the aim is to separate some complex data into different categories. For example, in pattern recognition, we may need to simply separate circles from squares. That is to label them into two different classes. In other applications, we have to answer a yes–no question, which is a binary classification. If there are $k$ different classes, we can in principle first classify them into two classes: (say) class 1 and non-class 1. We then focus on the non-class 1 and divide them into two different classes, and so on.

Mathematically speaking, for a given set but scattered data, the objective is to separate them into different regions/domains or types. In the simplest case, the outputs are just class either $A$ or $B$; in other words, that is, either +1 or −1.

### 9.4.1 Statistical Learning Theory

For the case of two-class classifications, we have the learning examples or data as $(\boldsymbol{x}_i, y_i)$ where $i = 1, 2, \dots, n$ and $y_i \in \{-1, +1\}$. The aim of such learning is to find a function $f_\beta(\boldsymbol{x})$ from allowable functions $\{f_\beta : \beta \in \Omega\}$ in the parameter space $\Omega$ such that

$$f_\beta(\boldsymbol{x}_i) \mapsto y_i \qquad (i = 1, 2, \dots, n), \tag{9.52}$$

and that the expected risk $E(\beta)$ is minimal. That is, the minimization of the risk

$$E(\beta) = \frac{1}{2} \int |f_\beta(x) - y| \mathrm{d}P(\boldsymbol{x}, y), \tag{9.53}$$

where $P(\boldsymbol{x}, y)$ is an unknown probability distribution, which makes it impossible to calculate $E(\beta)$ directly. A simple approach is to use the so-called empirical risk

$$E_p(\beta) \approx \frac{1}{n} \sum_{i=1}^{n} \frac{1}{2} \big|f_\beta(\boldsymbol{x}_i) - y_i\big|. \tag{9.54}$$

A main drawback of this approach is that a small risk or error on the training set does not necessarily guarantee a small error on prediction if the number $n$ of training data is small.

In the framework of structural risk minimization and statistical learning theory, there exists an upper bound for such errors. For a given probability of at least $1 - p$, the Vapnik bound for the errors can be written as

$$E(\beta) \leq E_p(\beta) + \phi \left[ \frac{h}{n}, \frac{\log(p)}{n} \right], \tag{9.55}$$

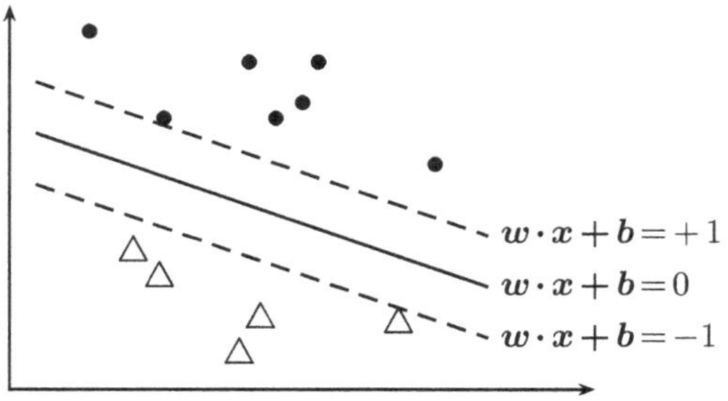


**Figure 9.4** Hyperplane, maximum margins, and a linear support vector machine.

where

$$\phi\left[\frac{h}{n}, \frac{\log(p)}{n}\right] = \sqrt{\frac{1}{n}\left[h\left(\log\frac{2n}{h}+1\right) - \log\left(\frac{p}{4}\right)\right]}. \tag{9.56}$$

Here $h$ is a parameter, often referred to as the Vapnik–Chervonenskis dimension (or simply VC-dimension). This dimension describes the capacity for prediction of the function set $f_\beta$. In the simplest binary classification with only two values of $+1$ and $-1$, $h$ is essentially the maximum number of points which can be classified into two distinct classes in all possible $2^h$ combinations.

### 9.4.2 Linear Support Vector Machine

The basic idea of classification is to try to separate different samples into different classes. For binary classifications such as the triangles and spheres (or solid dots) as shown in Figure 9.4, we intend to construct a hyperplane

$$\boldsymbol{w} \cdot \boldsymbol{x} + \boldsymbol{b} = 0, \tag{9.57}$$

so that these samples can be divided into classes with triangles on one side and the spheres on the other side. Here, the normal vector $\boldsymbol{w}$ and $\boldsymbol{b}$ have the same size as $\boldsymbol{x}$, and they can be determined using the data, though the method of determining them is not straightforward. This requires the existence of a hyperplane; otherwise, this approach will not work. In this case, we have to use other methods.

It is worth pointing out that $\boldsymbol{w} \cdot \boldsymbol{x} + \boldsymbol{b} = 0$ can also be written as

$$\boldsymbol{w}^{\mathrm{T}}\boldsymbol{x} + \boldsymbol{b} = 0, \tag{9.58}$$

but the dot product form of $\boldsymbol{w} \cdot \boldsymbol{x}$ explicitly highlights the nature of a hyperplane governed by the normal direction $\boldsymbol{w}$.

In essence, if we can construct such a hyperplane, we should construct two hyperplanes (shown as dashed lines in Figure 9.4) so that the two hyperplanes should be as far away as possible and no samples should be between these two planes. Mathematically, this is equivalent to two equations

$$\boldsymbol{w} \cdot \boldsymbol{x} + \boldsymbol{b} = +1 \tag{9.59}$$

and

$$\boldsymbol{w} \cdot \boldsymbol{x} + \boldsymbol{b} = -1. \tag{9.60}$$

From these two equations, it is straightforward to verify that the normal (perpendicular) distance between these two hyperplanes is related to the norm $||\boldsymbol{w}||$ via

$$d = \frac{2}{||\boldsymbol{w}||}. \tag{9.61}$$

A main objective of constructing these two hyperplanes is to maximize the distance or the margin between the two planes. The maximization of $d$ is equivalent to the minimization of $||w||$ or more conveniently $||w||^2/2$. Here, $||w||$ is the standard $L_2$-norm defined in earlier chapters. From the optimization point of view, the maximization of margins can be written as

$$\text{minimize} \quad \frac{1}{2}||\boldsymbol{w}||^2 = \frac{1}{2}||\boldsymbol{w}||_2^2 = \frac{1}{2}(\boldsymbol{w} \cdot \boldsymbol{w}). \tag{9.62}$$

If we can classify all the samples completely, for any sample $(\boldsymbol{x}_i, y_i)$ where $i = 1, 2, \ldots, n$, we have

$$\boldsymbol{w} \cdot \boldsymbol{x}_i + \boldsymbol{b} \geq +1, \qquad \text{if } (\boldsymbol{x}_i, y_i) \in \text{one class} \tag{9.63}$$

and

$$\boldsymbol{w} \cdot \boldsymbol{x}_i + \boldsymbol{b} \leq -1, \qquad \text{if } (\boldsymbol{x}_i, y_i) \in \text{the other class.} \tag{9.64}$$

As $y_i \in \{+1, -1\}$, the above two equations can be combined as

$$y_i(\boldsymbol{w} \cdot \boldsymbol{x}_i + \boldsymbol{b}) \geq 1 \qquad (i = 1, 2, \ldots, n). \tag{9.65}$$

However, in reality, it is not always possible to construct such a separating hyperplane. A very useful approach is to use nonnegative slack variables

$$\eta_i \geq 0 \qquad (i = 1, 2, \ldots, n), \tag{9.66}$$

so that

$$y_i(\boldsymbol{w} \cdot \boldsymbol{x}_i + \boldsymbol{b}) \geq 1 - \eta_i \qquad (i = 1, 2, \ldots, n). \tag{9.67}$$

Now the optimization problem for the SVM becomes

$$\text{minimize } \Psi = \frac{1}{2}||\boldsymbol{w}||^2 + \lambda \sum_{i=1}^{n} \eta_i, \tag{9.68}$$

subject to

$$y_i(\boldsymbol{w} \cdot \boldsymbol{x}_i + \boldsymbol{b}) \geq 1 - \eta_i, \tag{9.69}$$

$$\eta_i \geq 0 \qquad (i = 1, 2, \dots, n), \tag{9.70}$$

where $\lambda > 0$ is a penalty parameter to be chosen appropriately. Here, the term $\sum_{i=1}^{n} \eta_i$ is essentially a measure of the upper bound of the number of misclassifications on the training data.

By using Lagrange multipliers $\alpha_i \geq 0$, we can rewrite the above constrained optimization into an unconstrained version, and we have

$$L = \frac{1}{2}||\boldsymbol{w}||^2 + \lambda \sum_{i=1}^{n} \eta_i - \sum_{i=1}^{n} \alpha_i[y_i(\boldsymbol{w} \cdot \boldsymbol{x}_i + \boldsymbol{b}) - (1 - \eta_i)]. \tag{9.71}$$

From this, we can write the Karush–Kuhn–Tucker (KKT) conditions as

$$\frac{\partial L}{\partial \boldsymbol{w}} = \boldsymbol{w} - \sum_{i=1}^{n} \alpha_i y_i \boldsymbol{x}_i = 0, \tag{9.72}$$

$$\frac{\partial L}{\partial \boldsymbol{b}} = -\sum_{i=1}^{n} \alpha_i y_i = 0, \qquad y_i(\boldsymbol{w} \cdot \boldsymbol{x}_i + \boldsymbol{b}) - (1 - \eta_i) \geq 0, \tag{9.73}$$

$$\alpha_i[y_i(\boldsymbol{w} \cdot \boldsymbol{x}_i + \boldsymbol{b}) - (1 - \eta_i)] = 0 \qquad (i = 1, 2, \dots, n), \tag{9.74}$$

$$\alpha_i \geq 0, \qquad \eta_i \geq 0 \qquad (i = 1, 2, \dots, n). \tag{9.75}$$

From the first KKT condition, we get

$$\boldsymbol{w} = \sum_{i=1}^{n} y_i \alpha_i \boldsymbol{x}_i. \tag{9.76}$$

It is worth pointing out here that only the nonzero coefficients $\alpha_i$ contribute to the overall solution. This comes from the KKT condition (9.74), which implies that when $\alpha_i \neq 0$, the inequality (9.69) must be satisfied exactly, while $\alpha_i = 0$ means the inequality is automatically met. Therefore, only the corresponding training data $(\boldsymbol{x}_i, y_i)$ with $\alpha_i > 0$ can contribute to the solution, and thus such $\boldsymbol{x}_i$ form the support vectors (hence, the name SVM). All the other data with $\alpha_i = 0$ become irrelevant.

There is a dual problem for this SVM optimization problem, and it can be shown that the solution for $\alpha_i$ can be found by solving the following quadratic programming:

$$\text{maximize} \sum_{i=1}^{n} \alpha_i - \frac{1}{2} \sum_{i,j=1}^{n} \alpha_i \alpha_j y_i y_j (\boldsymbol{x}_i \cdot \boldsymbol{x}_j), \tag{9.77}$$

subject to

$$\sum_{i=1}^{n} \alpha_i y_i = 0, \tag{9.78}$$

$$0 \leq \alpha_i \leq \lambda \qquad (i = 1, 2, \ldots, n). \tag{9.79}$$

From the coefficients $\alpha_i$, we can write the final classification or decision function as

$$f(\boldsymbol{x}) = \text{sign}\left[\sum_{i=1}^{n} \alpha_i y_i (\boldsymbol{x} \cdot \boldsymbol{x}_i) + \boldsymbol{b}\right], \tag{9.80}$$

where sign($x$) is the classic sign function, which is sign($x$) = +1 if $x > 0$, −1 if $x < 0$, and 0 if $x = 0$.

### 9.4.3 Kernel Functions and Nonlinear SVM

In reality, most problems are nonlinear, and the above linear SVM cannot be used. Ideally, we should find some nonlinear transformation $\phi$ so that the data can be mapped onto a high-dimensional space where the classification becomes linear (see Figure 9.5). The transformation should be chosen in a certain way so that their dot product leads to a kernel-style function

$$K(\boldsymbol{x}, \boldsymbol{x}_i) = \phi(\boldsymbol{x}) \cdot \phi(\boldsymbol{x}_i), \tag{9.81}$$

which enables us to write our decision function as

$$f(\boldsymbol{x}) = \text{sign}\left[\sum_{i=1}^{n} \alpha_i y_i K(\boldsymbol{x}, \boldsymbol{x}_i) + \boldsymbol{b}\right]. \tag{9.82}$$

From the theory of eigenfunctions, we know that it is possible to expand functions in terms of eigenfunctions. In fact, we do not need to know such transformations, we can directly use kernel functions $K(\boldsymbol{x}, \boldsymbol{x}_i)$ to complete this task. This is the so-called kernel function trick. Now the main task is to chose a suitable kernel function for a given problem.

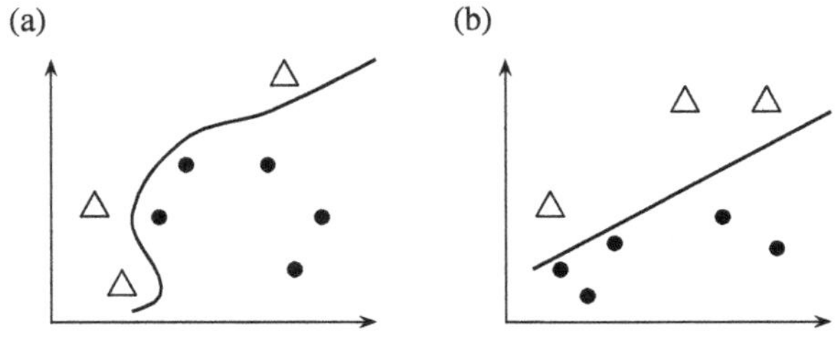

**Figure 9.5** Kernel functions by nonlinear transformation. (a) Input space and (b) feature space.

For most problems concerning a nonlinear SVM, we can use

$$K(\boldsymbol{x}, \boldsymbol{x}_i) = (\boldsymbol{x} \cdot x_i)^d \tag{9.83}$$

for polynomial classifiers, and

$$K(\boldsymbol{x}, \boldsymbol{x}_i) = \tanh[k(\boldsymbol{x} \cdot \boldsymbol{x}_i) + \Theta] \tag{9.84}$$

for neural networks. The most widely used kernel is the Gaussian radial basis function (RBF)

$$K(\boldsymbol{x}, \boldsymbol{x}_i) = \exp\left[-\frac{||\boldsymbol{x} - \boldsymbol{x}_i||^2}{(2\sigma^2)}\right]$$
$$= \exp\left[-\gamma||\boldsymbol{x} - \boldsymbol{x}_i||^2\right] = \exp[-\gamma r^2], \tag{9.85}$$

for nonlinear classifiers. Here, $r = ||\boldsymbol{x} - \boldsymbol{x}_i||$. This kernel can easily be extended to any high dimensions. Here, $\sigma^2$ is the variance and $\gamma = 1/2\sigma^2$ is a constant. In fact, $\gamma$ is a hyper-parameter, which needs to be tuned for each SVM.

Following a similar procedure as discussed earlier for linear SVMs, we can obtain the coefficients $\alpha_i$ by solving the following optimization problem:

$$\text{maximize} \sum_{i=1}^{n} \alpha_i - \frac{1}{2}\alpha_i\alpha_j y_i y_j K(\boldsymbol{x}_i, \boldsymbol{x}_j). \tag{9.86}$$

It is worth pointing out that, when matrix $\boldsymbol{A} = y_i y_j K(\boldsymbol{x}_i, \boldsymbol{x}_j)$ is a symmetric positive definite matrix, the above maximization problem becomes a quadratic programming problem, and can thus be solved efficiently by standard quadratic programming techniques.

There are many software packages (commercial or open source) which are easily available, so we will not provide any discussion of the implementation. In addition, some methods and their variants are still an area of active research. Interested readers can refer to more advanced literature.

## 9.5 Deep Learning

Deep learning is not just a buzz word nowadays, it is a very powerful tool for many applications related to artificial intelligence. Image processing, pattern recognition, and speech recognition become much more accurate because of the effective use of deep learning techniques such as deep convolutionary neural networks (CNN) and RBMs.

### 9.5.1 Learning

In general, learning in the broad context of machine learning can be divided into three categories: supervised learning, unsupervised learning, and reinforcement learning.

Supervised learning use data with known labels or classes. Regression is an example of supervised learning where the target outputs $y$ are real numbers via a model or model class $y = f(\boldsymbol{x}, \boldsymbol{\alpha})$ where $\boldsymbol{\alpha}$ is a model parameter vector. In addition, classification is also supervised learning. For example, classification using the SVM belongs to this category.

Clustering is a good example of unsupervised learning. There is no need to have labeled data, and learning is to figure out the internal structures and representation of the data.

Reinforcement learning is based on some scalar reward objective to be maximized, based on the data and some future rewards. However, this type of learning is limited to deal with cases when the number of key parameters is not huge (e.g. a few dozens).

It is worth pointing out that the learning in neural networks can be supervised or unsupervised, depending on the type of data. To use labeled data with target outputs to train neural networks, it is supervised learning. Otherwise, if there are no target outputs, the learning becomes unsupervised.

In addition to the traditional categories of learning, other forms and new types of learning exist in the literature. For example, semi-supervised learning is the learning of largely using a lot of unlabeled data with a few labeled data where both labeled data and unlabeled are assumed to be drawn from the same probability distributions. In essence, semi-supervised learning implicitly uses the assumption that unlabeled data can be labeled with the same labels used by labeled data for classification tasks.

Transfer learning transfers knowledge from one supervised learning task to another, which requires additional labeled data from a different (but related) task. The requirement of such extra labeled may be expensive.

Self-taught teaching is a two-stage learning approach where learning is carried out first on the unlabeled data to learn a representation, and then this learned represented is applied to labeled data for classification tasks. There is no assumption that the labeled data were drawn from the same distributions for the unlabeled data, thus the data can be in different classes and from different distributions. In other words, the self-taught learning can be considered as transfer learning from unlabeled data, or unsupervised transfer.

### 9.5.2 Deep Neural Nets

The essence of deep learning is the feedforward deep neural network (i.e. multilayer perceptron) with different levels of abstraction, and training of such multilayer perceptron networks (MLPN) is done by the backpropagation algorithm. Here, "deep" means multiple layers, which can range from a few layers to several hundred layers or even more. Such learning can be done by one layer at a time.

Deep learning intends to learn representations and feature hierarchies from higher-level features or complicated concepts formed by lower-level features or simpler concepts, and such a system consists of multiple levels of abstraction without human-crafted feature creation.

In order for the deep networks to work well, it requires both a large amount of training data (with a wide range of diversity) and computing power to realize the training practically. For example, for face recognition and image classifications, the number of images can vary from a few million to hundreds of million. Therefore, it is no surprise that big companies such as Google, Microsoft, IBM, and Amazon are among those who have the most successful deep learning systems. AlexNet and TensorFlow are good examples of deep learning software packages. There are also many good tutorials on these vast topics. For example, the Stanford website udldl.stanford.edu/wiki/index.php can be a very helpful starting point.

### 9.5.3 Tuning of Hyper-Parameters

Many techniques we have discussed in this chapter have algorithm-dependent hyper-parameters. For example, $k$ in the $k$-means, $\lambda$ in the regularization methods, the learning rate $\eta$, and the number of layers in deep learning are all hyper-parameters. At the moment, the choices of such hyper-parameters are mainly by researchers, based on parametric studies, the type of problems, and empirical studies, even personal expertise and experience. The initial choice can be an educated guess, then fine-tuning is then attempted. Some researchers use uniform, grid-style search, while others use various approaches. In fact, the optimal choice of such hyper-parameters is a challenging task, it is the optimization of an optimization problem, and the choice is largely problem-dependent.

Currently, there still lacks rigorous theory for choosing such parameters. Recent studies suggested random search, heuristic methods, and metaheuristic algorithms such as genetic algorithms and the firefly algorithm can be very useful to optimize hyper-parameters. We will introduce such metaheuristic algorithms such as particle swarm optimization, firefly algorithm, and cuckoo search in future chapters.

## Exercises

**9.1** For five data points, discuss the challenge of finding the right number of clusters?

**9.2** Carry out $k$-means clustering using R or Python.

**9.3** Use any software to carry out a simple deep learning task.

## Bibliography

Aggarwal, C.C. (2015). *Data Mining: The Textbook*. New York: Springer.

Bengio, Y. (2009). Learning deep architecture for AI. *Foundations and Trends in Machine Learning* 2 (1): 1–127.

Bengio, Y. (2012). Deep learning of representations for unsupervised and transfer learning. In: *JMLR Workshop and Conference Proceedings* (Workshop on Unsupervised and Transfer Learning) 27: 17–37.

Bishop, C. (2007). *Pattern Recognition and Machine Learning*. New York: Springer.

Bradley, P.S., Fayyad, U.M., and Reina, C. (1998). Scaling clustering algorithms to large databases. In: *Proceedings of the 4th International Conference on Knowledge Discovery and Data Mining*, 9–15. Menlo Park, CA: AAAI Press.

Buduma, N. (2015). *Fundamentals of Deep Learning: Designing Next-Generation Artificial Intelligence Algorithms*. Sebstopol, CA: O'Reilly Media.

Changhau, I. (2017). *Loss Functions in Artificial Neural Networks*. Github.io online mates. https://isaacchanghau.github.io/post/loss_functions/ (accessed 26 April 2018).

Cortes, C. and Vapnik, V. (1995). Support-vector networks. *Machine Learning* 20 (3): 273–297.

Cristianini, N. and Shawe-Taylor, J. (2000). *An Introduction to Support Vector Machines and Other Kernel-Based Learning Methods*. Cambridge, UK: Cambridge University Press.

Dhaenens, C. and Jourdan, L. (2016). *Metaheuristics for Big Data*. Hoboken, NJ: Wiley.

Fischer, A. and Igel, C. (2014). Training restricted Boltzmann machines: an introduction. *Pattern Recognition* 47 (1): 25–39.

Goodfellow, I., Bengio, Y., and Courville, A. (2017). *Deep Learning*. Cambridge, MA: The MIT Press.

Guha, S., Rastogi, R., and Shim, K. (1998). CURE: an efficient clustering algorithm for large databases. In: *Proceedings of the 1998 ACM SIGMOD International Conference on Management of Data*, 73–84. New York: ACM.

Haykin, S.S. (1999). *Neural Networks: A Comprehensive Foundation*. Upper Saddle River, NJ: Prentice Hall.

Heaton, J. (2015). *Artificial Intelligence for Humans, Volume 3: Deep Learning and Neural Networks*. Chesterfield, MO: CreateSpace Independent Publishing.

Hilbert, M. (2016). Big data for development: a review of promises and challenges. *Development Policy Review* 34 (1): 135–174.

Hinton, G.E. (2007). Learning multiple layers of representation. *Trends in Cognitive Sciences* 11: 428–434.

Hinton, G.E. (2009). Deep belief networks. *Scholarpedia* 4 (5): 5947.

Hinton, G.E. (2010). *A Practical Guide to Training Restricted Boltzmann Machines*. UTML TR 2010-003, Technical Report. University of Toronto.

Hinton, G.E., Deng, L., Yu, D. et al. (2012). Deep neural networks for acoustic modeling in speech recognition – the shared views of four research groups. *IEEE Signal Processing Magazine* 29 (6): 82–97.

Hinton, G.E., Osindero, S., and Teh, Y.W. (2006). A fast learning algorithm for deep belief nets. *Neural Computation* 18 (7): 1527–1554.

Hinton, G.E. and Salakhutdinov, R. (2006). Reducing the dimensionality of data with neural networks. *Science* 313 (5786): 504–507.

Holmes, D.E. (2017). *Big Data: A Very Short Introduction*. Oxford: Oxford University Press.

Hopfield, J.J. (1982). Neural networks and physical systems with emergent collective computational abilities. *Proceedings of the National Academy of Sciences* 79: 2554–2558.

Hurwitz, J., Nugent, A., Halper, F. et al. (2013). *Big Data for Dummies*. Hoboken, NJ: Wiley.

Janocha, K. and Czarnecki, W.M. (2017). On loss function for deep neural networks in classification. *Theoretical Foundations of Machine Learning (TFML 2017)*. https://arxiv.org/abs/1702.05659 (28 February 2018).

Kecman, V. (2001). *Learning and Soft Computing – Support Vector Machines, Neural Networks, Fuzzy Logic Systems*. Cambridge, MA: The MIT Press.

Larochelle, H. and Bengio, Y. (2008). Classification using discriminative restricted boltzmann machines. In: *Proceedings of the 25th International Conference on Machine Learning (ICML2008)*, 536. New York: ACM.

LeCun, Y., Bengio Y., and Hinton, G.E. (2015). Deep learning. *Nature* 521: 436–444.

LeCun, Y., Bottou, L., Orr, G.B. et al. (2012). Efficient backdrop. In: *Neural Networks: Tricks of the Trade*, 9–48. Berlin: Springer.

Leskovec, J., Rajaraman, A., and Ullman, J.D. (2014). *Mining of Massive Datasets*, 2e. Cambridge, UK: Cambridge University Press.

Marr, B. (2015). *Big Data: Using Smart Big Data, Analytics and Metrics to Make Better Decisions and Improve Performance*. Hoboken, NJ: Wiley.

Mayer-Schönberger, V. and Cukier, K. (2013). *Big Data: A Revolution That Will Transform How We Live, Work, and Think*. London: John Murray Publishers.

McCulloch, W. and Pitts, W. (1943). A logical calculus of the ideas immanent in nervous activity. *The Bulletin of Mathematical Biophysics* 5 (4): 115–133.

Raina, R., Battle, A., Lee, H. et al. (2007). Self-taught learning: transfer learning from unlabelled data. *Proceedings of 24th International Conference on Machine Learning (ICML)* (20–24 June, 2007), Corvalis, OR, 759–766.

Rumelhart, D.E., Hinton, G.E., and Williams, R.J. (1986). Learning representations by back-propagating errors. *Nature* 323 (6088): 533–536.

Smolensky, P. (1986). Information processing in dynamical systems: foundations of harmony theory, Chapter 6. In: *Parallel Distributed Processing: Explorations in the Microstructure of Cognition, Volume 1: Foundation* 194–281. Cambridge, MA: MIT Press.

Vapnik, V.N. (1995). *The Nature of Statistical Learning Theory*. Berlin: Springer-Verlag.
Winston, P.H. (1992). *Artificial Intelligence*, 3e. Reading, MA: Addison-Wesley Publishing.
Witten, I.H., Frank, E., Hall, M.A. et al. (2016). *Data Mining: Practical Machine Learning Tools and Techniques*, 4e. Cambridge, MA: Morgan Kaufmann.
Yang, X.S. (2010). *Engineering Optimization: An Introduction with Metaheuristic Applications*. Hoboken, NJ: Wiley.
Zaki, M.J. and Meira Jr., W. (2014). *Data Mining and Analysis: Fundamental Concepts and Algorithms*. Cambridge, UK: Cambridge University Press.

# 10

# Queueing Theory and Simulation

Queueing is an unavoidable part of modern life, and it seems that queueing is everywhere in banks, restaurants, airports, call centers, and leisure parks. Queueing also applies to internet queries, email message deliveries, telephone conversations, and many other applications. Thus, modeling queueing behavior can be of both theoretical interest and practical importance. The Danish engineer, A.K. Erlang, was the first to study the queueing process in the telephone system, and the relevant key ideas can be applied to many other queueing systems. This chapter introduces the fundamentals of queueing models.

## 10.1 Introduction

### 10.1.1 Components of Queueing

Apart from gaining insight into the queueing characteristics, one of the main aims of modeling queues is to improve the efficiency of queue management and to minimize the overall waiting costs and service costs. In order to model queueing systems properly, we have to identify their common components such as the rates of arrival, service, and departure.

From our experience and empirical observations, we can analyze the basic components or factors in a queue, and they are:

- Arrival rate ($\lambda$): This can be represented as the rate or the number of arrivals per unit time in a queue.
- Service rate ($\mu$): This is how quickly a service (e.g. serving a cup of coffee) is provided, which can be represented as the number of customers served per unit time or the time taken to serve a customer.
- Number of servers ($s$): This can be represented as the number of counters in a supermarket or the number of cashiers in a bank. If each queue requires one server only, the number of servers can also be considered as the number of queues.

*Optimization Techniques and Applications with Examples*, First Edition. Xin-She Yang.

The above components are most important. However, in practice, factors such as capacity in a restaurant or a bank is also important. In this case, we have to consider three additional factors or components:

- Maximum queueing length ($L_{max}$): The capacity of a popular restaurant or a coffee shop may be limited by the number of tables, thus there is a maximum queuing length in practice.
- Population size ($N$): This can be considered as the pool for potential customers that can arrive at the queueing system. In practice, the population size is always finite, but in the simplest case of queue modeling, we can assume that the population size is infinite.
- Queueing rule ($R$): The simplest queueing rule can be first-come and first-serviced, that is, a first-in first-out (FIFO) rule. Other rules can also apply when appropriate. For example, LIFO means last in and first out, and SIRO means service in random order.

In terms of practical queue settings and managements, there are usually four possibilities:

1) Single phase and single server: This is the simplest case where there is one queue with one server. Examples include queues at a photocopier machine, ATM, or a small corner shop.
2) Single phase and multiservers: This corresponds to a queue with multiple servers. For example, queueing in banks and some information centers seems to use this popular system. In this case, each customer in the queue is given a number and once a server is available, the next customer in the system is called to be served. For example, many information desk systems belong to this category.
3) Multiphases and single server: Sometimes, the service can be more complicated and thus can be divided into different stages or phases. For example, a drive-in McDonald, a customer in the driving queue orders first, then drives onto the next window to wait and collect the meal. This also applies to many services such as restaurants, student applications where tasks may take more than one stage or require two or more servicing facilities.
4) Multiphases and multi-servers: This represents a very generalized case where multiple service stages are need to serve each customer and there are multiple servers available for the service. A good example is at the airport where customers check in first, then go through the security, and then go to the boarding gates. At each phase, there are queues and each phase has multiple servers.

### 10.1.2 Notations

In order to model a queueing system properly, the standard notations introduced by D. G. Kendall are usually used.

In essence, Kendall's notation consists of six parts, in general, which can be written as

$$A/B/s/L/P/R. \tag{10.1}$$

Here, the first part A denotes the arrival model. For example, if the arrival probability obeys a Poisson's distribution, which is also Markovian, we can write "A" as "M." Here, the main characteristics of a Markovian process is that it is a memoryless process and thus the arrivals should be independent of each other.

The second part B denotes the service time distribution. For example, the most commonly used distribution for service time is exponential, which is also Markovian. The third part "s" denotes the number of servers, which is just a positive integer. The fourth part "L" denotes the queueing capacity or the maximum queueing length. If not given explicitly, it is usually assumed that the capacity is infinite. In addition, the fifth part "P" denotes the population size, and it can be assumed to be infinite if not stated explicitly.

The final part is the queueing rule "R" or queueing discipline. The default rule is FIFO. In case of $L = \infty$, $P = \infty$, and D = FIFO, Kendall's notation can be often simplified to only the first three letters. Thus, the simplest queue model is

$$M/M/1, \tag{10.2}$$

which corresponds to the case of a single server with both arrival and service being Markovian. The implicit assumptions are that both the capacity and population size are infinite, while the queueing rule is FIFO.

To model a queueing process with Poisson arrivals and exponential service time for 5 servers with a population size of 100 and each server has a capacity of 15 with a queue discipline of FIFO, the system model can be written compactly as M/M/5/15/100/FIFO or simply as M/M/5/15/100.

Kendall's notation can be used to describe almost all queueing processes. However, in our present discussions here, we will focus on a single-phased queue with a single server. Thus, the following additional assumptions are made:

- The arrival of each individual is independent, thus bulk arrivals (e.g. a coach of tourists, a group of students) are not allowed.
- The service time is also independent. That is, the service time of the previous customer is independent of the service time of the next customer.
- The whole queuing system is stable in the sense that the arrival and service probability distributions remain unchanged.
- The queueing discipline is simply "first come, first served" (i.e. FIFO).

With these assumptions, the model can be simplified and we will not explicitly discuss these assumptions any further in the rest of this chapter.

## 10.2 Arrival Model

The most widely used model for arrivals is the Poisson process. Let us first review the Poisson distribution before modeling queues.

### 10.2.1 Poisson Distribution

The Poisson distribution can be thought of as the limit of the binomial distribution when the number of trial is very large and the probability is sufficiently small. Typically, it is concerned with the number of events that occur in a certain time interval (e.g. number of telephone calls in an hour) or spatial area. The Poisson distribution can be written as

$$P(X = x) = \frac{\lambda^x e^{-\lambda}}{x!} \qquad (\lambda > 0), \tag{10.3}$$

where $x = 0, 1, 2, \ldots, n$ and $\lambda$ is the mean of the distribution.

Obviously, the sum of all the probabilities must be equal to one. That is,

$$\begin{aligned}\sum_{x=0}^{\infty} \frac{\lambda^x e^{-\lambda}}{x!} &= \frac{\lambda^0 e^{-\lambda}}{0!} + \frac{\lambda^1 e^{-\lambda}}{1!} + \frac{\lambda^2 e^{-\lambda}}{2!} + \frac{\lambda^3 e^{-\lambda}}{3!} + \cdots \\ &= e^{-\lambda}\left[1 + \lambda + \frac{\lambda^2}{2!} + \frac{\lambda^3}{3!} + \cdots\right] \\ &= e^{-\lambda} e^{\lambda} = e^{-\lambda+\lambda} = e^0 = 1.\end{aligned} \tag{10.4}$$

Many stochastic processes such as the number of phone calls in a call center, number of earthquakes in a given period, and the number of cars passing through a junction obey the Poisson distribution. Let us look at an example.

**Example 10.1** Suppose you receive one email per hour on average. If you attend a 1-hour lesson, what is the probability of receiving exactly two emails after the lesson? What is the probability of no email at all?

Since the distribution is Poisson with $\lambda = 1$, the probability of receiving two emails is

$$P(X = 2) = \frac{\lambda^2 e^{-\lambda}}{2!} = \frac{1^2 e^{-1}}{2!} \approx 0.184.$$

The probability of no email is

$$P(X = 0) = \frac{\lambda^0 e^{-\lambda}}{0!} = \frac{1^0 e^{-1}}{0!} \approx 0.368.$$

The probability of receiving exactly one email is

$$P(X = 1) = \frac{\lambda^1 e^{-\lambda}}{1!} = \frac{1^1 e^{-1}}{1} \approx 0.368.$$

On the other hand, what is the probability of receiving two or more emails? This means $X = 2, 3, 4, \ldots$, which have an infinite number of terms. That is,

$$P(X \geq 2) = P(X = 2) + P(X = 3) + P(X = 4) + \cdots,$$

but how do we calculate this probability? Since the total probability is one or

$$\begin{aligned} 1 &= P(X = 0) + P(X = 1) + P(X = 2) + P(X = 3) + P(X = 4) + \cdots \\ &= P(X = 0) + P(X = 1) + P(X \geq 2), \end{aligned}$$

we have

$$P(X \geq 2) = 1 - P(X = 0) - P(X = 1) = 1 - 0.184 - 0.368 \approx 0.264.$$

That is, the probability of receiving two or more emails is about 0.264.

Furthermore, what is the probability of receiving exactly one email in a 15-minute interval?

We already know that $\lambda = 1$ for one hour, so $\lambda_* = 1/4 = 0.25$ for a 15-minute period. The probability of receiving exactly one email in a 15-minute period is

$$P(X = 1) = \frac{\lambda_*^1 \mathrm{e}^{-\lambda_*}}{1!} = \frac{0.25^1 \mathrm{e}^{-0.25}}{1} \approx 0.195.$$

In this example, we have used $\lambda_* = \lambda t$ where $t$ is the time interval in the same time unit when defining $\lambda$. In general, we should use $\lambda t$ to replace $t$ when dealing with the arrivals in a fixed period $t$. Thus, the Poisson distribution becomes

$$P(X = n) = \frac{(\lambda t)^n e^{-\lambda t}}{n!}. \tag{10.5}$$

Therefore, in the above example, if the lesson is a two-hour session, the probability of getting exactly one email after the two-hour session is

$$P(X = 1) = \frac{(1 \times 2)^1 \mathrm{e}^{-1\times 2}}{1!} \approx 0.271.$$

Using the definitions of mean and variance, it is straightforward to prove that $E(X) = \lambda$ and $\sigma^2 = \lambda$ for the Poisson distribution. For example, the mean or expectation $E(X)$ can be calculated by

$$\begin{aligned} E(X) &= \sum_{x=0}^{\infty} x P(X = x) = \sum_{x=0}^{\infty} x \frac{\lambda^x \mathrm{e}^{-\lambda}}{x!} \\ &= 0 \times \mathrm{e}^{-\lambda} + 1 \times (\lambda \mathrm{e}^{-\lambda}) + 2 \times \left(\frac{\lambda^2 \mathrm{e}^{-\lambda}}{2!}\right) + 3 \times \left(\frac{\lambda^3 \mathrm{e}^{-\lambda}}{3!} + \cdots\right) \\ &= \lambda \mathrm{e}^{-\lambda} \left[1 + \lambda + \frac{\lambda^2}{2!} + \frac{\lambda^3}{3!} + \cdots\right] = \lambda \mathrm{e}^{-\lambda} \mathrm{e}^{+\lambda} = \lambda. \end{aligned}$$

The parameter $\lambda$ controls the location of the peak and the spread (or standard deviation). In essence, $\lambda$ describes a Poisson distribution uniquely, so some textbooks use Poisson($\lambda$) to denote a Poisson distribution with parameter $\lambda$.

Poisson distributions have an interesting property. For two independent random variables $U$ and $V$ that obey Poisson distributions: Poisson($\lambda_1$) and Poisson($\lambda_2$), respectively, $S = U + V$ obeys Poisson($\lambda_1 + \lambda_2$). From the basic Poisson distribution, we know that

$$\begin{aligned} P(S=n) &= \sum_{k=0}^{n} P(U=k, V=n-k) = \sum_{k=1}^{n} P(U=k)P(V=n-k) \\ &= \sum_{k=0}^{n} \frac{\lambda_1^k e^{-\lambda_1}}{k!} \cdot \frac{\lambda_2^{n-k} e^{-\lambda_2}}{(n-k)!} = e^{-(\lambda_1+\lambda_2)} \frac{1}{n!} \sum_{k=0}^{n} \frac{n!}{k!(n-k)!} \lambda_1^k \lambda_2^{n-k} \\ &= e^{-(\lambda_1+\lambda_2)} \frac{(\lambda_1+\lambda_2)^n}{n!} = e^{-\lambda} \frac{\lambda^n}{n!}, \end{aligned} \tag{10.6}$$

where $\lambda = \lambda_1 + \lambda_2$. In the above calculations, we have used the fact that $U$ and $V$ are independent (and thus the joint probability is the product of their probabilities). We have also used the binomial expansions

$$(\lambda_1 + \lambda_2)^n = \sum_{k=0}^{n} \frac{n!}{k!(n-k)!} \lambda_1^k \lambda_2^{n-k}. \tag{10.7}$$

Let us look at an example.

**Example 10.2** For two students A and B, Student A receives on average 1 email per hour and Student B receives on average 1.5 emails per hour. What is the probability of receiving a total of exactly 4 emails in 1 hour?

Since $\lambda_1 = 1$ and $\lambda_2 = 1.5$, we have that A obeys Poisson(1) and B obeys Poisson(1.5), so A + B obeys Poisson($\lambda_1 + \lambda_2$) = Poisson(2.5). The probability of receiving exactly 4 emails is

$$P(A+B=4) = \frac{(\lambda_1+\lambda_2)^4 e^{-(\lambda_1+\lambda_2)}}{4!} = \frac{2.5^4 e^{-2.5}}{4!} \approx 0.134.$$

In addition, the probability of none of them receiving any email is

$$P(A+B=0) = \frac{2.5^0 e^{-2.5}}{0!} \approx 0.082.$$

The arrival model of a Poisson distribution means that the inter-arrival times obey an exponential distribution

$$p(x) = \lambda e^{-\lambda x} \quad (x > 0), \tag{10.8}$$

with a mean of $1/\lambda$.

### 10.2.2 Inter-arrival Time

For the ease of our discussion here, let us define the basic concepts of Poisson processes first. Loosely speaking, a Poisson process is an arrival process in which the inter-arrival times are independent and identically distributed (iid) random variables, and such inter-arrivals are exponentially distributed $f(t) = \lambda e^{-\lambda t}$.

An interesting property of the Poisson process is the memoryless properties of the inter-arrival time derived from the exponential distribution.

For a Poisson arrival process with an arrival sequence $(A_1, A_2, \dots, A_k, \dots)$ with the arrival time $T_k$ for the $k$th arrival, the inter-arrival time is $\Delta T_k = T_{k+1} - T_k$ between arrival $A_k$ and $A_{k+1}$.

From the Poisson's distribution, the first arrival at $t$ means that there is no arrival in the interval $[0, t]$, thus its probability is

$$\text{Poisson}(T_1 > t) = P(n = 0) = \frac{(\lambda t)^0 e^{-\lambda t}}{0!} = e^{-\lambda t}. \tag{10.9}$$

This is also true that

$$P(\Delta T_k > t) = P[n(T_k + t) - n(T_k) = 0] = P(n = 0) = e^{-\lambda t}, \tag{10.10}$$

which means that it has a memoryless property. Therefore, the inter-arrival time obeys an exponential distribution with parameter $\lambda$.

As the total probability of all inter-arrival times must be one, the probability density function $f(t)$ should be divided by a scaling factor or normalization factor $\int_0^\infty e^{-\lambda t} dt = 1/\lambda$. Thus, we have

$$f(t) = \lambda e^{-\lambda t}. \tag{10.11}$$

Here, the simultaneous multiple arrivals are not allowed. For a higher arrival rate, as long as the time can be subdivided into sufficiently many small intervals, arrivals can always occur in sequence and thus the above results still hold.

## 10.3 Service Model

The most widely used service time model is the exponential distribution.

### 10.3.1 Exponential Distribution

The exponential distribution has the following probability density function

$$f(x) = \mu e^{-\mu x} \quad (\mu > 0, \quad x > 0), \tag{10.12}$$

and $f(x) = 0$ for $x \leq 0$. Its mean and variance are

$$\frac{1}{\mu}, \qquad \sigma^2 = \frac{1}{\mu^2}. \tag{10.13}$$

The expectation $E(X)$ of an exponential distribution is

$$E(X) = \int_{-\infty}^{\infty} x\mu e^{-\mu x} dx = \int_{0}^{\infty} x\mu e^{-\mu x} dx$$
$$= \left[-xe^{-\mu x} - \frac{1}{\mu} e^{-\mu x}\right]_0^{\infty} = \frac{1}{\mu}. \tag{10.14}$$

For $E(X^2)$, we have

$$E(X^2) = \int_{0}^{\infty} x^2 \mu e^{-\mu x} dx = [-x^2 e^{-\mu x}]_0^{\infty} + 2\int_{0}^{\infty} xe^{-\mu x} dx$$
$$= [-x^2 e^{-\mu x}]_0^{\infty} + \left[-\frac{2x}{\mu} e^{-\mu x} - \frac{2}{\mu^2} e^{-\mu x}\right]_0^{\infty} = \frac{2}{\mu^2}.$$

Here, we have used the fact that $x$ and $x^2$ grow slower than $\exp(-\mu x)$ decreases. That is, $x\exp(-\mu x) \to 0$ and $x^2 \exp(-\mu x) \to 0$ when $x \to \infty$.

From

$$E(X^2) = [E(X)]^2 + \sigma^2 = \frac{1}{\mu^2} + \mathrm{Var}(X), \tag{10.15}$$

we have

$$\mathrm{Var}(X) = \frac{2}{\mu^2} - \left(\frac{1}{\mu}\right)^2 = \frac{1}{\mu^2}. \tag{10.16}$$

Exponential distributions are widely used in queuing theory and simulating discrete events. As we have seen earlier, the arrival process of customers in a bank is a Poisson process and the time interval between arrivals (or inter-arrival time) obeys an exponential distribution.

The service time of a queue typically obeys an exponential distribution

$$P(t) = \begin{cases} \mu e^{-\mu t}, & t \geq 0, \\ 0, & t < 0, \end{cases} \tag{10.17}$$

where $\mu$ is the average number of customer served per unit time, and thus $\tau = 1/\mu$ is the mean service time. Thus, the service time as a random variable $X$ less than some time $(t)$ is the cumulative distribution

$$P(X \leq t) = \int_{-\infty}^{t} \mu e^{-\mu\tau} d\tau = \int_{0}^{t} \mu e^{-\mu\tau} d\tau = -e^{-\mu\tau}\Big|_0^t = 1 - e^{-\mu t}. \tag{10.18}$$

Obviously, as the total probability sum must be one, the probability of service time longer than a certain time is

$$P(X \geq t) = 1 - P(X \leq t) = e^{-\mu t}. \tag{10.19}$$

**Example 10.3** If you are in a queue in a bank, you observe that it takes 2 minutes on average to service a customer. The service time obeys a cumulative distribution function (CDF)

$$P(X \leq t) = 1 - e^{-\mu t},$$

what is the probability of taking less than 1 minute to the next customer?

We know that $\mu = 1/2 = 0.5$ (2 minutes per customer or 0.5 customer per minute), so the probability is thus

$$P(t \leq 1) = 1 - e^{-0.5\times 1} \approx 0.393.$$

The probability of taking longer than 5 minutes is

$$P(t \geq 5) = e^{-\mu t} = e^{-0.5\times 5} = e^{-2.5} \approx 0.008.$$

The arrival processes of many services can be approximated by this model.

### 10.3.2 Service Time Model

Since the service time $T$ is exponentially distributed, it has a memoryless property. From the conditional probability

$$P(T > \tau + t) = P(T > \tau)P(T > t), \tag{10.20}$$

we have

$$P(T > \tau)P(T > t) = e^{-\mu\tau}e^{-\mu t} = e^{-\mu(\tau+t)} = P(T > \tau + t), \tag{10.21}$$

where we have not considered the scaling/normaliation factors for simplicity. The above equation means that

$$P(T > \tau + t | T > t) = P(T > \tau), \tag{10.22}$$

which means that the time difference $\tau$ is independent of the time $t$, so it does not matter when the time starts (thus it is memoryless). This also means that the past has no effect on the future.

It is worth pointing out that the exponential distribution is the only memoryless distribution for continuous random variables.

### 10.3.3 Erlang Distribution

As the service time is a random variable $S$, which is exponentially distributed

$$h(t) = \begin{cases} \mu e^{-\mu t}, & t \geq 0, \\ 0, & t < 0. \end{cases} \tag{10.23}$$

the joint service time distribution of two customers or two services with two iid random variables $S_1$ and $S_2$ can be calculated by the convolution integral

$$(h * h)(t) = \int_{-\infty}^{\infty} h(t - \tau)h(\tau)d\tau. \tag{10.24}$$

Thus, the joint probability density of $S_1 + S_2$ is

$$h_2 = \begin{cases} \mu^2 t e^{-\mu t}, & t \geq 0, \\ 0, & t < 0. \end{cases} \tag{10.25}$$

Similarly, the sum of random variables $S_1 + S_2 + S_3$ of the same iid is the convolution of $h$ with $h_2$. Following the same line of thinking, we can conclude that the joint probability density of

$$T_s = \sum_{k=1}^{n} S_k = S_1 + S_2 + \cdots + S_n \tag{10.26}$$

can be calculated by

$$h_n = \begin{cases} \frac{\mu^n t^{n-1}}{(n-1)!} e^{-\mu t}, & t \geq 0, \\ 0, & t < 0, \end{cases} \tag{10.27}$$

which is the Erlang distribution for $t \geq 0$ and $\mu \geq 0$. This distribution is also a special case of the gamma distribution.

## 10.4 Basic Queueing Model

### 10.4.1 M/M/1 Queue

Now let us focus on the M/M/1 queue model with an arrival rate $\lambda$ (obeying the Poisson process) and the service time $\mu$ obeying the exponential distribution (see Figure 10.1).

Obviously, if $\lambda > \mu$, the system will not be stable because the service is slower than the arrival and the queueing length will thus grow unboundedly (leading to infinity). Thus, in order to analyze the stable characteristics of a proper queueing system, it requires that the ratio $\rho = \lambda/\mu$ is less than 1. That is,

$$\rho = \frac{\lambda}{\mu} < 1, \tag{10.28}$$

which is a performance measure for the queueing system. In fact, $\rho$ can be considered as the average server utilization because $\rho = 1$ means that the server is almost 100% busy (on average). Alternatively, $\rho$ can be considered as the number of customers or the probability in service for the server. In other words, the probability of no customer at all is $p_0 = 1 - \rho$.

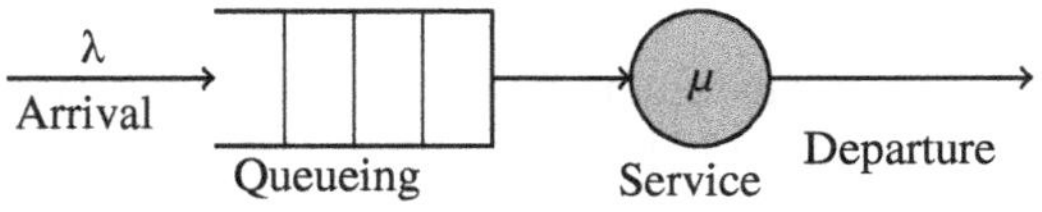


**Figure 10.1** A simple queue representation of M/M/1.

The key idea here is to look at the long-time limiting behavior of the system (not the short-time transient behavior), which allows us to identify the main characteristics of the stable, steady-state queueing system. Thinking along this line, the probability of $n$ customers is

$$p_n = \rho^n p_0 = \rho^n(1-\rho), \tag{10.29}$$

which is essentially a geometric distribution. It is worth pointing out that this probability is the limiting probability.

From the definition of the mean for a probability mass function, we can estimate the mean number of customers in the system as

$$\begin{aligned} E(n) &= \sum_{n=0}^{\infty} n p_n = \sum_{n=0}^{\infty} n \rho^n p_0 = p_0 \sum_{n=0}^{\infty} n \rho^n \\ &= p_0[1+\rho+2\rho^2+\cdots+n\rho^n+\cdots] = (1-\rho)\cdot\frac{\rho}{(1-\rho)^2} = \frac{\rho}{1-\rho}, \end{aligned} \tag{10.30}$$

where we have used

$$\sum_{n=0}^{\infty} n\rho^n = \frac{\rho}{(1-\rho)^2}, \qquad p_0 = 1-\rho. \tag{10.31}$$

Since $\rho$ is a constant, the above equality can be proved by the following steps:

$$\begin{aligned} \sum_{n=0}^{\infty} n\rho^n &= \sum_{n=0}^{\infty} \rho \frac{d\rho^n}{d\rho} = \rho \frac{d}{d\rho}\left[\sum_{n=0}^{\infty} \rho^n\right] \\ &= \rho\frac{d}{d\rho}\left[\frac{1}{1-\rho}\right] = \rho\cdot\frac{1}{(1-\rho)^2} = \frac{\rho}{(1-\rho)^2}. \end{aligned} \tag{10.32}$$

Thus, the average number of customers in the system can be estimated by

$$L = E(n) = \frac{\rho}{1-\rho} = \frac{\lambda}{\mu-\lambda}. \tag{10.33}$$

It is worth pointing out that $L$ will become infinite as $\rho \to 1$ or $\mu \to \lambda$. In fact, many quantities many diverge when $\rho \to 1$.

Since the number of customers in service is $\rho = \lambda/\mu$, the number of customers ($L_q$) in the queue (not served yet) can be estimated as

$$\begin{aligned} L_q &= L - \rho = \frac{\rho}{1-\rho} - \rho = \frac{\rho}{1-\rho} - \frac{\rho(1-\rho)}{1-\rho} \\ &= \frac{\rho^2}{1-\rho} = \frac{\lambda^2}{\mu(\mu-\lambda)}. \end{aligned} \tag{10.34}$$

However, we have to be careful in order to calculate the average time ($W$) of a customer spent in the system correctly. One naive (but incorrect) approach is to use $W = L\tau = \lambda/\mu(\mu-\lambda)$, where $\tau = 1/\mu$ is the average service time.

The correct approach is as follows: since the service time is exponentially distributed with parameter $\mu$, for a queueing system with $n$ customers, its total service time obeys the Erlang distribution as discussed earlier:

$$q_n(t) = \frac{\mu^n}{(n-1)!} t^{n-1} e^{-\mu t}. \tag{10.35}$$

Let $T_w$ be the waiting time and distribution of the waiting time $Q(t) = p(T_w \leq t)$, so the no waiting at all means that

$$Q(0) = p_0 = 1 - \rho. \tag{10.36}$$

Since the service time is exponentially distributed and memoryless, the remaining service time of the customer being served should obey the same exponential distribution. For a detailed discussion on this issue, readers can refer to the book by Bhat and Miller (2002). For a small time increment $\delta t$ for any time $t > 0$, the probability increment of $T_w$ in the interval $[t, t + \delta t]$ (or $t < T_w \leq t + \delta t$) is given by

$$\begin{aligned}
\delta Q(t) &= \sum_{n=1}^{\infty} p_n q_n(t) \delta t = \sum_{n=1}^{\infty} (1-\rho)\rho^n \frac{\mu^n}{(n-1)!} t^{n-1} e^{-\mu t} \delta t \\
&= (1-\rho) e^{-\mu t} \left[ \sum_{n=1}^{\infty} \frac{(\rho\mu)^n t^{n-1}}{(n-1)!} \right] \delta t = (1-\rho) e^{-\mu t} \sum_{n=1}^{\infty} \frac{\lambda^n t^{n-1}}{(n-1)!} \delta t \\
&= (1-\rho) e^{-\mu t} \left[ \lambda \sum_{n=1}^{\infty} \frac{(\lambda t)^{n-1}}{(n-1)!} \right] \delta t \\
&= (1-\rho) e^{-\mu t} \left[ \lambda e^{\lambda t} \right] \delta t = (1-\rho) \lambda e^{-(1-\rho)\mu t} \delta t,
\end{aligned} \tag{10.37}$$

where we have used $\lambda = \rho\mu$ and the Taylor series of $\exp(\lambda t)$,

$$\exp(\lambda t) = \sum_{k=0}^{\infty} \frac{(\lambda t)^k}{k!} = \sum_{n=1}^{\infty} \frac{(\lambda t)^{n-1}}{(n-1)!}, \quad k = n - 1. \tag{10.38}$$

Thus, the probability density function of the waiting time in queue ($T_w$) is

$$P_w(t) = (1-\rho)\lambda e^{-\mu(1-\rho)t} = \lambda(1-\rho) e^{-(\mu-\lambda)t} \quad (t > 0). \tag{10.39}$$

Using $\delta t = dt$, the distribution $Q(t)$ becomes

$$\begin{aligned}
Q(T_w \leq t) &= p_0 + \int_0^t [\delta Q] dt = 1 - \rho + \lambda(1-\rho) \int_0^t e^{-\mu(1-\rho)t} dt \\
&= 1 - \rho e^{-\mu(1-\rho)t} = 1 - \frac{\lambda}{\mu} e^{-(\mu-\lambda)t},
\end{aligned} \tag{10.40}$$

which is the cumulative distribution function for waiting time $T_w < t$. Thus, the probability of waiting time $T_w > t$ is the complementary part, which means

$$H(T_w > t) = 1 - Q(T_w \leq t) = \rho e^{-\mu(1-\rho)t} = \frac{\lambda}{\mu} e^{-(\mu-\lambda)t}. \tag{10.41}$$

Finally, the mean waiting time can be calculated by

$$W_q = E(T_w) = \int_0^\infty tP_w(t)dt = \int_0^\infty t\lambda(1-\rho)e^{-\mu(1-\rho)t}dt$$
$$= \lambda(1-\rho)\left[\int_0^\infty te^{-\mu(1-\rho)t}dt\right]$$
$$= \lambda(1-\rho)\left[\frac{1}{\mu^2(1-\rho)^2}\right] = \frac{\lambda}{\mu^2(1-\rho)} = \frac{\rho}{\mu(1-\rho)}. \quad (10.42)$$

Therefore, the average time of a customer spent in the queueing system (waiting time $W_q$ plus the service time $\tau$) is thus

$$W = W_q + \tau = \frac{\rho}{\mu(1-\rho)} + \frac{1}{\mu}$$
$$= \frac{1}{\mu}\left[\frac{\rho}{1-\rho} + 1\right] = \frac{1}{\mu(1-\rho)} = \frac{1}{(\mu-\lambda)}. \quad (10.43)$$

**Example 10.4** A small shop usually has an arrival rate of about 20 customers per hour, while the counter can typically serve a customer every 2 minutes. Assuming the arrivals are Poisson and the service time is exponential, calculate the relevant quantities in the shop system.

The arrival rate is $\lambda = 20\text{h}^{-1}$ and the service rate is $\mu = 60/2 = 30\text{h}^{-1}$, so we have

$$\rho = \frac{20}{30} = \frac{2}{3} \approx 0.66667.$$

The expected queue length is

$$L = \frac{\rho^2}{1-\rho} = \frac{(2/3)^2}{1-\frac{2}{3}} = 1.333.$$

The mean waiting time is

$$W_q = \frac{\lambda}{\mu(\mu-\lambda)} = \frac{20}{30(30-20)} = \frac{1}{15} \text{ (hour)} = 4 \text{ (minute)}.$$

The probability of no waiting at all is

$$P_0 = 1 - \rho = \frac{1}{3}.$$

The probability of waiting longer than $t$ is given by

$$P(T_w > t) = \rho \exp[-(\mu-\lambda)t] \quad (t > 0).$$

For example, the probability of waiting longer than 5 minute is

$$P\left(T_w > \frac{5}{60}\right) = \frac{2}{3}\exp\left[-(30-20)\times\frac{5}{60}\right] = 0.29.$$

Similarly, the probability of waiting longer than 15 minute is

$$P\left(T_{\rm w} > \frac{15}{60}\right) = \frac{2}{3}\exp\left[-(30-20)\times\frac{15}{60}\right] \approx 0.05.$$

It is worth pointing out that, from Eqs. (10.34) and (10.43), we have

$$W = \frac{1}{(\mu-\lambda)} = \frac{1}{\lambda}\frac{\lambda}{(\mu-\lambda)} = \frac{L}{\lambda} \tag{10.44}$$

or

$$L = \lambda W, \tag{10.45}$$

which is the well-known Little's law.

In addition, from $L_{\rm q} = \rho^2/(1-\rho)$ and $W_{\rm q} = \rho/[\mu(1-\rho)]$, we have

$$L_{\rm q} = \frac{\rho^2}{(1-\rho)} = \frac{\lambda\rho}{\mu(1-\rho)} = \lambda W_{\rm q}, \tag{10.46}$$

which means that Little's law also applies to the waiting time in queue and the number of customers in queue. Little's law is also valid for a queueing system with $s$ servers (i.e. M/M/s model).

### 10.4.2 M/M/s Queue

For the extension of the results about the M/M/1 queue model with a single server to $s \geq 1$ servers, formal derivations usually require the full Markovian model, limiting probability and balance equations. Such derivations are beyond the scope of this book. Therefore, we will state the main results without formal proofs.

For a queueing system with $s$ servers and $n$ customers, it is obvious that there is no need to queue if $n \leq s$. The average utility measure is

$$U_\rho = \frac{\rho}{s} = \frac{\lambda}{s\mu} < 1. \tag{10.47}$$

It is worth pointing out that it requires $U_\rho = \rho/s < 1$ (not $\rho < 1$) here.

The probability of zero customer is given by

$$p_0 = \left[\sum_{k=0}^{s-1}\frac{\rho^k}{k!} + \frac{\rho^s}{s!}\left(\frac{s\mu}{s\mu-\lambda}\right)\right]^{-1} = \frac{1}{\rho^s/s!(1-U) + \sum_{k=0}^{s-1}\rho^k/k!}. \tag{10.48}$$

The probability of $n$ customers is

$$p_n = \begin{cases} \frac{\rho^n}{n!}p_0 & (n \leq s), \\ \frac{\rho^n}{s!\, s^{n-s}}p_0 & (n > s). \end{cases} \tag{10.49}$$

The average number of customers in the system is

$$L = \sum_{n=0}^{\infty} np_n = \sum_{n=0}^{s-1} np_n + \sum_{n=s}^{\infty} np_n = p_0 \left[ \sum_{n=0}^{s-1} n\frac{\rho^n}{n!} + \sum_{n=s}^{\infty} s \frac{\rho^n}{s!s^{n-s}} \right]$$

$$= \frac{p_0}{s!} \left[ \sum_{n=0}^{s-1} n\rho^n + s^s \sum_{n=s}^{\infty} n\frac{\rho^n}{s^n} \right] = \rho + p_0\rho \left[ \frac{\rho^s}{(s-1)!(s-\rho)^2} \right]$$

$$= \rho + p_0 \left[ \frac{\rho^{s+1}}{(s-1)!(s-\rho)^2} \right] = \frac{\lambda}{\mu} + p_0 \left[ \frac{(\lambda/\mu)^s(\lambda\mu)}{(s-1)!(s\mu-\lambda)^2} \right]. \quad (10.50)$$

This means that the average number of customers in queue is

$$L_q = L - \rho = \frac{(\lambda/\mu)^s(\lambda\mu)}{(s-1)!\,(s\mu-\lambda)}. \quad (10.51)$$

The probability of waiting in the queue when $n > s$ is

$$P_{n>s} = p_0 \frac{\rho^{s+1}}{s!s(1-\rho/s)}. \quad (10.52)$$

The waiting time can be calculated by

$$W_q = \frac{\rho^s p_0}{s!(s\mu)(1-\rho/s)^2} = \frac{\rho^s p_0}{s!(s\mu)} \frac{s^2}{(s-\rho)^2}. \quad (10.53)$$

In addition, the waiting time probability distribution $T_w < t$ is given by

$$Q(T_w < t) = 1 - \frac{\rho^s p_0}{s!(1-\rho/s)} e^{-(s\mu-\lambda)t} = 1 - \frac{\rho^s p_0}{s!(1-\rho/s)} e^{-\mu(s-\rho)t}, \quad (10.54)$$

thus the probability of waiting longer than $t$ is given by

$$Q_q(T_w > t) = 1 - Q(T_w < t) = \frac{\rho^s p_0}{s!(1-\rho/s)} e^{-\mu(s-\rho)t}. \quad (10.55)$$

It is straightforward to verify that the above formulae will become the results for M/M/1 when $s = 1$ if we use $0! = 1$.

The average time of a customer spent in the system can be obtained using Little's law, and we have

$$W = \frac{L}{\lambda}. \quad (10.56)$$

Let us revisit the small shop example discussed earlier.

**Example 10.5** Suppose the shop is getting busier at weekends, the arrival rate becomes 50 per hour, the shop has to open 2 counters at the same time and each counter can still serve (on average) one customer every 2 minutes. Assuming the other conditions remain the same (Poisson arrival, exponential service time), what are the new performance measures for the shop at weekends?

Now we have $\lambda = 50\text{h}^{-1}$, and $\mu = 60/2 = 30/\text{h}^{-1}$, and $s = 2$. The utility measure is

$$U = \frac{\lambda}{s\mu} = \frac{50}{2 \times 30} = \frac{5}{6} < 1,$$

with

$$\rho = \frac{50}{30} = \frac{5}{3}.$$

The probability of no customer is

$$p_0 = \left[ \frac{\rho^s}{s!} \frac{s}{(s-\rho)} + \sum_{k=0}^{1} \frac{\rho^k}{k!} \right]^{-1} = \frac{1}{11} \approx 0.090\,909.$$

The expected average number of customers in the shop is

$$L = \rho + p_0 \left[ \frac{\rho^{s+1}}{(s-1)!(s-\rho)^2} \right] = \frac{5}{3} + \frac{1}{11} \left[ \frac{(5/3)^3}{(2-1)!(2-5/3)^2} \right] \approx 5.45$$

The expected waiting time in the shop is

$$W_q = \frac{\rho^s p_0}{s!(s\mu)(1-\rho/s)^2} = \frac{(5/3)^2 \times 1/11}{2!(2 \times 30)(1-5)/(3 \times 2)^2} = 0.075\,76 \text{ (hour)},$$

which is about 4.5 minute. The probability of no waiting at all $t = 0$ is

$$Q(0) = 1 - \frac{\rho^s p_0}{s!(1-\rho/s)} = 1 - \frac{\rho^s p_0}{s!(1-U)}$$
$$= 1 - \frac{(5/3)^2 \times 1/11}{2!(1-5/6)} \approx 0.2424.$$

The probability of waiting longer than 10 minutes (or 10/60 hours) is

$$Q(T_w > 10 \text{ minute}) = \frac{\rho^s p_0}{s!(1-\rho/s)} e^{-\mu(s-\rho)t}$$
$$= \frac{(5/3)^2 \times 1/11}{2!(1-5/3 \times 2)} \exp\left[-30\left(2-\frac{5}{3}\right)\frac{10}{60}\right] \approx 0.14.$$

Now let us discuss an important property about the queuing systems.

## 10.5 Little's Law

As we have seen earlier, Little's law of a queuing system states that the average number $L$ of items in the system is equal to the average arrival rate $\lambda$ multiplied by the average time $W$ that an item spends in the queuing system. That is,

$$L = \lambda W.$$

This simple law provides some good insight into queueing systems, and relevant quantities can be estimated without any detailed knowledge of a particular queueing system.

**Example 10.6** For example, a system has an arrival rate of 2 items per minute, and the average queue length is 8. What is the average waiting time for an item?

We know that $\lambda = 2$ and $L = 8$, so

$$W = \frac{L}{\lambda} = \frac{8}{2} = 4,$$

which means that an item usually waits 4 minutes before being processed.

The above results obtained for an M/M/1 or M/M/s system provide some useful insight into the queueing systems. However, real-world queues are more complicated because the assumptions we made may not be true. For example, real-world queues at a restaurant can be time-dependent because there are more customers at lunch time and in the evening. Therefore, the stable assumptions may not be true at all. More generalized queueing models should be considered. Interested readers can refer to more advanced literature on queueing theory and applications, listed at the end of this chapter.

## 10.6 Queue Management and Optimization

Queue management is crucially important to the success of many organizations and applications, from retail business and event management to call centers and the Internet routing. The conditions in real-world queues are dynamic, time-varying with uncertainty. Even though the mathematical models may no longer be valid, queues still have to be managed, and optimization still have to be carried out whenever appropriate.

The management of queues may include many aspects, from physical settings and structures to the estimation of key parameters and predictability of various quantities. For example, a business should observe and estimate the number of customers, queue length, waiting time, service time, and other quantities so as to be prepared for queueing management. Then, the number of servers should be able to vary so as to reduce the waiting time and queue length. Efficient queue management should aim to serve a majority (say 95%) within a fixed time limit (for example, 3 or 5 minutes). Customers' rating and satisfaction can be largely influenced by the waiting time and ease of exiting the queues. Amazon's online one-click checkout is a primary example for efficient queueing management.

In order to provide sufficiently accurate estimates of key parameters, it requires a multidisciplinary approach to use a variety of data and methodologies, including historical information, current arrival information, mobile sensors and cameras, statistical analysis, forecasting, appointment systems, classification and clustering (sort queries from customers), machine learning, and artificial intelligence. Monitoring of queues and communications about the queue status, comfortable waiting environment, engagement and interactions with customers, and effective service mechanism are all part of queue management systems. A successful queue management system should be able to optimize customer experience so as to maintain a successful business in the long run.

Optimization can be carried out when estimating key parameters, dynamic allocation of servers, mining historical data, and predicting future trends.

**Example 10.7** From the M/M/s queue theory discussed earlier, if the aim is to minimize the customer waiting $W_q$ for given $\lambda$, $\mu$ and other parameters, we can adjust $s$ so that

$$\text{minimize} \quad \frac{\rho^s p_0}{s!(s\mu)} \frac{s^2}{(s-\rho)^2}, \tag{10.57}$$

where $s$ should be a positive integer $s \geq 1$. This is a nonlinear optimization problem, but it is not difficult to solve in principle because it is just a function of a single variable.

The main issue is that even with a good solution of $s$, it may be less useful in practice because the actual queue setting can be different such that the model is just a very crude approximation. However, in many applications, a simple estimate can be sufficient to provide enough information for proper queue management.

There is the queueing rule of thumb to estimate the servers needed for a particular setting. For a queue system with multiple servers $s$, if $N$ is the total number of customers to be served during a total period of $T$, then we can use $N/T$ to approximate the arrival rate $\lambda$. That is $\lambda = N/T$. In addition, the average service time $\tau$ can be estimated as $\tau = 1/\mu$. From the above discussion, we know that the queue system is valid and stable if

$$U = \frac{\lambda}{s\mu} < 1, \tag{10.58}$$

which becomes

$$U = \frac{N/T}{s\mu} = \frac{N\tau}{sT} < 1. \tag{10.59}$$

Based on this, Teknomo derived a queueing rule of thumb for the number of servers

$$s \geq \left\lceil \frac{N\tau}{T} \right\rceil, \tag{10.60}$$

which can be a handy estimate.

**Example 10.8** For example, a supermarket can become very busy during lunch time. Suppose that there may be 500 customers for a 2-hour lunch period. If each customer can be served within 2 minutes at checkout, how many checkout counters should be available?

We know $N = 500$, $T = 2 \times 3600 = 7200$ seconds, and $\tau = 2 \times 60 = 120$ seconds, so we have

$$s = \left\lceil \frac{500 \times 120}{7200} \right\rceil = \left\lceil \frac{25}{3} \right\rceil = 9.$$

However, this simplified model does not give enough information about the queue length and waiting time. For such information, we need to use the complicated formulae discussed earlier in the chapter. Obviously, the dynamic, noisy nature of real-world settings requires an effective queue management system using real-time data and service management.

## Exercises

**10.1** An IT help desk typically receives 20 queries per hour and the help desk team can handle at most 30 queries per hour. Assuming this process obeys a Poisson model, write down the queue model for this process using Kendall's notation.

**10.2** For the previous question, what is the average waiting time in the queue? What is the probability of no wait at all? What is the probability of waiting longer than 10 minutes?

**10.3** A shop has the maximum of five counters and each counter can serve a customer every 2 minutes. The shop has 100 customers per hour, what is the average queue length if 3 counters are open? If the total number of customers in all the queues should not be more than 5, how many counters should be used?

**10.4** A busy road has a traffic volume of about $Q = 250$ vehicles per hour. If a Poisson distribution is used, show that

$$P(t \geq h) = \mathrm{e}^{-h/T}, \quad P(t < h) = 1 - \mathrm{e}^{-h/T},$$

with $\lambda = Q/3600$ (cars per second) and the mean headway $h$ (between successive cars) is $T = 1/\lambda = 3600/Q$. If it takes about 15 seconds to cross the road, what is the probability of no waiting at all? If some one walks slower and may take 20 seconds to cross the road, what is the probability of find a gap between 15 and 20 seconds?

**10.5** A small shop has 5 parking spaces, and each customer takes on average about 10 minutes to shop. If there are 18 customers per hour driving to the shop, what is the probability of not finding a parking space upon arrival?

## Bibliography


Bertsekas, D. and Gallager, R. (1992). *Data Networks*, 2e. Englewood Cliffs, NJ: Prentice Hall.

Bhat, U.N. (2008). *An Introduction to Queueing Theory: Modelling and Analysis in Applications*. Boston: Birkhäuser.

Bhat, U.N. and Miller, G.K. (2002). *Elements of Applied Stochastic Processes*, 3e. New York: Wiley.

Buzen, J.P. (1973). Computational algorithms for closed queueing networks with exponential servers. *Communications of the ACM* 16 (9): 527–531.

Daigle, J.N. (2010). *Queueing Theory and Applications to Packet Telecommunication*. New York: Springer.

Daley, D.J. and Servi, L.D. (1998). Idle and busy periods in stable M/M/k queues. *Journal of Applied Probability* 35 (4): 950–962.

Erlang, A.K. (1909). The theory of probabilities and telephone conversations. *Nyt Tidsskrift for Matematik B* 20 (1): 33–39.

Gross, D. and Harris, C.M. (1998). *Fundamentals of Queueing Theory*. New York: Wiley.

Halfin, S. and Whitt, W. (1981). Heavy-traffic limits for queues with many exponential servers. *Operations Research* 29 (3): 567–588.

Harchol-Balter, M. (2013). *Performance Modeling and Design of Computer Systems: Queueing Theory in Action*. Cambridge, UK: Cambridge University Press.

Jackson, J.R. (1957). Networks of waiting lines. *Operations Research* 5 (4): 518–521.

Kelly, F.P. (1975). Networks of queues with customers of different types. *Journal of Applied Probability* 12 (3), 542–554.

Kendall, D.G. (1953). Stochastic processes occurring in the theory of queues and their analysis by the method of the imbedded Markov chain. *Annals of Mathematical Statistics* 24 (3): 338–354.

Kingman, J.F.C. (2009). The first erlang century – and the next. *Queueing Systems* 63 (1): 3–4.

Lester, L. (2010). *Queueing Theory: A Linear Algebraic Approach*, 2e. New York: Springer.

Little, J.D.C. (1961). A proof for the queueing formula: $L = \lambda W$. *Operations Research* 9 (3): 383–387.

Mannering, F.L., Washburn, S.S., and Kilareshi, W.P. (2008). *Principles of Highway Engineering and Traffic Analysis*. Hoboken, NJ: Wiley.
Murdoch, J. (1978). *Queueing Theory: Worked Examples and Problems*. London UK: Palgrave Macmillan.
Newell, G.F. (1971). *Applications of Queueing Theory*. London: Chapman and Hall.
Saaty, T.L. (1961). *Elements of Queueing Theory*. New York: McGraw-Hill.
Simchi-Levi, D. and Trick, M.A. (2011). Introduction to Little's law as viewed on its 50th anniversary. *Operations Research* 59 (3): 535.
Teknomo, K. (2012). Queuing rule of thumb based on M/M/s queuing theory with applications in construction management. *Civil Engineering Dimension* 14 (3): 139–146.

# Part IV

# Advanced Topics

# 11

# Multiobjective Optimization

All the optimization problems we have discussed so far have only a single objective. In reality, we often have to optimize multiple objectives simultaneously. For example, we may want to improve the performance of a product while trying to minimize the cost at the same time. In this case, we are dealing with multiobjective optimization problems. Many new concepts are required for solving multiobjective optimization. Furthermore, these multiobjectives can be conflicting, and thus some trade-offs are needed. As a result, a set of Pareto-optimal solutions have to be found, rather than a single solution. This often requires multiple runs of solution algorithms.

## 11.1 Introduction

The optimization problem with a single objective discussed so far can be considered as a scalar optimization problem because the objective function always reaches a single global optimal value or a scalar. For multiobjective optimization, the multiple objective functions form a vector, and thus it is also called vector optimization.

Any multiobjective optimization problem can generally be written as

$$\underset{\boldsymbol{x}\in\mathbb{R}^D}{\text{minimize}}\ \boldsymbol{f}(\boldsymbol{x}) = [f_1(\boldsymbol{x}), f_2(\boldsymbol{x}), \ldots, f_M(\boldsymbol{x})],$$

subject to

$$g_j(\boldsymbol{x}) \le 0 \quad (j = 1, 2, \ldots, J), \tag{11.1}$$

$$h_k(\boldsymbol{x}) = 0 \quad (k = 1, 2, \ldots, K), \tag{11.2}$$

where $\boldsymbol{x} = (x_1, x_2, \ldots, x_D)^T$ is the vector of decision variables. In some formulations used in the optimization literature, inequalities $g_j (j = 1, \ldots, J)$ can also include any equalities because an equality $\phi(\boldsymbol{x}) = 0$ can be converted into two inequalities $\phi(\boldsymbol{x}) \le 0$ and $\phi(\boldsymbol{x}) \ge 0$. However, for clarity, we list here the equalities and inequalities separately.

*Optimization Techniques and Applications with Examples*, First Edition. Xin-She Yang.

The space $\mathcal{F} = \mathbb{R}^D$ spanned by the vectors of decision variables $\boldsymbol{x}$ is called the search space. The space $\mathcal{S} = \mathbb{R}^M$ formed by all the possible values of objective functions is called the response space or objective space. Comparing with the single objective function whose objective space is (at most) $\mathbb{R}$, the objective space for multiobjective optimization is considerably much larger and higher. In addition, as we know that we are dealing with multiobjectives $\boldsymbol{f}(\boldsymbol{x}) = [f_i]$ where $i = 1, 2, \ldots, M$, for simplicity, we can write $f_i$ as $\boldsymbol{f}(\boldsymbol{x})$ without causing any confusion in certain context.

Multiobjective optimization problems, unlike a single objective optimization problem, do not necessarily have an optimal solution that minimizes all the multiobjective functions simultaneously. Often, different objectives may conflict each other and the optimal parameters of some objectives usually do not lead to the optimality of other objectives (sometimes even make them worse). For example, we want the first-class quality service on our holidays and at the same time we want to pay as little as possible. The high-quality service (one objective) will inevitably cost much more, and this is in conflict with the other objective (to minimize cost).

Therefore, among these often conflicting objectives, we have to choose some trade-off or a certain balance of objectives. If none of these are possible, we must choose a list of preferences so that which objectives should be achieved first. More importantly, we have to compare different objectives and make a compromise. This usually requires a reformulation, and one of the most popular approaches is to find a scalar-valued function that represents a weighted combination or preference order of all objectives. Such a scalar function is often referred to as the preference function or utility function. A simple way to construct this scalar function is to use the weighted sum

$$\Phi\left(f_1(\boldsymbol{x}), \ldots, f_M(\boldsymbol{x})\right) = \sum_{i=1}^{M} w_i f_i(\boldsymbol{x}), \tag{11.3}$$

where $w_i \geq 0$ are the weighting coefficients.

Naively, some may think what happens if one tries to optimize each objective individually so that each will achieve the best (the minimum for a minimization problem), then we have

$$F^* = (f_1^*, f_2^*, \ldots, f_M^*), \tag{11.4}$$

which is called the ideal objective vector. However, there is no solution that corresponds to this ideal vector. That is to say, it is a nonexistent solution. The only exception is when all the objectives correspond to the same solution, and in this case, these multiobjectives are not conflicting, leading to the case when the Pareto front typically collapses into a single point.

For multiobjective optimization, we have to introduce some new concepts related to Pareto optimality.

## 11.2 Pareto Front and Pareto Optimality

A vector $\boldsymbol{u} = (u_1, \dots, u_D)^T \in \mathcal{F}$ is said to dominate another vector $\boldsymbol{v} = (v_1, \dots, v_D)^T$ if and only if $u_i \leq v_i$ for $\forall i \in \{1, \dots, D\}$ and $\exists i \in \{1, \dots, D\} : u_i < v_i$. This "partial less" or component-wise relationship is denoted by

$$\boldsymbol{u} \prec \boldsymbol{v}, \tag{11.5}$$

which is equivalent to

$$\forall i \in \{1, \dots, D\} \, : \, u_i \leq v_i \wedge \exists i \in \{1, \dots, D\} \, : \, u_i < v_i. \tag{11.6}$$

Here, $\wedge$ means the logical "and." In other words, no component of $\boldsymbol{u}$ is larger than the corresponding component of $\boldsymbol{v}$, and at least one component is smaller. Similarly, we can define another dominance relationship $\preceq$ by

$$\boldsymbol{u} \preceq \boldsymbol{v} \Longleftrightarrow \boldsymbol{u} \prec \boldsymbol{v} \vee \boldsymbol{u} = \boldsymbol{v}. \tag{11.7}$$

Here, $\vee$ means "or." It is worth pointing out that for maximization problems, the dominance can be defined by replacing $\prec$ with $\succ$.

A point or a solution $\boldsymbol{x}_* \in \mathbb{R}^D$ is called a Pareto optimal solution or non-inferior solution to the optimization problem if there is no $\boldsymbol{x} \in \mathbb{R}^D$ satisfying $f_i(\boldsymbol{x}) \leq f_i(\boldsymbol{x}_*)$, $(i = 1, 2, \dots, M)$. In other words, $\boldsymbol{x}_*$ is Pareto optimal if there exists no feasible vector (of decision variables in the search space) which would decrease some objectives without causing an increase in at least one other objective simultaneously. That is to say, optimal solutions are solutions which are not dominated by any other solutions. When mapping to objective vectors, they represent different trade-offs between multiple objectives.

Furthermore, a point $\boldsymbol{x}_* \in \mathcal{F}$ is called a non-dominated solution if no solution can be found that dominates it. A vector is called ideal if it contains the decision variables that correspond to the optima of objectives when each objective is considered separately.

Unlike the single objective optimization with often a single optimal solution, multiobjective optimization will lead to a set of solutions, called the Pareto optimal set $\mathcal{P}^*$, and the decision vectors $\boldsymbol{x}_*$ for this solution set are thus called non-dominated. That is to say, the set of optimal solutions in the decision space forms the Pareto (optimal) set. The image of this Pareto set in the objective or response space is called the Pareto front. In the literature, the set $\boldsymbol{x}_*$ in the decision space that corresponds to the Pareto optimal solutions is also called an efficient set. The set (or plot) of the objective functions of these non-dominated decision vectors in the Pareto optimal set forms the so-called Pareto front $\mathcal{P}$ or Pareto frontier.

In short, $\boldsymbol{u} \preceq \boldsymbol{v}$ means that $\boldsymbol{u}$ dominates $\boldsymbol{v}$, or $\boldsymbol{v}$ is dominated by $\boldsymbol{u}$. This definition may be too theoretical. To put in the practical terms, $\boldsymbol{u}$ is non-inferior to $\boldsymbol{v}$ (i.e. $\boldsymbol{u}$ is better or no worse than $\boldsymbol{v}$). Intuitively, when $\boldsymbol{u}$ dominates $\boldsymbol{v}$, we can loosely say $\boldsymbol{u}$ is better than $\boldsymbol{v}$. The domination concept provides a good way to compare solutions for multiobjective optimization, and the aim of multiobjective optimization is to find such non-dominated solutions. For any two solution vectors $\boldsymbol{x}_1$ and $\boldsymbol{x}_2$, there are only three possibilities: $\boldsymbol{x}_1$ dominates $\boldsymbol{x}_2$, or $\boldsymbol{x}_2$ dominates $\boldsymbol{x}_1$, or $\boldsymbol{x}_1$ and $\boldsymbol{x}_2$ do not dominate each other. Among many interesting properties of domination, transitivity still holds. That is, if $\boldsymbol{x}_1$ dominates $\boldsymbol{x}_2$, and $\boldsymbol{x}_2$ dominates $\boldsymbol{x}_3$, then $\boldsymbol{x}_1$ dominates $\boldsymbol{x}_3$.

Using the above notation, the Pareto front $\mathcal{P}$ can be defined as the set of non-dominated solutions so that

$$\mathcal{P} = \{\boldsymbol{s} \in \mathcal{S} \big| \nexists\, \boldsymbol{s}' \in \mathcal{S} : \boldsymbol{s}' \prec \boldsymbol{s}\}, \tag{11.8}$$

or in terms of the Pareto optimal set in the search space

$$\mathcal{P}^* = \{\boldsymbol{x} \in \mathcal{F} \big| \nexists\, \boldsymbol{x}' \in \mathcal{F} : \boldsymbol{f}(\boldsymbol{x}') \prec \boldsymbol{f}(\boldsymbol{x})\}. \tag{11.9}$$

All the non-dominated solutions in the whole feasible search space form the so-called globally Pareto-optimal set, which is simply referred to as the Pareto front.

The identification of the Pareto front is not an easy task, and it often requires a parametric analysis, say, by treating all but one objective, say, $f_i$ in an $M$-objective optimization problem so that $f_i$ is a function of $f_1, \dots, f_{i-1}, f_{i+1}, \dots,$ and $f_M$. By maximizing the $f_i$ when varying the values of the other $M-1$ objectives so that there can be enough solution points that will trace out the Pareto front properly.

**Example 11.1** For example, we have four Internet service providers $A$, $B$, $C$, and $D$. We have two objectives to choose their service (i) as cheap as possible and (ii) higher bandwidth. Their details are listed below:

| IP provider | Cost (£mo$^{-1}$) | Bandwidth (Mb) |
|---|---|---|
| $A$ | 20 | 80 |
| $B$ | 25 | 112 |
| $C$ | 30 | 56 |
| $D$ | 40 | 112 |

From the table, we know that option $C$ is dominated by $A$ and $B$ because both objectives are improved (low cost and faster). Option $D$ is dominated by $B$. Thus, solution $C$ is an inferior solution, and so is $D$. Both solutions $A$ and $B$ are non-inferior solutions or non-dominated solutions. However, which solution ($A$ or $B$) to choose is not easy, as provider $A$ outperforms $B$ on the first objective (cheaper) while $B$ outperforms $A$ on another objective (faster). In this case, we say these two solutions are incomparable. The set of the non-dominated solutions $A$ and $B$ forms the Pareto front which is a mutually incomparable set.

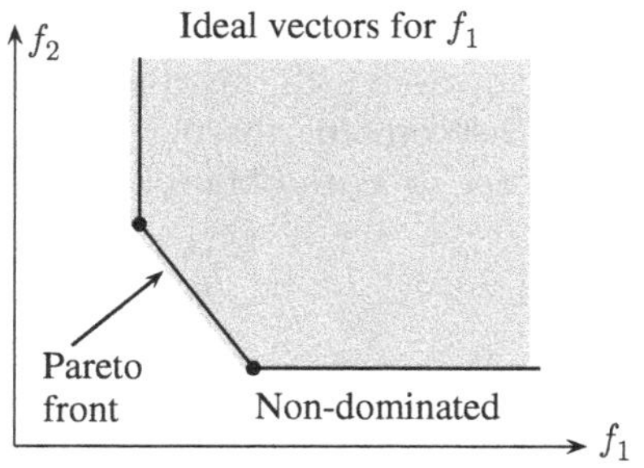


**Figure 11.1** The non-dominated set, Pareto front and ideal vectors in a minimization problem with two objectives $f_1$ and $f_2$.

Obviously, if we combine these two into a single composite objective, we can compare, for example, the cost per unit Mb. In this case, we essentially reformulate the problem as a scalar optimization problem. For choice $A$, each Mb costs £0.25, while it costs about £0.22 for choice $B$. So we should choose $B$. However, in reality, we usually have many incomparable solutions, and it is often impossible to in some way. In addition, the real choice depends on our preference and emphasis on objectives.

For a minimization problem with two objectives, the basic concepts of the non-dominated set, Pareto front, and ideal vectors are shown in Figure 11.1. It is worth pointing out that the Pareto frontiers are sketched here as straight lines, but in reality Pareto frontiers can be very complex curves for two objectives, while they can be complex surfaces for three objectives.

## 11.3 Choice and Challenges

Even we have produced high-quality multiple Pareto-optimal solutions, the final choice of a point on the Pareto front is usually up to the decision makers who have higher-level information or rules to make the decision. Such higher-level information is typically nontechnical, inaccurate, highly subjective, and often not part of the optimization problem.

Multiobjective optimization is usually difficult to solve, partly due to the lack of efficient tools, and partly due to complexity of these types of problems. Loosely speaking, there are three ways to deal with multiobjective problems: direct approach, aggregation or transformation, and Pareto set approximation. However, the current trends tend to be evolutionary approaches for approximating Pareto fronts.

Direct approach is difficult, especially in the case when multiple objectives seem conflicting. Therefore, we often use aggregation or transformation by combining multiple objectives into a single composite objective so that the standard methods for optimization discussed in this book can be used. We will focus on this approach in the rest of the chapter. However, with this approach, the solutions typically depend on the way how we combine the objectives. A third way is to try to approximate the Pareto set so as to obtain a set of mutually non-dominated solutions.

## 11.4 Transformation to Single Objective Optimization

To transform a multiobjective optimization problem into a single objective, we can often use the method of weighted sum, and utility methods. We can also choose the most important objective of our interest as the only objective, while rewriting other objectives as constraints with imposed limits.

### 11.4.1 Weighted Sum Method

The weighted sum method combines all the multi-objective functions into one scalar, composite objective function using the weighted sum

$$F(\boldsymbol{x}) = w_1 f_1(\boldsymbol{x}) + w_2 f_2(\boldsymbol{x}) + \cdots + w_M f_M(\boldsymbol{x}). \tag{11.10}$$

An issue arises in assigning the weighting coefficients $\boldsymbol{w} = (w_1, w_2, \ldots, w_M)$ because the solution strongly depends on the chosen weighting coefficients. Obviously, these weights have to be nonnegative, satisfying

$$\sum_{i=1}^{M} w_i = 1, \quad w_i \in [0, 1]. \tag{11.11}$$

Let us first look at an example.

**Example 11.2** The classical three-objective functions are commonly used for testing multi-objective optimization algorithms. These functions are

$$f_1(x, y) = x^2 + (y - 1)^2, \tag{11.12}$$

$$f_2(x, y) = (x - 1)^2 + y^2 + 2, \tag{11.13}$$

$$f_3(x, y) = x^2 + (y + 1)^2 + 1, \tag{11.14}$$

where $(x, y) \in [-2, 2] \times [-2, 2]$.

If we combine all the three functions into a single function $f(x, y)$ using the weighted sum, we have

$$f(x, y) = \alpha f_1 + \beta f_2 + \gamma f_3, \qquad \alpha + \beta + \gamma = 1. \tag{11.15}$$

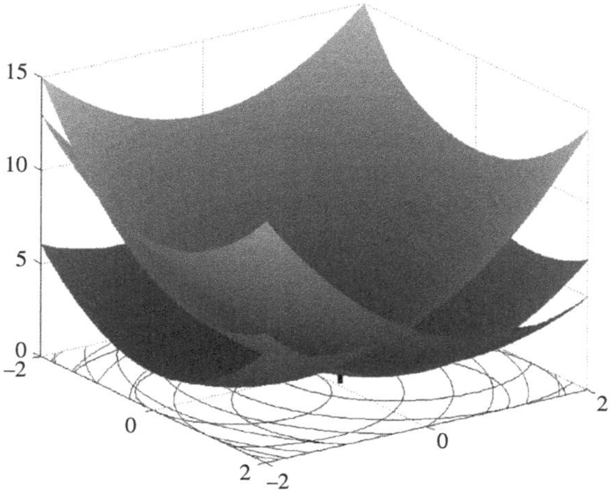


**Figure 11.2** Three functions reach the global minimum at $x_* = \beta, y_* = \alpha - \gamma$.

The stationary point is determined by

$$\frac{\partial f}{\partial x} = 0, \qquad \frac{\partial f}{\partial y} = 0, \tag{11.16}$$

which lead to

$$2\alpha + 2\beta(x - 1) + 2\gamma = 0 \tag{11.17}$$

and

$$2\alpha(y - 1) + 2\beta y + 2\gamma(y + 1) = 0. \tag{11.18}$$

The solutions are

$$x_* = \beta, \qquad y_* = \alpha - \gamma. \tag{11.19}$$

This implies that $x_* \in [0, 1]$ and $y_* \in [-1, 1]$. Consequently, $f_1 \in [0, 5]$,$f_2 \in [2, 4]$, and $f_3 \in [1, 6]$. In addition, the solution or the optimal location varies with the weighting coefficients $\alpha$, $\beta$, and $\gamma$. In the simplest case $\alpha = \beta = \gamma = 1/3$, we have

$$x_* = \frac{1}{3}, \qquad y_* = 0. \tag{11.20}$$

This location is marked with a short thick line in Figure 11.2.

Now the original multiobjective optimization problem has been transformed into a single objective optimization problem. Thus, the solution methods for solving single objective problems are all valid.

However, there is an important issue here. The combined weighted sum transforms the optimization problem into a single objective, this is not necessarily equivalent to the original multiobjective problem because the extra weighting coefficients could be arbitrary, while the final solutions still depend on these coefficients. Furthermore, there are so many ways to construct the weighted sum function and there is no easy guideline to choose which form is the best for a given problem. When there is no rule to follow, the simplest choice obviously is to use the linear form. But, there is no reason why the weighted sum should be linear. In fact, we can use other combinations such as the following quadratic weighted sum

$$\Pi(\boldsymbol{x}) = \sum_{i=1}^{M} w_i f_i^2(\boldsymbol{x}) = w_1 f_1^2(\boldsymbol{x}) + \cdots + w_M f_M^2(\boldsymbol{x}), \tag{11.21}$$

and other forms.

Another important issue is that how to choose the weighting coefficients as the solutions depend on these coefficients. The choice of weighting coefficients is essentially to assign a preference order by the decision maker to the multiobjectives. This leads to a more general concept of the utility function (or preference function) which reflects the preference of the decision maker(s).

Ideally, a different weight vector should result in a different trade-off point on the Pareto front; however, in reality, this is usually not the case. Different combinations of weight coefficients can lead to the same point or points very close to each other, and consequently the points are not uniformly distributed on the Pareto front. In fact, a single trade-off solution on the Pareto front just represents one sampling point, and there is no technique to ensure uniform sampling on the front. All the issues still form an active area of research.

It is worth pointing out that the linear weighted sum method

$$\Pi(\boldsymbol{x}) = \sum_{i=1}^{M} w_i f_i(\boldsymbol{x}), \quad \sum_{i=1}^{M} w_i = 1, \quad w_i \geq 0, \tag{11.22}$$

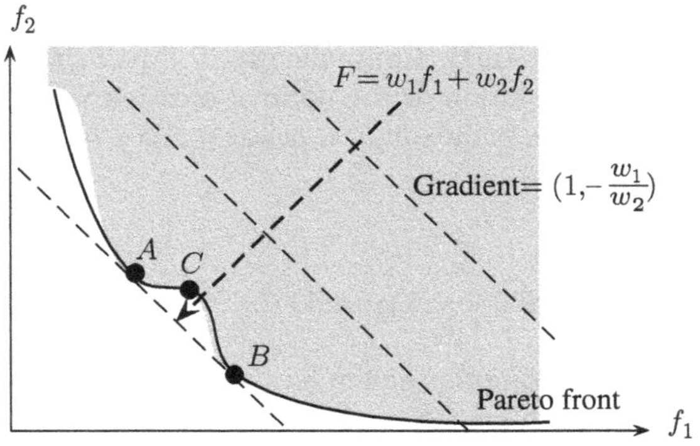


Figure 11.3 Weighted sum method for two objectives $f_1$ and $f_2$ and $w_1 + w_2 = 1$.

only works for problems with convex Pareto fronts. As we can see from Figure 11.3 where two objectives are combined into one for a given set of $w_1 + w_2 = 1$, the composite function $F$ is minimized. For any given set $(w_1, w_2)$, a (dashed) line has a gradient $(1, -w_1/w_2)$ which will become tangent to the Pareto front when moving downward to the left, and that touching point is the minimum of $F$. However, at the non-convex segment, if the aim is point $C$, however, the weighted sum method will usually lead to either point $A$ or point $B$, depending on the values of $w_1$ (since $w_2 = 1 - w_1$).

Though the weighted sum method is one of the most widely used, due to its simplicity, however, it is usually difficult to generate a good set of points that are uniformly distributed on the Pareto front. Furthermore, proper scalings or normalization of the objectives are often needed so that the ranges/values of each objective should be comparable; otherwise, the weight coefficients are not well distributed and thus leading to biased sampling on the Pareto front.

For more complex multiobjective optimization problems, another widely used and yet more robust method is the $\epsilon$-constraint method. Before, we proceed, let us discuss the utility method which can be considered as the different ways of forming composite objectives.

### 11.4.2 Utility Function

The weighted sum method is essentially a deterministic value method if we consider the weighting coefficients as the ranking coefficients. This implicitly assumes that the consequence of each ranking alternative can be characterized with certainty. This method can be used to explore the implications of alternative value judgement. The utility method, on the other hand, considers uncertainty in the criteria values for each alternative, which is a more realistic method because there is always some degree of uncertainty about the outcome of a particular alternative.

Utility (or preference) functions can be associated with the risk attitude or preferences. For example, if you are offered a choice between a guaranteed £500 and a 50/50 chance of zero and £1000. How much are you willing to pay to take the gamble? The expected payoff of each choice is £500 and thus it is fair to pay $0.5 \times 1000 + (1 - 0.5) \times 0 =$ £500 for such a gamble. A risk-seeking decision maker would risk a lower payoff in order to have a chance to win a higher prize, while a risk-averse decision maker would be happy with the safe choice of £500. For a risk-neutral decision maker, the choice is indifferent between a guaranteed £500 and the 50/50 gamble since both choices have the same expected value of £500.

In reality, the risk preference can vary from person to person and may depend on the type of problem. The utility function can have many forms, and one of the simplest is the exponential utility (of representing preference)

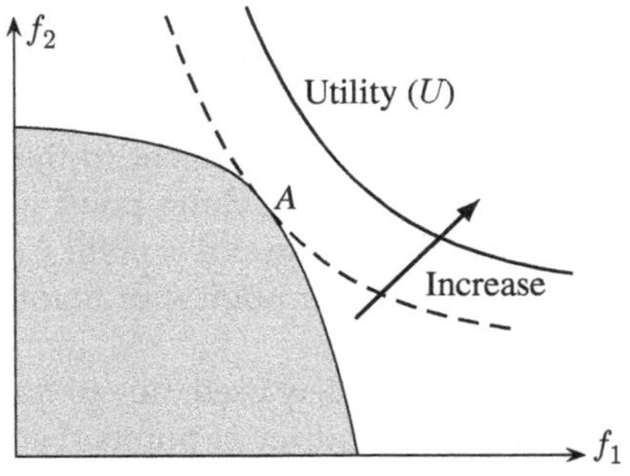


**Figure 11.4** Finding the Pareto solution with maximum utility in a maximization problem with two objectives.

$$u(x) = \frac{1 - e^{-(x-x_a)/\rho}}{1 - e^{-(x_b-x_a)/\rho}}, \tag{11.23}$$

where $x_a$ and $x_b$ are the lowest and highest level of $x$, and $\rho$ is called the risk tolerance of the decision maker.

The utility function defines combinations of objective values $f_1, \dots, f_M$ which a decision maker finds equally acceptable or indifferent. So the contours of the constant utility are referred to as the indifference curves. The optimization now becomes the maximization of the utility. For a maximization problem with two objectives $f_1$ and $f_2$, the idea of the utility contours (indifference curves), Pareto front, and the Pareto solution with maximum utility (point $A$) are shown in Figure 11.4. When the utility function touches the Pareto front in the feasible region, it then provides a maximum utility Pareto solution (marked with $A$).

For two objectives $f_1$ and $f_2$, the utility function can be constructed in different ways. For example, the combined product takes the following form

$$U(f_1, f_2) = k f_1^{\alpha} f_2^{\beta}, \tag{11.24}$$

where $\alpha$ and $\beta$ are nonnegative exponents and $k$ a scaling factor. On the other hand, the aggregated utility function for the same two objectives can be defined as

$$U(f_1, f_2) = \alpha f_1 + \beta f_2 + [1 - (\alpha + \beta)] f_1 f_2. \tag{11.25}$$

There are many other forms. The aim of constructing utility functions by a decision maker is to form a mapping $U: \mathbb{R}^M \mapsto \mathbb{R}$ so that the total utility function has a monotonic and/or convexity properties for easy analysis. It will also improve the quality of the Pareto solution(s) with maximum utility. Let us look at a simple example.

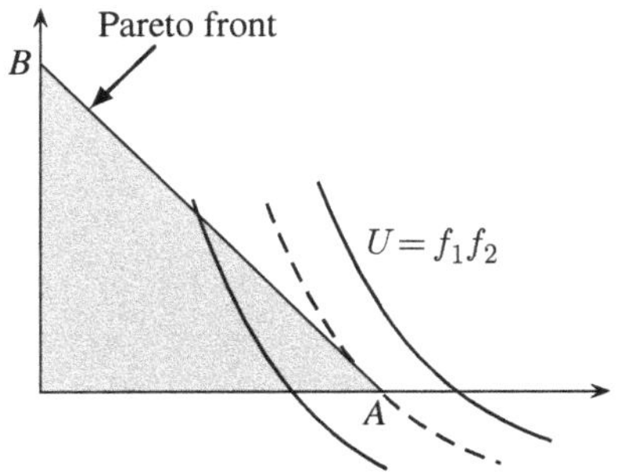


**Figure 11.5** Pareto front is the line connecting $A(5, 0)$ and $B(0, 5/\alpha)$. The Pareto solution with maximum utility is $U_* = 25$ at point $A$.

**Example 11.3** We now try to solve the simple two-objective optimization problem:

$$\underset{(x,y)\in\Re^2}{\text{maximize}} f_1(x, y) = x + y, \qquad f_2(x, y) = x,$$

subject to

$$x + \alpha y \leq 5 \qquad (x \geq 0, \ \ y \geq 0),$$

where $0 < \alpha < 1$. Let us use the simple utility function

$$U = f_1 f_2,$$

which combines the two objectives. The line connecting the two corner points $(5, 0)$ and $(0, 5/\alpha)$ forms the Pareto front (see Figure 11.5). It is easy to check that the Pareto solution with the maximum utility is $U = 25$ at $A(5, 0)$ when the utility contours touch the Pareto front with the maximum possible utility.

The complexity of multiobjective optimization makes the construction of utility functions a difficult task as there are many ways to construct such functions.

## 11.5 The $\epsilon$-Constraint Method

An interesting way of dealing with multiobjective optimization is to write objectives except one as constraints. Let us try to rewrite the following unconstrained optimization as a single objective constrained optimization problem

$$\text{minimize} f_1(\boldsymbol{x}), f_2(\boldsymbol{x}), \dots , f_M(\boldsymbol{x}).$$

To achieve this goal, we often choose the most important objective of our preference, say, $f_k(\boldsymbol{x})$ as the main objective, while imposing limits on the other objectives. That is,

$$\text{minimize} \quad f_k(\boldsymbol{x}),$$
$$\text{subject to} \quad f_i \leq \epsilon_i \quad (i = 1, 2, k-1, k+1, \dots, M),$$

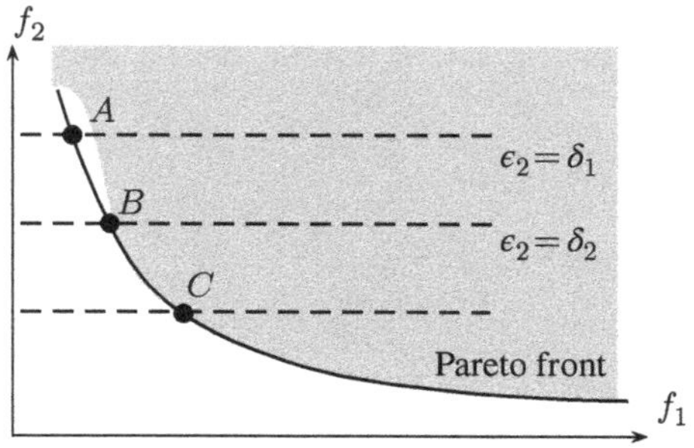


**Figure 11.6** Slicing the objective domain in the $\epsilon$-constraint method.

where the limits $\epsilon_i$ are given. In the simplest case, we can choose $k = 1$. Here, for simplicity and ease of discussion, we have omitted the constraints in this formulation, but we can always add them into the formulation when necessary.

In principle, the problem can be solved using the standard optimization algorithms for single objective optimization. In essence, this is a slicing method which splits the objective domain into different subdomains. For example, in the case of a bi-objective problem as shown in Figure 11.6, we take $f_2$ as a constraint. This problem becomes

$$\text{Minimize } f_1(\boldsymbol{x}), \tag{11.26}$$

subject to

$$f_2(\boldsymbol{x}) \leq \epsilon_2, \tag{11.27}$$

where $\epsilon_2$ is a number, not necessarily small. For any given value of $\epsilon_2$, the objective domain is split into two subdomains: $f_2 \leq \epsilon_2 = \delta_1$ (feasible) and $f_2 > \epsilon_2 = \delta_1$ (infeasible). The minimization of $f_1$ in the feasible domain leads to the globally optimal point $A$. Similarly, for a different value of $\epsilon_2 = \delta_2$, the minimum of $f_1$ gives point $B$.

**Example 11.4** Let us look at a bi-objective optimization example, called Schaffer's min–min function

$$\text{Minimize} f_1(x) = x^2, \quad f_2(x) = (x-2)^2, \quad x \in [-10^3, 10^3]. \tag{11.28}$$

If we use $f_1$ as the objective and $f_2 \leq \epsilon_2$ as the constraint, we can set $\epsilon_2 \in [0, 4]$ with 20 different values. Then, we can solve it using a single objective optimization technique. The 20 points of approximated Pareto optimal solutions and the true Pareto front are shown in Figure 11.7. However, if we use $f_2$ as the objective and $f_1$ as the constraint, we follow exactly the same procedure and the results are shown in Figure 11.8. As we can see from both figures, the distributions of the approximate Pareto points are different, though they look similar.

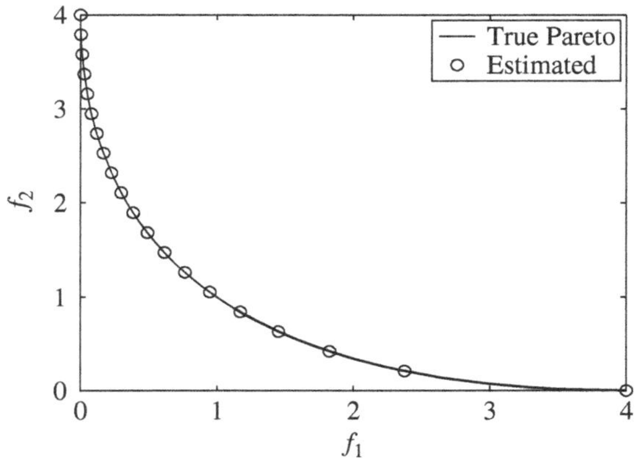


**Figure 11.7** True Pareto front and the estimated front when setting $f_1$ as the objective and $f_2$ as the constraint.

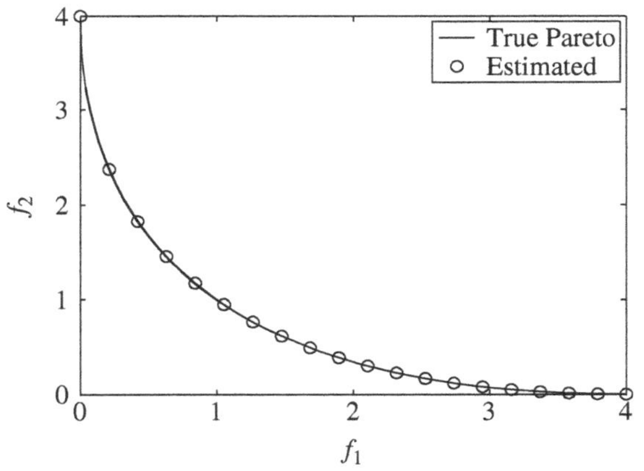


**Figure 11.8** True Pareto front and the estimated front when setting $f_2$ as the objective and $f_1$ as the constraint.

As this example has demonstrated, the distributions of the sampling points on the Pareto front may depend on the actual formulation and the order of choosing the main objective.

The advantage of this method is that it works well for complex problems with either convex or non-convex Pareto fronts. However, it does have some disadvantages. There could be many different formulations by choosing which objective as the main objective and the rest of objectives as constraints. Different formulations may lead to different computational efforts. In addition, there is no guarantee that the points generated on the Pareto front are uniformly distributed as seen in the previous example.

Furthermore, it is difficult to impose the right range of $\epsilon_i$. In the previous example, if we set $\epsilon_2$ too small, say, $\epsilon_2 \to 0$, there may not be a feasible solution. On the other hand, if we set $\epsilon$ too high, it will be difficult to find the minimum of $f_1$ even if it exists, because the number of evaluations for this single objective optimization problem may increase. In practice, some prior knowledge is required to impose the correct limits. Otherwise, the solutions obtained may not be the solution to the original problem.

The good news is that recent trends tend to use evolutionary approaches such as genetic algorithms (GAs) and cuckoo search. We will briefly introduce some of these metaheuristic methods in the rest of this chapter and in Chapter 13 and 14 of this book.

## 11.6 Evolutionary Approaches

### 11.6.1 Metaheuristics

So far, we have seen that finding solutions to a multiobjective optimization problem is usually difficult, even by the simple weighted sum method and utility function. However, there are other promising methods that work well for multiobjective optimization problems, especially the metaheuristic methods such as GAs, simulated annealing, particle swarm, optimization (PSO), cuckoo search, and firefly algorithm. We will introduce some of these algorithms in later chapters of this book. For the moment, we just highlight their key ideas.

There are many potential ways to extend the standard simulated annealing to solve multiobjective optimization problems. A direct and yet simple extension is to modify the acceptance probability as the joint probability

$$p_a = \prod_{j=1}^{m} p_j = \prod_{j=1}^{m} e^{-\Delta f_j / k_B T_j},$$

where $k_B$ is Boltzmann constant which can be taken as $k_B = 1$, and $\Delta f_j = f_j(\boldsymbol{x}_n) - f_j(\boldsymbol{x}_{n-1})$ is the change of each individual objective. For details, please refer to the article by Suppapitnarm et al.

In 1985, Schaffer was probably the first to use vector evaluated genetic algorithms (VEGA) to solve multiobjective optimization without using any composite aggregation by combining all objectives into a single objective. Since then, many metaheuristic algorithms such as PSO, simulated annealing, and the firefly algorithm have been extended to solve multiobjective optimization problems successfully.

The main difficulty to extend any single-objective optimization algorithm to accommodate the multiobjectives is that we have to consider the dominance of each candidate solution set. Among many widely used methods for multiobjective optimization, we will introduce the elitist non-dominated sorting genetic algorithm (NSGA-II) in the rest of this chapter.

### 11.6.2 Non-Dominated Sorting Genetic Algorithm

The elitist non-dominated sorting GA, or NSGA-II for short, was developed by Deb et al., which has become a popular method for solving multiobjective optimization problems by finding multiple Pareto solutions. Its key features include the use of elitist, diversity-preserving mechanism, and emphasis on non-dominated solutions.

The main step in NSGA-II is that the $n$ offsprings $S_t$ are created at iteration $t$ from the $n$ parents $P_t$ using the standard GAs. The whole population $W_t$ of size $2n$ is formed by joining $S_t$ and $P_t$, i.e. $W_t = P_t \cup S_t$. Then, the non-dominated sorting is applied to $W_t$. Afterward, the new population is filled, one at a time, by different non-dominated solutions. Because the population of $W_t$ is $2n$, only half of it will be put into the new population, selecting the non-dominated solutions for the Pareto front with the higher diversity, while discarding the rest of the solutions.

The crowding distance $D_i$ used in the NSGA-II is essentially based on the cardinality of the solution sets and their distance to the solution boundaries. The crowded tournament selection is based on ranking and distance. In other words, if a solution $\boldsymbol{x}_i$ has a better rank than $\boldsymbol{x}_j$, we select $\boldsymbol{x}_i$. If the ranks are the same but $D_i > D_j$, we select $\boldsymbol{x}_i$ based on its crowding distance. For details, please refer to Deb et al. (2002). The main steps of NSGA-II can be summarized as follows:

1) Create a new population $W_t = P_t \cup S_t$ by combining $P_t$ and applying non-dominated sorting.
2) Identify different fronts $PF_i$ $(i = 1, 2, \ldots)$.
3) Generate $P_{t+1}$ from $P_{t+1} = \emptyset$ with $i = 1$ and fill $P_{t+1} = P_{t+1} \cup PF_i$ until size $n$.
4) Carry out crowding-sort by using the crowd-distance to sort some $PF_i$ to $P_{t+1}$.
5) Generate new offsprings $S_{t+1}$ from $P_{t+1}$ via crowded tournament-based genetic operators: crossover, mutation, and selection.

There are many other methods for solving multiobjective optimization problems, including differential evolution for multiobjective optimization (DEMO), strength pareto evolutionary algorithm (SPEA), multiobjective cuckoo search (MOCS), multiobjective firefly algorithm (MOFA), and multiobjective flower pollination algorithm (MOFPA). We will introduce some of these nature-inspired algorithms in Chapters 13 and 14 of this book.

## Exercises

**11.1** Find the Pareto front analytically of the bi-objective problem

$$\min f_1(x) = x^2, \quad \min f_2(x) = (x-1)^2.$$

**11.2** Find the Pareto front for the ZDT2 test benchmark

$$\min f_1 = x_1, \quad \min f_2 = g(x)\left[1 - \left(\frac{x_1}{g(x)}\right)^2\right],$$

where

$$g(x) = 1 + \frac{9}{D-1}\sum_{i=1}^{D} x_i \quad (0 \le x_i \le 1),$$

for $i = 1, 2, \dots, D$ and $D \ge 2$.

## Bibliography

Boyd, S. and Vandenberge, L. (2004). *Convex Optimization*. Cambridge, UK: Cambridge University Press.

Coello, C.A.C., Pulido, G.T., and Lechuga, M.S. (2004). Handling multiple objectives with particle swarm optimization. *IEEE Transactions on Evolutionary Computation* 8 (3): 256–279.

Deb, K. (1995). *Optimization for Engineering Design: Algorithms and Examples*. New Delhi: Prentice-Hall.

Deb, K. (2001). *Multi-objective Optimization Using Evolutionary Algorithms*. Chichester: Wiley.

Deb, K., Agrawal, S., Pratap, A. et al. (2002). A fast and elitist multi-objective genetic algorithm: NSGA-II. *IEEE Transactions on Evolutionary Computation* 6 (2): 182–197.

Fonseca, C.M. and Fleming, P.J. (1995). An overview of evolutionary algorithms in multiobjective optimization. *Evolutionary Computation* 3 (1): 1–16.

Knowles, J.D. and Corne, D.W. (2000). Approximating the non-dominated front using the pareto archived evolution strategy. *Evolutionary Computation* 8 (2): 149–172.

Konak, A., Coit, D.W., and Smith, A.E. (2006). Multi-objective optimization using genetic algorithms: a tutorial. *Reliability Engineering and System Safety* 91 (9): 992–1007.

Pareto, V. (1972). *Manuale di Economica Politica*. London: Macmillan.

Sawaragi, Y., Nakayama, H., and Tanino, T. (1985). *Theory of Multiobjective Optimization*. London: Academic Press.

Schaffer, J.D. (1985). Multiple objective optimization with vector evaluated genetic algorithms. In: *Proceedings of the First International Conference on Genetic Algorithms and Their Applications* (ed. J. Grefenstette), 93–100. Hillsdale, NJ: L. Erlbaum Associates Inc.

Srinivas, N. and Deb, K. (1994). Multiple objective optimization using nondominated sorting in genetic algorithms. *Evolutionary Computation* 2 (3): 221–248.

Suppapitnarm, A., Seffen, K.A., Parks, G.T. et al. (2000). A simulated annealing algorithm for multiobjective optimization. *Engineering Optimization* 33 (1): 59–85.

Talbi, E.G. (2009). *Metaheuristics: From Design to Implementation*. Hoboken, NJ: Wiley.

Yang, X.S. (2013). Multi-objective firefly algorithm for continuous optimization. *Engineering with Computers* 29 (2): 175–184.

Yang, X.S. and Deb, S. (2013). Multiobjective cuckoo search for design optimization. *Computers & Operations Research* 40 (6): 1616–1624.

Yang, X.S., Karamanoglu, M., and He, X. (2013). Multi-objective flower algorithm for optimization. *Procedia Computer Science* 18, 861–868.

Zitzler, E., Deb, K., and Thiele, L. (2000). Comparison of multiobjective evolutionary algorithms: empirical results. *Evolutionary Computation* 8 (2): 173–195.

Zitzler, E., Laumanns, M., and Bleuler, S. (2004). A tutorial on evolutionary multiobjective optimization. In: *Metaheuristics for Multiobjective Optimization* (ed. X. Gandibleux, M. Sevaux, K. Sörensen et al.), Lecture Notes in Economics and Mathematical Systems, vol. 535, 3–37. Berlin: Springer.

# 12

# Constraint-Handling Techniques

When we discussed the optimization techniques for solving constrained optimization in earlier chapters, we have introduced a few ways of handling constraints, including the method of Lagrange multipliers, slack variables, and penalty methods. There are many ways of dealing with constraints. In fact, such constraint-handling techniques can form important topics of many books and comprehensive review articles.

We now briefly summarize the methods we have covered and introduce some additional methods for handling constraints.

## 12.1 Introduction and Overview

Numerous constraint-handling techniques can be classified in a few ways. Loosely speaking, we can divide them into two major categories: classic methods and recent methods. Classic/traditional methods are still widely used in many applications, and new recent developments have been largely based on the hybrid of evolutionary ideas with these traditional methods. Therefore, the differences between the old and new are relatively arbitrary and purely for the purpose of arguments here.

Traditional methods include the penalty methods, transformation methods, and special representation, and separation of objective and constraints.

Penalty methods try to convert a constrained optimization problem into an unconstrained one by incorporating its constraints in the revised objective. However, this introduces more parameters into the problem, but if proper values are used, the converted unconstrained problem can often be solved by many algorithms relatively effectively.

The separation of objective function $f(\boldsymbol{x})$ and constraints is another class of methods and has gained attention in recent years. For example, the Powell and

*Optimization Techniques and Applications with Examples*, First Edition. Xin-She Yang.

Skolnick (1993) approach uses a fitness function $\rho(x)$ to incorporate $M$ equalities $\phi_i(\boldsymbol{x}) = 0$ and $N$ inequalities $\psi_j(\boldsymbol{x}) \leq 0$ in the following way:

$$\rho(\boldsymbol{x}) = \begin{cases} f(\boldsymbol{x}) & \text{if feasible,} \\ 1 + \mu\left[\sum_{j=1}^{N} \max\{0, \psi_j(\boldsymbol{x})\} + \sum_{i=1}^{M} |\phi_i(\boldsymbol{x})|\right] & \text{otherwise,} \end{cases} \tag{12.1}$$

where $\mu > 0$ is a constant. This form essentially maps unfeasible solutions to ranks in the range of $1 - \infty$, which ensures that a feasible solution should always have a better fitness value than an infeasible solution. It is worth pointing out that the fitness can be considered as ranks, and, for minimization problems, lower numbers mean better fitness.

Other methods, or different methods known with other names, can also be put into these categories. For example, direct methods can be classified into the separation of objective function and constraints, while the Lagrange multiplier method can be considered as the special representation or even penalty methods, depending on the perspectives we may have.

Direct approaches intend to find the feasible regions enclosed by the constraints. This is often difficult, except for a few special cases. Numerically, we can generate a potential solution, and check if all the constraints are satisfied. If all the constraints are met, then it is a feasible solution, and the evaluation of the objective function can be carried out. If one or more constraints are not satisfied, this potential solution is discarded, and a new solution should be generated. We then proceed in a similar manner. As we can expect, this process is slow and inefficient. Better approaches are to incorporate the constraints so as to formulate the problem as an unconstrained one, including the method of Lagrange multipliers and penalty methods.

Other traditional methods that we have covered in this book include the use of slack variables to turn inequality constraints to equality constraints (then use the Lagrange multiplier), and generalized reduced gradient (GRG) methods.

Recent methods reflect some new trends in constraint-handling techniques, including feasibility methods, stochastic ranking (SR), adaptive penalty methods and new special operator methods, $\epsilon$-constrained method, multiobjective approach, and hybrid or ensemble methods.

## 12.2 Method of Lagrange Multipliers

The method of Lagrange multipliers has a rigorous mathematical basis, while the penalty method is simple to implement in practice. So let us review these methods first.

The method of Lagrange multipliers converts a constrained problem to an unconstrained one. For example, if we want to minimize a function

$$\underset{\boldsymbol{x}\in\mathbb{R}^n}{\text{minimize}} f(\boldsymbol{x}), \qquad \boldsymbol{x} = (x_1, \dots, x_n)^{\mathrm{T}} \in \mathbb{R}^n, \tag{12.2}$$

subject to multiple nonlinear equality constraints

$$h_j(\boldsymbol{x}) = 0 \qquad (j = 1, 2, \dots, M). \tag{12.3}$$

We can use $M$ Lagrange multipliers $\lambda_j (j = 1, \dots, M)$ to reformulate the above problem as the minimization of the following function:

$$L(\boldsymbol{x}, \lambda_j) = f(\boldsymbol{x}) + \sum_{j=1}^{M} \lambda_j h_j(\boldsymbol{x}). \tag{12.4}$$

The optimality requires the following stationary conditions:

$$\frac{\partial L}{\partial x_i} = \frac{\partial f}{\partial x_i} + \sum_{j=1}^{M} \lambda_j \frac{\partial h_j}{\partial x_i} \quad (i = 1, \dots, n) \tag{12.5}$$

and

$$\frac{\partial L}{\partial \lambda_j} = h_j = 0 \;\; (j = 1, \dots, M). \tag{12.6}$$

These $M + n$ equations will determine the $n$ components of $\boldsymbol{x}$ and $M$ Lagrange multipliers. As $\partial L/\partial h_j = \lambda_j$, we can consider $\lambda_j$ as the rate of the change of the quantity $L(\boldsymbol{x}, \lambda_j)$ as a functional of $h_j$.

**Example 12.1** Now let us look at a simple example

$$\underset{u,v}{\text{maximize}} f = u^{2/3}v^{1/3},$$

subject to

$$3u + v = 9.$$

First, we write it as an unconstrained problem using a Lagrange multiplier $\lambda$, and we have

$$L = u^{2/3}v^{1/3} + \lambda(3u + v - 9).$$

The conditions for optimality are

$$\frac{\partial L}{\partial u} = \frac{2}{3}u^{-1/3}v^{1/3} + 3\lambda = 0, \qquad \frac{\partial L}{\partial v} = \frac{1}{3}u^{2/3}v^{-2/3} + \lambda = 0,$$

and

$$\frac{\partial L}{\partial \lambda} = 3u + v - 9 = 0.$$

The first two conditions give $2v = 3u$, whose combination with the third condition leads to

$$u = 2, \qquad v = 3.$$

Thus, the maximum of $f_*$ is $\sqrt[3]{12}$. In addition, this gives

$$\lambda = -\frac{2}{9}\left(\frac{3}{2}\right)^{1/3}. \tag{12.7}$$

Here, we only discussed the equality constraints. For inequality constraints, things become more complicated. One way is to use the so-called slack variables to convert inequality constraints to equalities. We have already introduced this method in the earlier chapters.

## 12.3 Barrier Function Method

When we introduced the interior-point method, we have used logarithmic barrier functions to deal with inequalities. As we have introduced this method in greater detail, we just briefly highlight the main idea here.

For an inequality-constrained optimization problem

$$\text{minimize} \quad f(\boldsymbol{x}), \quad \boldsymbol{x} \in \mathbb{R}^n, \tag{12.8}$$

subject to

$$g_i(\boldsymbol{x}) \leq 0 \quad (i = 1, 2, \ldots, N), \tag{12.9}$$

we can use logarithmic barrier functions to write it as

$$\text{minimize} \quad L(x, \mu) = f(x) - \mu \sum_{i=1}^{N} \log\left[-g_i(\boldsymbol{x})\right], \tag{12.10}$$

which is an unconstrained optimization problem and can be solved by any appropriate method discussed in this book.

## 12.4 Penalty Method

For a nonlinear optimization problem with equality and inequality constraints, a common method of incorporating constraints is the penalty method. For the optimization problem

$$\begin{aligned} &\underset{\boldsymbol{x} \in \mathbb{R}^n}{\text{minimize}} f(\boldsymbol{x}), \quad \boldsymbol{x} = (x_1, \ldots, x_n)^{\mathrm{T}} \in \mathbb{R}^n, \\ &\text{subject to } \phi_i(\boldsymbol{x}) = 0 \qquad (i = 1, \ldots, M), \\ &\qquad\qquad \psi_j(\boldsymbol{x}) \leq 0 \qquad (j = 1, \ldots, N), \end{aligned} \tag{12.11}$$

the idea is to define a penalty function so that the constrained problem is transformed into an unconstrained problem. One commonly used penalty formulation is

$$g(\boldsymbol{x}) = f(\boldsymbol{x}) + P(\boldsymbol{x}), \tag{12.12}$$

where $P(\boldsymbol{x})$ the penalty term defined by

$$P(\boldsymbol{x}) = \sum_{j=1}^{N} \nu_j \max\{0, \psi_j(\boldsymbol{x})\} + \sum_{i=1}^{M} \mu_i |\phi_i(\boldsymbol{x})|. \tag{12.13}$$

Here, $\mu_i > 0, \mu_j > 0$ are penalty constants or penalty factors. The advantage of this method is to transform the constrained optimization problem into an unconstrained one. That is, all the constraints are incorporated into the new objective function. However, this introduces more free parameters whose values need to be defined so as to solve the problem appropriately.

Other forms of the penalty functions can be used. For example, we can use a smoother version

$$P(\boldsymbol{x}) = \sum_{j=1}^{N} \nu_j \max\{0, \psi_j(\boldsymbol{x})\}^2 + \sum_{i=1}^{M} \mu_i \phi_i^2(\boldsymbol{x}). \tag{12.14}$$

In addition, there is no need to fix the values of penalty parameters. In fact, the variations of penalty parameters with iterations may be advantageous. These techniques with time-dependent penalty parameters belong an active research area of dynamic penalty function methods. Furthermore, many traditional constraint-handling techniques now have been combined with evolutionary algorithms, and these approaches themselves also become evolutionary. This becomes a major trend that constraint-handling takes on an evolutionary approach.

## 12.5 Equality Constraints via Tolerance

Good ways of handling equality constraints are to incorporate them into the objective by using the method of Lagrange multiplier, the reduced gradient method, and the ways used in the interior-point method.

However, equality constraints can be handled directly with some modifications. Sometimes, it might be easier to change an equality constraint into two inequality constraints, so that we only have to deal with inequalities in the implementation, but this rarely works in practice because the search volume can become extremely small.

Naively, $h(\boldsymbol{x}) = 0$ is always equivalent to $h(\boldsymbol{x}) \leq 0$ and $h(\boldsymbol{x}) \geq 0$ (or $-h(\boldsymbol{x}) \leq 0$), but it will not work in practice. Because the feasibility volume for $h(x) = 0$ and therefore, a randomly-sampled solution has almost zero probability of satisfying this equality.

One remedy is to use some approximation techniques, and a widely used one is to use a tolerance $\epsilon > 0$:

$$|h(\boldsymbol{x})| - \epsilon \leq 0, \tag{12.15}$$

which is equivalent to two inequalities:

$$h(\boldsymbol{x}) - \epsilon \leq 0 \tag{12.16}$$

and

$$-h(\boldsymbol{x}) - \epsilon \leq 0. \tag{12.17}$$

Obviously, the accuracy is controlled by $\epsilon$. To get better results, there is no need to fix $\epsilon$. At the beginning of the iterations, a large value $\epsilon$ can be used, and then as the iterations converge, a smaller $\epsilon$ can be used.

## 12.6 Feasibility Criteria

An effective and yet popular constraint-handling technique in this category was proposed by K. Deb (2000), which is combined with genetic algorithms. In this method, three feasible criteria are used in terms of a binary tournament selection mechanism.

1) For one feasible solution and one infeasible solution, the feasible one is chosen first.
2) For two feasible solutions, the one with the better objective value is preferred.
3) For two infeasible solutions, the one with the lower degree of constraint violation is chosen first.

Within this method, the constraint violation is the penalty term, and that is

$$P(\boldsymbol{x}) = \sum_{j=1}^{N} \max \left[0, \psi_j(x)\right]^2 + \sum_{i=1}^{M} |\phi_i(\boldsymbol{x})|, \tag{12.18}$$

which includes both inequalities and equalities.

Such feasibility-based methods have been extended and applied in many algorithms and applications, often in combination with evolutionary algorithms. In fact, the feasibility rules can be considered as some sort of fitness related to the problem, and such fitness-based evolutionary methods aim to select solutions that are the most fittest in the sense that they are feasible with the lowest objective values (for minimization problems). Feasibility rules can be absolute or relative, and Mezura-Montes and Coello (2011) provided a comprehensive review on this constraint-handling topic.

## 12.7 Stochastic Ranking

Another constraint-handling technique is called SR, originally developped by Runarsson and Yao in 2000. One of the advantages of this method is to avoid the under or over penalty associated with the penalty methods. In SR a user-defined parameter $\lambda_u$ is used to control and compare infeasible solutions. The swap conditions between two individuals in the population are based on the sum of constraint violation and their objective function values. Ranking of the solutions is carried out by a bubble-sort-style process. In essence, a uniformly-distributed random number $u$ is used to control the switch in the form of $u < \lambda_u$, and therefore, dynamic parameter control methods can be used. Again, SR has been used in combination with evolutionary algorithms such as differential evolution in the literature.

A related approach for handling constraints is the so-called $\epsilon$-constrained method, developed by Takahama and Sakai (2006). The $\epsilon$-constrained method essentially uses two parts: the relaxation limits to consider the feasibility of solutions, in terms of the sum of constraint violation and comparison of objective function values, and a lexicographical ordering mechanism such that the objective minimization is preceded by the minimization of constraint violation. Two sets of solutions $\boldsymbol{x}_1$ and $\boldsymbol{x}_2$ can be compared and ranked by objective $f(\boldsymbol{x})$ and constraint violation $P(\boldsymbol{x})$ as follows:

$$\{f(\boldsymbol{x}_1), P(\boldsymbol{x}_1)\} \leq \epsilon \{f(\boldsymbol{x}_2), P(\boldsymbol{x}_2)\}, \tag{12.19}$$

which is equivalent to

$$\begin{cases} f(\boldsymbol{x}_1) \leq f(\boldsymbol{x}_2), & \text{if } P(\boldsymbol{x}_1), P(\boldsymbol{x}_2) \leq \epsilon \\ f(\boldsymbol{x}_1) \leq f(\boldsymbol{x}_2), & \text{if } P(\boldsymbol{x}_1) = P(\boldsymbol{x}_2), \\ P(\boldsymbol{x}_1) \leq P(\boldsymbol{x}_2), & \text{otherwise.} \end{cases} \tag{12.20}$$

Here, $\epsilon \geq 0$ controls the level of comparison. Obviously, we have two special cases: $\epsilon = \infty$ and $\epsilon = 0$. The former is equivalent to the comparison of objective function values only, while the latter $\epsilon = 0$ provides a lexicographical ordering mechanism where the objective minimization is preceded by the minimization of the constraint violation.

It should be noted that this $\epsilon$-constrained method should not be confused with the $\epsilon$-constraint method for multiobjective optimization. They are two different methods and for different purposes.

All these methods, novel/new penalty methods, and other evolutionary decoder methods can be classified into the evolutionary approach of constraint-handling techniques.

## 12.8 Multiobjective Constraint-Handling and Ranking

In many cases, multiobjective optimization problems can be converted into single objective optimization by methods such as weighted sum methods. It seems that the multiobjective approach to constraint-handling tends to do the opposite. Naively, one may think it may not be a good approach; however, some studies show that such multiobjective approaches can be highly competitive.

For example, one of the multiobjective approaches to constraint-handling was the so-called infeasibility driven evolutionary algorithm (IDEA) proposed by Ray et al. (2009), which uses an extra objective in addition to the original objective function. This extra objective measures the constraint violation and the ranking of solutions. If a constraint is satisfied by a solution, a zero rank is assigned to that solution for that constraint, and the total rank of a solution as a measure of the constraint violation is the sum of all the ranks for all the constraints. For evolutionary algorithms, new generations of solutions are sorted into two sets: a feasible set and infeasible set, and non-dominated sorting is applied. The results are quite competitive.

There are quite a few other multiobjective approaches. Furthermore, constraints can be approximated and even dynamic constraint-handling techniques can be used, and a brief survey can be found in more advanced literature.

## Exercises

**12.1** Solve the constrained optimization problem $f = x^2 + y^2 + 2xy$ subject to $y - x^2 + 2 = 0$.

**12.2** Solve the constrained optimization by the method of Lagrange multipliers

$$\text{minimize} \quad f(x, y) = x^3 - 3xy^2,$$

subject to

$$h(x, y) = x - y^2 = 1.$$

**12.3** Use a logarithm barrier to solve the constrained problem $\min f(x) = (x - 1)^2 + 1$ subject to $x^2 \geq 4$.

## Bibliography

Boyd, S. and Vandenberge, L. (2004). *Convex Optimization*. Cambridge, UK: Cambridge University Press.

Cagnina, L.C., Esquivel, S.C., and Coello, C.A. (2008). Solving engineering optimization problems with the simple constrained particle swarm optimizer. *Informatica* 32 (3): 319–326.

Coello, C.A.C. (2000). Use of a self-adaptive penalty approach for engineering optimization problems. *Computers in Industry* 41 (2): 113–127.

Deb, K. (1995). *Optimization for Engineering Design: Algorithms and Examples*. New Delhi: Prentice-Hall.

Deb, K. (2000). An efficient constraint handling method for genetic algorithms. *Computer Methods in Applied Mechanics and Engineering* 186 (2–4): 311–338.

Gill, P.E., Murray, W., and Wright, M.H. (1982). *Practical Optimization*. Bingley: Emerald Group Publishing Ltd.

Konak, A., Coit, D.W., and Smith, A.E. (2006). Multi-objective optimization using genetic algorithms: a tutorial. *Reliability Engineering and System Safety* 91 (9): 992–1007.

Koziel, S. and Michalewicz, Z. (1999). Evolutionary algorithms, homomorphous mappings, and constrained parameter optimization. *Evolutionary Computation* 7 (1): 19–44.

Mezura-Montes, E. (2009). *Constraint-Handling in Evolutionary Optimization*, Studies in Computational Intelligence, vol. 198. Berlin: Springer.

Mezura-Montes, E. and Coello, C.A.C. (2011). Constraint-handling in nature-inspired numerical optimization: past, present and future. *Swarm and Evolutionary Computation* 1 (4): 173–194.

Powell, D. and Skolnick, M.M. (1993). Using genetic algorithms in engineering design optimization with non-linear constraints. In: *Proceedings of the Fifth International Conference on Genetic Algorithms* (ed. S. Forrest), ICGA-93, University of Illinois at Urbana-Champaign, 424–431. San Mateo, CA: Morgan Kaufmann Publishers.

Ray, T., Sing, H.K., Isaacs, A., and Smith, W. (2009). Infeasibility driven evolutionary algorithm for constrained optimization. In: *Constraint-Handling in Evolutionary Optimization* (ed. E. Mezura-Montes), Studies in Computational Intelligence, vol. 198, 145–165. Berlin: Springer-Verlag.

Runarsson, T.P. and Yao, X. (2000). Stochastic ranking for constrained evolutionary optimization. *IEEE Transactions on Evolutionary Computation* 4 (3): 284–294.

Takahama, T. and Sakai, S. (2006). Solving constrained optimization problems by the e-constrained particle swarm optimizer with adaptive velocity limit control. *Proceedings of the IEEE Conference on Cybernetics and Intelligent Systems (CIS'2006)*, Bangkok, Thailand (7–9 June 2006), 683–689. IEEE Publication.

Wright, S. (1997). *Primal-Dual Interior-Point Methods*. Philadelphia, PA: SIAM.
Yang, X.S. (2010). *Engineering Optimization: An Introduction with Metaheuristic Applications*. Hoboken, NJ: Wiley.
Yeniay, O. (2005). Penalty function methods for constrained optimization with genetic algorithms. *Mathematical and Computational Applications* 10 (1): 45–56.

# Part V

# Evolutionary Computation and Nature-Inspired Algorithms

# 13

# Evolutionary Algorithms

The optimization methods we discussed so far are deterministic in the sense that solutions can uniquely be determined by an iterative procedure starting from an initial point. The only randomness is the starting point which is either an educated guess or initilized randomly. In terms of algorithm structure, the only exception is the stochastic gradient method where the approximation to the true gradient is carried out with some randomness. For almost all analytical methods and search algorithms, we seem to try to avoid randomness, except for the part in probability and statistics. On the other hand, modern trends in solving tough optimization problems tend to use evolutionary algorithms and nature-inspired metaheuristic algorithms, especially those based on swarm intelligence (SI).

## 13.1 Evolutionary Computation

In reality, randomness is everywhere and there is no strong reason for not using randomness in developing algorithms. In fact, many modern search algorithms use randomness very effectively. For example, heuristic methods use a trial-and-error approach to find the solutions to many difficult problems. Modern stochastic search methods are often evolutionary algorithms, or more recently called "meta-heuristic." Here *meta* means "beyond" or "higher level," while *heuristic* means "to find" or "to discover by trial and error."

All these algorithms form part of the evolutionary computation. Loosely speaking, evolutionary computation is part of computational intelligence in the wide context of computer science and artificial intelligence. However, the definitions and boundary between different disciplines can be rather arbitrary.

Two major characteristics of modern metaheuristic methods are nature-inspired, and a balance between randomness and regularity. Almost all modern heuristic methods such as genetic algorithms (GA), particle swarm optimization (PSO), cuckoo search (CS), and firefly algorithm (FA) are

*Optimization Techniques and Applications with Examples*, First Edition. Xin-She Yang.

nature-inspired as they have been developed based on the study of natural phenomena, learning from the beauty and effectiveness of nature.

In addition, a balanced use of randomness with a proper combination with certain deterministic components is in fact the essence of making such algorithms so powerful and effective. If the randomness in an algorithm is too high, then the solutions generated by the algorithm do not converge easily as they could continue to "jump around" in the search space. If there is no randomness at all, then they can suffer the same disadvantages as those of deterministic methods (such as the gradient-based search). Therefore, a certain tradeoff is needed. Many NP-hard problems such as the traveling salesman problem can be solved using metaheuristic methods. In this chapter and in Chapter 14, we will introduce some of these metaheuristic methods.

## 13.2 Evolutionary Strategy

The evolution strategy (ES) was developed by I. Rechenberg, H.P. Schwefel and their collaborators in the 1960s and early 1970s. The basic idea was to use representations (that are problem-dependent) via mutation and selection to mimic the evolution of natural systems. Here, the mutation is a local random walk in the sense that the solution is modified by perturbing each component by a normally distributed random number. Mathematically speaking, the degree or step size of such mutation is controlled by the standard deviation of the distribution. In addition, selection is to compare the fitness of different solutions so as to mimic the survival of the fittest.

In the simplest case where there are only two solutions in the population. The current solution (a point or a vector in the parameter design space) is considered as the parent solution, and this solution is modified or mutated to create a new (child) solution. The new solution is evaluated for its fitness in terms of the design objective. The child solution will become the new parent solution only if it is better or fitter than the old parent; otherwise, the solution is discarded and another mutated solution will be generated. This scheme is often referred to as the $(1+1)$-ES.

A better variant is the $(1+\lambda)$-ES where $\lambda \geq 1$ child solutions are generated by mutation. The fittest solution among $\lambda$ new solutions will become the new parent solution so as to pass onto the next generation.

ES, together with the GA, paved the ways for many evolutionary algorithms in the sense that almost all evolutionary algorithms share some similarity in their basic procedure:

- The population consists of $n$ individual solution vectors. Their fitness values are evaluated and ranked.
- New solutions are generated by modification such as mutation and crossover (mixing of two different solutions).

- The fitness of the new population is evaluated and the fittest will be selected to pass onto next generations.

Though the exact details of mutation, crossover, selection, and solution representations can be different for different algorithms, their essence is about the same. For example, selection for reproducing new solutions can be based on their fitness, while representations of solutions can be either as real numbers or as binary strings. In addition, mutation can be carried out component-wise or at a few fixed locations of the representations. There are a vast literature on these topics, and we will introduce some of the fundamentals of these algorithms, including the well-known GA.

## 13.3 Genetic Algorithms

The GA is an evolutionary algorithm and probably the most widely used. It is becoming a conventional and classic method. However, it does have fundamental genetic operators that have inspired many later algorithms, so we will introduce it in detail.

The GA, developed by John Holland and his collaborators in the 1960s and 1970s, is a model or abstraction of biological evolution based on Charles Darwin's theory of natural selection. Holland was the first to use crossover and recombination, together with mutation and selection, in the study of adaptive and artificial systems. These genetic operators form the essential part of the GA as a problem-solving strategy. Since then, many variants of genetic algorithms have been developed and applied to a wide range of optimization problems, from graph coloring to pattern recognition, from discrete systems (such as the traveling salesman problem) to continuous systems (e.g. the efficient design of airfoils in aerospace engineering), and from the financial market to multiobjective engineering optimization.

There are many advantages of metaheuristic algorithms such as GA over traditional optimization algorithms, the two most noticeable advantages are the ability to deal with complex problems, and parallelism. GA can deal with various types of optimization whether the objective (fitness) function is stationary or nonstationary (change with time), linear or nonlinear, continuous or discontinuous, or with random noise. As multiple offsprings in a population act like independent agents, the population (or any subgroup) can explore the search space in many directions simultaneously. This feature makes it ideal to parallelize the algorithm for implementation. Different parameters and even different groups of strings can be manipulated at the same time. Such advantages also map onto the algorithms based on SI and thus SI-based algorithms such as PSO and FA to be introduced later also possess such good advantages.

However, GA also have some disadvantages. The formulation of the fitness function, the population size, the choice of the important parameters such as the rate of mutation and crossover, and the selection criteria of new populations should be carried out carefully. Any inappropriate choice will make it difficult for the algorithm to converge, or it simply produces meaningless results.

### 13.3.1 Basic Procedure

There are many variants of the GA, and they now form a class of GA. The essence of GA involves the encoding of an objective function as arrays of bits or character strings to represent the chromosomes, the manipulation operations of strings by genetic operators, and the selection according to their fitness in the aim of finding a solution to the problem concerned. This is often done by the following procedure: (i) encoding of solutions into strings; (ii) defining a fitness function and selection criterion; (iii) creating a population of individuals and evaluating their fitness; (iv) evolving the population by generating new solutions using crossover, mutation, fitness-proportionate reproduction; (v) selecting new solutions according to their fitness and replacing the old population by better individuals; (vi) decoding the results to obtain the solution(s) to the problem.

These steps can be represented schematically as the pseudocode of GA shown in Algorithm 13.1. One iteration of creating a new population is called a generation. The fixed-length character strings are used in most GA during each generation although there is substantial research on the variable-length strings and coding structures. The coding of the objective function is usually in the form of binary arrays or real-valued arrays in the adaptive GA. For simplicity, we use binary strings for encoding and decoding. The genetic operators include crossover, mutation, and selection from the population.

**Algorithm 13.1** Pseudocode of genetic algorithms.

Objective function $f(\boldsymbol{x})$, $\boldsymbol{x} = (x_1, \dots, x_D)^{\mathrm{T}}$
Encode the solution into chromosomes (binary strings)
Define fitness $F$ (e.g. $F \propto f(\boldsymbol{x})$ for maximization)
Generate the initial population
Initialize probabilities of crossover ($p_c$) and mutation ($p_m$)
**while** ($t <$ Max number of generations)
    Generate new solution by crossover and mutation
    **if** $p_c >$ rand, Crossover; **end if**
    **if** $p_m >$ rand, Mutate; **end if**
    Accept the new solutions if their fitness increase
    Select the current best for new generation (elitism)
**end while**
Decode the results and visualization

| | | | | | | | |
|---|---|---|---|---|---|---|---|
| 1 | 0 | 1 | 1 | 0 | 1 | 0 | 1 |
| 0 | 1 | 1 | 0 | 0 | 0 | 1 | 1 |

Parent gene pair ⇓ (before crossover)

| | | | | | | | |
|---|---|---|---|---|---|---|---|
| 1 | 0 | 1 | 1 | 0 | 0 | 1 | 1 |
| 0 | 1 | 1 | 0 | 0 | 1 | 0 | 1 |

Child gene pair (after crossover)

**Figure 13.1** Diagram of crossover at a random crossover point.

| | | | | | | | |
|---|---|---|---|---|---|---|---|
| 1 | 1 | 0 | 1 | 1 | 0 | 0 | 1 |

Original gene ⇓ (before mutation)

| | | | | | | | |
|---|---|---|---|---|---|---|---|
| 1 | 1 | 0 | **0** | 1 | 0 | 0 | 1 |

New gene (after crossover)

**Figure 13.2** Schematic representation of mutation at a single site by flipping a randomly selected bit ($1 \to 0$).

The crossover of two parent strings is the main operator with a higher probability $p_c$ and is carried out by swapping one segment of one chromosome with the corresponding segment on another chromosome at a random position (see Figure 13.1). The crossover carried out in this way is a single-point crossover. Crossover at multiple points is also used in many GA to increase the efficiency of the algorithms.

The mutation operation is achieved by flopping the randomly selected bits (see Figure 13.2), and the mutation probability $p_m$ is usually small. The selection of an individual in a population is carried out by the evaluation of its fitness, and it can remain in the new generation if a certain threshold of the fitness is reached or the reproduction of a population is fitness-proportionate. That is to say, the individuals with higher fitness are more likely to reproduce. Alternatively, the best solutions with the highest fitness can be ensured to be passed onto the next generation, which is essentially a form of elitism.

### 13.3.2 Choice of Parameters

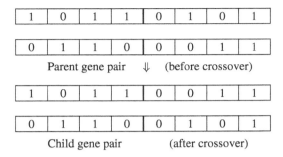

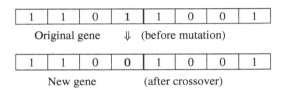

An important issue is the formulation or choice of an appropriate fitness function that determines the selection criterion in a particular problem. For the minimization of $f(\boldsymbol{x})$ using GA, one simple way of constructing a fitness function is to use the simplest form $F(\boldsymbol{x}) = A - f(\boldsymbol{x})$ with $A$ being a large constant (though $A = 0$ will do), thus the objective is to maximize the fitness function. However, there are many different ways of defining a fitness function. For example, we can use the individual fitness assignment relative to the whole population

$$F(\boldsymbol{x}_i) = \frac{f(\boldsymbol{x}_i)}{\sum_{i=1}^{N} f(\boldsymbol{x}_i)}, \tag{13.1}$$

where $N$ is the population size. The appropriate form of the fitness function will ensure that the solutions with higher fitness should be selected efficiently. Poorly defined fitness functions may result in incorrect or meaningless solutions.

Another important issue is the choice of various parameters. The crossover probability $p_c$ is usually very high, typically in the range of $0.7 - 0.99$. On the other hand, the mutation probability $p_m$ is usually small (usually $0.001 - 0.05$). If $p_c$ is too small, then the crossover occurs sparsely, which is not efficient for evolution. If the mutation probability is too high, the diversity of the population may be too high, which makes it harder for the system to converge.

The selection criterion is also important; how to select the current population so that the best individuals with higher fitness are preserved and passed on to the next generation. That is often carried out in association with a certain elitism. The basic elitism is to select the most fit individual (in each generation) which will be carried over to the new generation without being modified by genetic operators. This ensures that the best solution is achieved more quickly.

Other issues include multiple sites for mutation and crossover. Mutation at a single site is not very efficient; mutation at multiple sites will increase the evolution efficiency. However, too many mutants will make it difficult for the system to converge, or even make the system go astray to the wrong solutions. In reality, if the mutation is too high under high selection pressure, then the whole population might go extinct. Similarly, crossover can also be carried out at multiple parts, which can increase the mixing ability of the population and increase the efficiency of crossover.

In addition, the choice of the right population size is also very important. If the population size is too small, there is not enough evolution going on, and there is a risk that the whole population may go extinct. In the real world, for a species with a small population, ecological theory suggests that there is a real danger of extinction. Even though the system carries on, there is still a danger of premature convergence. In a small population, if a significantly more fit individual appears too early, it may reproduce enough offspring to overwhelm the whole (small) population. This will eventually drive the system to a local optimum (not the global optimum). On the other hand, if the population is too large, more evaluations of the objective function are needed, which will require an extensive computing time.

Using the basic procedure described here, we can implement the GA in any programming language. In fact, there is no need to do any programming (if you prefer) because there are many software packages (either freeware or commercial) about GA. For example, Matlab itself has an optimization toolbox including this algorithm.

## 13.4 Simulated Annealing

Simulated annealing (SA) is a random search technique for global optimization problems, and it mimics the annealing process in materials processing when a metal cools and freezes into a crystalline state with the minimum energy and larger crystal size so as to reduce the defects in metallic structures. The annealing process involves the careful control of temperature and cooling rate (often called annealing schedule).

The application of SA into optimization problems was pioneered by Kirkpatrick et al. in 1983. Since then, there have been extensive studies. Unlike the gradient-based methods and other deterministic search methods which have the disadvantage of becoming trapped in local minima, the main advantage of SA is its ability to avoid being trapped in local minima. In fact, it has been proved that SA will converge to its global optimality if enough randomness is used in combination with very slow cooling.

Metaphorically speaking, the iterations in SA are equivalent to dropping some bouncing balls over a landscape. As the balls bounce and lose energy, they will settle down to some local minima. If the balls are allowed to bounce enough times and lose energy slowly enough, some of the balls will eventually fall into the lowest global locations, hence the global minimum will be reached.

The basic idea of the simulated annealing algorithm is to use random search which not only accepts changes that improve the objective function, but also keeps some changes that are not ideal. In a minimization problem, for example, any better moves or changes that decrease the cost (or the value) of the objective function $f$ will be accepted; however, some changes that increase $f$ will also be accepted with a probability $p$. This probability $p$, also called the transition probability, is determined by

$$p = \exp\left(-\frac{\delta E}{k_B T}\right), \tag{13.2}$$

where $k_B$ is the Boltzmann constant, and $T$ is the temperature for controlling the annealing process. $\delta E$ is the change of the energy level. This transition probability is based on the Boltzmann distribution in physics. The simplest way to link $\delta E$ with the change of the objective function $\delta f$ is to use $\delta E = \gamma \delta f$ where $\gamma$ is a real constant. For simplicity without losing generality, we can use $k_B = 1$ and $\gamma = 1$. Thus, the probability $p$ simply becomes

$$p(\delta f, T) = e^{-\delta f/T}. \tag{13.3}$$

Whether or not to accept a change, we usually use a random number $r$ (drawn from a uniform distribution in [0,1]) as a threshold. Thus, if $p > r$ or

$$p = e^{-\delta f/T} > r, \tag{13.4}$$

it is accepted.

Here the choice of the right temperature is crucially important. For a given change $\delta f$, if $T$ is too high ($T \to \infty$), then $p \to 1$, which means almost all changes will be accepted. If $T$ is too low ($T \to 0$), then any $\delta f > 0$ (worse solution) will rarely be accepted as $p \to 0$ and thus the diversity of the solution is limited, but any improvement $\delta f$ will almost always be accepted. In fact, the special case $T \to 0$ corresponds to the gradient-based method because only better solutions are accepted, and the system is essentially climbing up or descending a hill. Therefore, if $T$ is too high, the system is at a high energy state on the topological landscape, and the minima are not easily reached. If $T$ is too low, the system may be trapped in a local minimum (not necessarily the global minimum), and there is not enough energy for the system to jump out of the local minimum to explore other potential global minima. So a proper, initial temperature should be calculated.

Another important issue is how to control the cooling process so that the system cools down gradually from a higher temperature to ultimately freeze to a global minimum state. There are many ways to control the cooling rate or the decrease in temperature.

Two commonly used cooling schedules are: linear and geometric cooling. For a linear cooling process, we have

$$T = T_0 - \beta t,$$

(or $T \to T - \delta T$ with a temperature increment $\delta T$). Here, $T_0$ is the initial temperature, and $t$ is the pseudo time for iterations. $\beta$ is the cooling rate, and it should be chosen in such a way that $T \to 0$ when $t \to t_f$ (maximum number of iterations), which usually gives $\beta = T_0/t_f$.

The geometric cooling essentially decreases the temperature by a cooling factor $0 < \alpha < 1$ so that $T$ is replaced by $\alpha T$ or

$$T(t) = T_0 \alpha^t \qquad (t = 1, 2, \dots, t_f). \tag{13.5}$$

The advantage of the second method is that $T \to 0$ when $t \to \infty$, and thus there is no need to specify the maximum number of iterations $t_f$. For this reason, we will use this geometric cooling schedule. The cooling process should be slow enough to allow the system to stabilize easily. In practise, $\alpha = 0.7–0.99$ is commonly used.

In addition, for a given temperature, multiple evaluations of the objective function are needed. If there are too few evaluations, there is a danger that the system will not stabilize and subsequently will not converge to its global optimality. If there are too many evaluations, it is time-consuming, and the system

**Algorithm 13.2** Simulated annealing algorithm.

Objective function $f(\boldsymbol{x})$, $\boldsymbol{x} = (x_1, \dots, x_D)^{\mathrm{T}}$
Initialize initial temperature $T_0$ and initial guess $\boldsymbol{x}_0$
Set final temperature $T_f$ and max number of iterations $N$
Define cooling schedule $T \mapsto \alpha T$, $(0 < \alpha < 1)$
**while** ( $T > T_f$ or $t < t_f$ )
  Move randomly to new location $\boldsymbol{x}_{t+1}$
  Calculate $\delta f = f(\boldsymbol{x}_{t+1}) - f(\boldsymbol{x}_t)$
  Accept the new solution if better
  **if** not improved
  Generate a random number $r$
  Accept if $p = \exp[-\delta f / T] > r$
  **end if**
  Update the best $\boldsymbol{x}_*$ and $f_*$
**end while**

will usually converge too slowly as the number of iterations to achieve stability may be exponential to the problem size.

Therefore, there is a balance between the number of evaluations and solution quality. We can either do many evaluations at a few temperature levels or do few evaluations at many temperature levels. There are two major ways to set the number of iterations: fixed or varied. The first uses a fixed number of iterations at each temperature, while the second is designed to increase the number of iterations at lower temperatures so that the local minima can be fully explored.

The basic procedure of the simulated annealing algorithm can be summarized as the pseudocode shown in Algorithm 13.2.

In order to find a suitable starting temperature $T_0$, we can use any information about the objective function. If we know the maximum change $\max(\delta f)$ of the objective function, we can use it to estimate an initial temperature $T_0$ for a given probability $p_0$. That is,

$$T_0 \approx -\frac{\max(\delta f)}{\ln p_0}. \tag{13.6}$$

If we do not know the possible maximum change of the objective function, we can use a heuristic approach. We can start evaluations from a very high temperature (so that almost all changes are accepted) and reduce the temperature quickly until about 50 or 60% of the worse moves are accepted, and then use this temperature as the new initial temperature $T_0$ for proper and relatively slow cooling.

For the final temperature, in theory it should be zero so that no worse move can be accepted. However, if $T_f \to 0$, more unnecessary evaluations are needed. In practise, we simply choose a very small value, say, $T_f = 10^{-10} - 10^{-5}$, depending on the required quality of the solutions and time constraints.

The implementation of SA to optimize the banana function can be demonstrated using a Matlab/Octave program. We have used the initial temperature $T_0 = 1.0$, the final temperature $T_f = 10^{-10}$, and a geometric cooling schedule with $\alpha = 0.9$. A demo code in Matlab can be found at the Mathworks.[1]

## 13.5 Differential Evolution

Differential evolution (DE) was developed by R. Storn and K. Price in 1997. It is a vector-based algorithm, which has some similarity to pattern search and GA due to its use of crossover and mutation. DE is a stochastic search algorithm with self-organizing tendency and does not use the information of derivatives. Thus, it is a population-based, derivative-free method. In addition, DE uses real-number as solution strings, thus no encoding and decoding is needed.

For a $D$-dimensional optimization problem with $D$ parameters, a population of $n$ solution vectors are initially generated, we have $\boldsymbol{x}_i$ where $i = 1, 2, \ldots, n$. For each solution $\boldsymbol{x}_i$ at any generation $t$, we use the conventional notation as

$$\boldsymbol{x}_i^t = (x_{1,i}^t, x_{2,i}^t, \ldots, x_{D,i}^t), \tag{13.7}$$

which consists of $D$ components in the $D$-dimensional space. This vector can be considered as the chromosomes or genomes.

DE consists of three main steps: mutation, crossover, and selection.

Mutation is carried out by the mutation scheme. For each vector $\boldsymbol{x}_i$ at any time or generation $t$, we first randomly choose three distinct vectors $\boldsymbol{x}_p$, $\boldsymbol{x}_q$, and $\boldsymbol{x}_r$ at $t$ (see Figure 13.3), and then generate a so-called donor vector by the mutation scheme

$$\boldsymbol{v}_i^{t+1} = \boldsymbol{x}_p^t + F(\boldsymbol{x}_q^t - \boldsymbol{x}_r^t), \tag{13.8}$$

where $F \in [0, 2]$ is a parameter, often referred to as the differential weight. This requires that the minimum number of population size is $n \geq 4$. In principle, $F \in [0, 2]$, but in practice, a scheme with $F \in [0, 1]$ is more efficient and stable. In fact, almost all the studies in the literature use $F \in (0, 1)$.

From Figure 13.3, we can see that the perturbation $\boldsymbol{\delta} = F(\boldsymbol{x}_q - \boldsymbol{x}_r)$ to the vector $\boldsymbol{x}_p$ is used to generate a donor vector $\boldsymbol{v}_i$, and such perturbation is directed.

Crossover is controlled by a crossover parameter $C_r \in [0, 1]$, controlling the rate or probability for crossover. The actual crossover can be carried out in two ways: binomial and exponential. The binomial scheme performs crossover on each of the $D$ components or variables/parameters. By generating

1 http://www.mathworks.co.uk/matlabcentral/fileexchange/29739-simulated-annealing-for-constrained-optimization

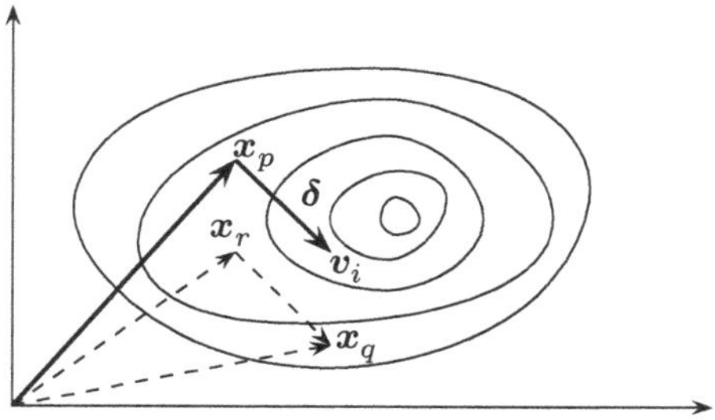


**Figure 13.3** Schematic representation of mutation vectors in differential evolution with movement $\boldsymbol{\delta} = F(\boldsymbol{x}_q - \boldsymbol{x}_r)$.

a uniformly distributed random number $r_i \in [0, 1]$, the $j$th component of $\boldsymbol{v}_i$ is manipulated as

$$u_{j,i}^{t+1} = \begin{cases} v_{j,i} & \text{if } r_i \le C_r, \\ x_{j,i}^t & \text{otherwise,} \end{cases} \qquad (j = 1, 2, \ldots, D). \tag{13.9}$$

This way, each component can be decided randomly whether or not to exchange with the counterpart of the donor vector.

In the exponential scheme, a segment of the donor vector is selected and this segment starts with a random integer $k$ with a random length $L$ which can include many components. Mathematically, this is to choose $k \in [0, D-1]$ and $L \in [1, D]$ randomly, and we have

$$u_{j,i}^{t+1} = \begin{cases} v_{j,i}^t & \text{for } j = k, \ldots, k - L + 1 \in [1, D], \\ x_{j,i}^t & \text{otherwise.} \end{cases} \tag{13.10}$$

As the binomial is simpler to implement, we will use the binomial crossover in our implementation.

Selection is essentially the same as that used in GA. It is to select the most fittest; that is, the minimum objective value for a minimization problem. Therefore, we have

$$x_i^{t+1} = \begin{cases} u_i^{t+1} & \text{if } f(u_i^{t+1}) \le f(x_i^t), \\ x_i^t & \text{otherwise.} \end{cases} \tag{13.11}$$

All the above three components can be summarized as the pseudocode shown in Algorithm 13.3. It is worth pointing out here that the use of $\boldsymbol{v}_i^{t+1} \neq \boldsymbol{x}_i^t$ may increase the evolutionary or exploratory efficiency. The overall search efficiency is controlled by two parameters: the differential weight $F$ and the crossover probability $C_r$.

Most studies have focused on the choice of $F$, $C_r$, and $n$ as well as the modifications of Eq. (13.8). In fact, when generating mutation vectors, we can use many different ways of formulating (13.8), and this leads to various schemes

**Algorithm 13.3** Pseudocode of differential evolution.

Initialize the population $\boldsymbol{x}_i$ randomly, set $F \in [0, 2]$ and $C_r \in [0, 1]$
**while** (stopping criterion)
  **for** $i = 1$ to $n$,
  For each $\boldsymbol{x}_i$, randomly choose three distinct vectors $\boldsymbol{x}_p$, $\boldsymbol{x}_r$ and $\boldsymbol{x}_r$
  Generate a new vector $\boldsymbol{v}$ by DE scheme (13.8)
  Generate a random index $J_r \in \{1, 2, \dots, D\}$ by permutation
  Generate a randomly distributed number $r_i \in [0, 1]$
  **for** $j = 1$ to $D$,
  For each parameter $\boldsymbol{v}_{j,i}$ ($j$th component of $\boldsymbol{v}_i$), update

$$\boldsymbol{u}_{j,i}^{t+1} = \begin{cases} \boldsymbol{v}_{j,i}^{t+1} & \text{if } r_i \leq C_r \text{ or } j = J_r \\ \boldsymbol{x}_{j,i}^{t} & \text{if } r_i > C_r \text{ and } j \neq J_r \end{cases}$$

  **end**
  Select and update the solution by Eq. (13.11)
  **end**
**end**
Post-process and output the best solution found

with the naming convention: DE/$x$/$y$/$z$ where $x$ is the mutation scheme (rand or best), $y$ is the number of difference vectors, and $z$ is the crossover scheme (binomial or exponential). So DE/Rand/1/* means the basic DE scheme using random mutation, one difference vector with either a binomial or exponential crossover scheme.

The basic DE/Rand/1/Bin scheme is given in Eq. (13.8). That is,

$$\boldsymbol{v}_i^{t+1} = \boldsymbol{x}_p^t + F(\boldsymbol{x}_q^t - \boldsymbol{x}_r^t). \tag{13.12}$$

If we replace the $\boldsymbol{x}_p^t$ by the current best $\boldsymbol{x}_{\text{best}}$ found so far, we have the so-called DE/Best/1/Bin scheme

$$\boldsymbol{v}_i^{t+1} = \boldsymbol{x}_{\text{best}}^t + F(\boldsymbol{x}_q^t - \boldsymbol{x}_r^t). \tag{13.13}$$

There is no reason that why we should not use more than thee distinct vectors. For example, if we use four different vectors plus the current best, we have the DE/Best/2/Bin scheme

$$\boldsymbol{v}_i^{t+1} = \boldsymbol{x}_{\text{best}}^t + F(\boldsymbol{x}_{k_1}^t + \boldsymbol{x}_{k_2}^t - \boldsymbol{x}_{k_3}^t - \boldsymbol{x}_{k_4}^t). \tag{13.14}$$

Furthermore, if we use five different vectors, we have the DE/Rand/2/Bin scheme

$$\boldsymbol{v}_i^{t+1} = \boldsymbol{x}_{k_1}^t + F_1(\boldsymbol{x}_{k_2}^t - \boldsymbol{x}_{k_3}^t) + F_2(\boldsymbol{x}_{k_4}^t - \boldsymbol{x}_{k_5}^t), \tag{13.15}$$

where $F_1$ and $F_2$ are differential weights in $[0, 1]$. Obviously, for simplicity, we can also take $F_1 = F_2 = F$. Following the similar strategy, we can design various schemes. For example, the above variants can be written in a generalized form

$$\boldsymbol{v}_i^{t+1} = \boldsymbol{x}_{k_1}^t + \sum_{s=1}^{m} F_s \cdot (\boldsymbol{x}_{k_2(s)}^t - \boldsymbol{x}_{k_3(s)}^t), \tag{13.16}$$

where $m = 1, 2, 3, \ldots$ and $F_s$ $(s = 1, \ldots, m)$ are the scale factors. The number of vectors involved on the right-hand side is $2m + 1$. In the above variants, $m = 1$ and $m = 2$ are used.

On the other hand, there is another type of variant, which uses an additional influence parameter $\lambda \in (0, 1)$. For example, the DE/rand-to-best/1/* variant can be written as

$$\boldsymbol{v}_i^{t+1} = \lambda \boldsymbol{x}_{\text{best}}^t + (1 - \lambda)\boldsymbol{x}_{k_1}^t + F(\boldsymbol{x}_{k_2}^t - \boldsymbol{x}_{k_3}^t), \tag{13.17}$$

which introduces an extra parameter $\lambda$. Again, these types of variants can be written in a generalized form

$$\boldsymbol{v}_i^{t+1} = \lambda \boldsymbol{x}_{\text{best}}^t + (1 - \lambda)\boldsymbol{x}_{k_1}^t + F \sum_{s=1}^{m} (\boldsymbol{x}_{k_2(s)}^t - \boldsymbol{x}_{k_3(s)}^t). \tag{13.18}$$

In fact, more than 10 different schemes have been formulated, and for details, readers can refer to the book by Price et al. (2005).

There are other good variants of DE, including self-adapting control parameter in differential evolution (jDE) by Brest et al. (2006), self-adaptive DE (SaDE) by Qin et al. (2009), and DE with the eagle strategy by Yang and Deb (2012).

## Exercises

**13.1** For GA, there are many implementations in the literature. Use any software package to find the minimum of $(1 - x)^2 + 100(y - x^2)^2$.

**13.2** Discuss the effects of mutation, crossover, and selection in GA.

**13.3** In SA, can the cooling schedule be a non-monotonically decreasing function?

**13.4** Discuss the choice of $F$ in DE.

## Bibliography

Auger A. (2005). Convergence results for the (1,$\lambda$)-SA-ES using the theory of $\phi$-irreducible markov chains. *Theoretical Computer Science* 334 (1–3): 35–69.

Blum, C. and Roli, A. (2003). Metaheuristics in combinatorial optimization: overview and conceptual comparison. *ACM Computing Surveys* 35 (2): 268–308.

Brest, J., Greiner, S., Boskovic, B. et al. (2006). Self-adapting control parameters in differential evolution: a comparative study on numerical benchmark functions. *IEEE Transactions on Evolutionary Computation* 10 (6): 646–657.

Brest, J. and Maucec, M. S. (2011). Self-adaptive differential evolution algorithm using population size reduction and three strategies. *Soft Computing* 15 (11): 2157–2174.

Dorigo, M. (1992). Opimization, learning and natural algorithms. PhD thesis. Politecnico di Milano, Italy.

Engelbrecht, A.P. (2005). *Fundamentals of Computational Swarm Intelligence*. Hoboken, NJ: Wiley.

Fogel, L.J., Owens, A.J., and Walsh, M.J. (1966). *Artificial Intelligence Through Simulated Evolution*. New York: Wiley.

Glover, F. (1986). Future paths for integer programming and links to artificial intelligence. *Computers & Operations Research* 13 (5): 533–549.

Goldberg, D.E. (1989). *Genetic Algorithms in Search, Optimisation and Machine Learning*. Reading, MA: Addison Wesley.

Holland, J. (1975). *Adaptation in Natural and Artificial Systems*. Ann Arbor: University of Michigan Press.

Judea, P. (1984). *Heuristics*. New York: Addison-Wesley.

Karaboga, D. (2005). *An Idea Based on Honeybee Swarm for Numerical Optimization*. Technical Report. Turkey: Erciyes University.

Kirkpatrick, S., Gellat, C.D., and Vecchi, M.P. (1983). Optimization by simulated annealing. *Science* 220 (4598): 671–680.

Koza, J.R. (1992). *Genetic Programming: On the Programming of Computers by Means of Natural Selection*. Cambridge, MA: MIT Press.

Price, K., Storn, R., and Lampinen, J. (2005). Differential Evolution: A Practical Approach to Global Optimization. Berlin: Springer.

Qin, A.K., Huang, V.L., and Suganthan, P.N. (2009). Differential evolution algorithm with strategy adaptation for global numerical optimization. *IEEE Transactions on Evolutionary Computation* 13 (2): 398–417.

Rechenberg, I. (1971). Evolutionsstrategie - Optimierung Technisher Systeme nach Prinzipien der Biogischen Evolution. PhD thesis. Technical University of Berlin.

Schwefel, H.P. (1974). Numerische Optimierung von Computer-Modellen. PhD thesis. Technical University of Berlin.

Schwefel, H.P. (1995). *Evolution and Optimum Seeking*. New York: Wiley.

Storn, R. and Price, K. (1997). Differential evolution: a simple and efficient heuristic for global optimization over continuous spaces. Journal of Global Optimization, 11 (4): 341–359.

Yang, X.S. (2008). *Nature-Inspired Metaheuristic Algorithms*. Bristol, UK: Luniver Press.

Yang, X.S. (2011). Metaheuristic optimization. *Scholarpedia* 6 (8): 11472.

Yang, X.S. and Deb, S. (2012). Two-stage eagle strategy with differential evolution. International Journal of Bio-Inspired Computation 4 (1): 1–5.

Yang, X.S., Cui, Z.H., Xiao, R.B. et al. (2013). *Swarm Intelligence and Bio-Inspired Computation: Theory and Applications*. London: Elsevier.

Yang, X.S. (2014). *Nature-Inspired Optimization Algorithms*. London: Elsevier Insight.

Yang, X.S., Chien, S.F., and Ting, T.O. (2015). *Bio-Inspired Computation in Telecommunications*. Waltham: Morgan Kaufmann.

Yang, X.S. and Papa, J.P. (2016). *Bio-Inspired Computation and Applications in Image Processing*. London: Academic Press.

Yang, X.S. (2017). *Nature-Inspired Algorithms and Applied Optimization*. Heidelberg: Springer.

# 14

# Nature-Inspired Algorithms

Many algorithms such as ant colony algorithms and firefly algorithm (FA) use the behavior of the so-called swarm intelligence (SI). All SI-based algorithms use the real-number randomness and some form of communication among agents or particles. These algorithms are usually easy to implement as there is no encoding or decoding of the parameters into binary strings as those in genetic algorithms where real-number strings can also be used. In addition, SI-based algorithms are very flexible and yet efficient to deal with a wide range of problems.

SI has attracted great attention in the last two decades and the literature has expanded significantly in the last few years. Many novel algorithms have appeared and new algorithms such as the FA and cuckoo search (CS) have been shown to be very efficient. In this chapter, we will briefly introduce some of the most recent algorithms.

## 14.1 Introduction to SI

Many new algorithms that are based on SI may have drawn inspiration from different sources, but they have some similarity to some of the components that are used in particle swarm optimization (PSO) and ant colony optimization (ACO). In this sense, the PSO and ACO pioneered the basic ideas of SI-based computation.

From the algorithm point of view, the aim of a swarming system is to let the system evolve and converge into some stable optimality. In this case, it has strong similarity to a self-organizing system. Such an iterative, self-organizing system can evolve, according to a set of rules or mathematical equations. As a result, such a complex system can interact and self-organize into certain converged states, showing some emergent characteristics of self-organization. In this sense, the proper design of an efficient optimization algorithm is equivalent to finding efficient ways to mimic the evolution of a self-organizing system.

*Optimization Techniques and Applications with Examples*, First Edition. Xin-She Yang.

In practice, all nature-inspired algorithms try to mimic some successful characteristics of biological, physical, or chemical systems in nature.

Among all evolutionary algorithms, algorithms based on SI dominate the landscape. There are many reasons for this dominance, though three obvious reasons are: (i) SI uses multiple agents as an evolving, interacting population, and thus provides good ways to mimic natural systems. (ii) Population-based approaches allow parallelization and vectorization implementations in practice, and are thus straightforward to implement. (iii) These SI-based algorithms are flexible and yet efficient enough to deal with a wide range of problems.

## 14.2 Ant and Bee Algorithms

Ant algorithms, especially the ACO developed by Dorigo in 1992, mimic the foraging behavior of social ants. In essence, all ant algorithms use pheromone as a chemical messenger and the pheromone concentration as the indicator of quality solutions to a problem of interest. From the implementation point of view, solutions are related to the pheromone concentration, leading to routes and paths marked by the higher pheromone concentrations as better solutions to problems such as discrete combinatorial problems.

Looking at the ACO closely, random route generation is mainly by mutation, while pheromone-based selection provides a mechanism for selecting shorter routes. There is no explicit crossover in ant algorithms. However, mutation is not a simple action as flipping digits as in genetic algorithms, the new solutions are essentially generated by fitness-proportional mutation. For example, the probability of ants in a network problem at a particular node $i$ to choose the route from node $i$ to node $j$ is given by

$$p_{ij} = \frac{\phi_{ij}^{\alpha} d_{ij}^{\beta}}{\sum_{i,j=1}^{n} \phi_{ij}^{\alpha} d_{ij}^{\beta}}, \tag{14.1}$$

where $\alpha > 0$ and $\beta > 0$ are the influence parameters, $\phi_{ij}$ is the pheromone concentration on the route between $i$ and $j$, and $d_{ij}$ the desirability of the same route. The selection is subtly related to some a priori knowledge about the route such as the distance $s_{ij}$ is often used so that $d_{ij} \propto 1/s_{ij}$.

On the other hand, bee algorithms do not usually use pheromone. For example, in the artificial bee colony (ABC) optimization algorithm, the bees in a colony are divided into three groups: employed bees (forager bees), onlooker bees (observer bees), and scouts. Randomization are carried out by scout bees and employed bees, and both are mainly mutation. Selection is related to honey or the objective. Again, there is no explicit crossover.

Both ACO and ABC only use mutation and fitness-related selection, and they can explore the search space relatively effectively, but convergence may be slow because it lacks crossover, and thus the subspace exploitation ability is very limited. In fact, the lack of crossover is very common in many meta-heuristic algorithms. In terms of exploration and exploitation, both ant and bee algorithms have strong exploration ability, but their exploitation ability is comparatively low. This may explain why they can perform reasonably well for some tough optimization, but the computational efforts such as the number of function evaluations can be very high.

As there is extensive literature about the ACO and relevant variants, we will not discuss them in detail here. Interested readers can refer to more advanced literature.

## 14.3 Particle Swarm Optimization

PSO was developed by Kennedy and Eberhart in 1995 and has become one of the most widely used SI-based algorithms. The PSO algorithm searches the space of an objective function by adjusting the trajectories of individual agents, called particles, as the piecewise paths formed by positional vectors in a quasi-stochastic manner. The movement of a swarming particle consists of two major components: a stochastic component and a deterministic component. Each particle is attracted toward the position of the current global best $\boldsymbol{g}^*$ and its own best location $\boldsymbol{x}_i^*$ in history, while at the same time it has a tendency to move randomly.

When a particle finds a location that is better than any previously found locations, then it updates it as the new current best for particle $i$. There is a current best for all $n$ particles at any time $t$ during iterations. The aim is to find the global best among all the current best solutions until the objective no longer improves or after a certain number of iterations. The movement of particles is schematically represented in Figure 14.1 where $\boldsymbol{x}_i^*$ is the current best for particle $i$, and $\boldsymbol{g}^* \approx \min\{f(\boldsymbol{x}_i)\}$ for $(i = 1, 2, \ldots, n)$ is the current global best at $t$.

The essential steps of the PSO can be summarized as the pseudocode shown in Algorithm 14.1.

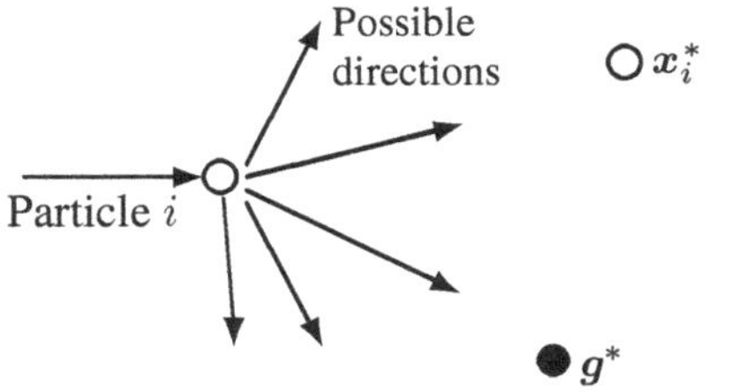


**Figure 14.1** Schematic representation of the motion of a particle in PSO, moving toward the global best $\boldsymbol{g}^*$ and the current best $\boldsymbol{x}_i^*$ for each particle $i$.

**Algorithm 14.1** Pseudocode of particle swarm optimization.

Objective function $f(\boldsymbol{x})$, $\boldsymbol{x} = (x_1, \dots, x_D)^{\mathrm{T}}$
Initialize locations $\boldsymbol{x}_i$ and velocity $\boldsymbol{v}_i$ of $n$ particles.
Find $\boldsymbol{g}^*$ from $\min\{f(\boldsymbol{x}_1), \dots, f(\boldsymbol{x}_n)\}$ (at $t = 0$)
**while** (criterion)
    **for** loop over all $n$ particles and all $D$ dimensions
    Generate new velocity $\boldsymbol{v}_i^{t+1}$ using Eq. (14.2)
    Calculate new locations $\boldsymbol{x}_i^{t+1} = \boldsymbol{x}_i^t + \boldsymbol{v}_i^{t+1}$
    Evaluate objective functions at new locations $\boldsymbol{x}_i^{t+1}$
    Find the current best for each particle $\boldsymbol{x}_i^*$
    **end for**
    Find the current global best $\boldsymbol{g}^*$
    Update $t = t + 1$ (pseudo time or iteration counter)
**end while**
Output the final results $\boldsymbol{x}_i^*$ and $\boldsymbol{g}^*$

Let $\boldsymbol{x}_i$ and $\boldsymbol{v}_i$ be the position vector and velocity for particle $i$, respectively. The new velocity vector is determined by the following formula

$$\boldsymbol{v}_i^{t+1} = \boldsymbol{v}_i^t + \alpha\epsilon_1[\boldsymbol{g}^* - \boldsymbol{x}_i^t] + \beta\epsilon_2[\boldsymbol{x}_i^{*(t)} - \boldsymbol{x}_i^t], \tag{14.2}$$

where $\epsilon_1$ and $\epsilon_2$ are two random vectors, and each entry taking the values between 0 and 1. The parameters $\alpha$ and $\beta$ are the learning parameters or acceleration constants, which can typically be taken as, say, $\alpha \approx \beta \approx 2$.

The initial locations of all particles should distribute relatively uniformly so that they can sample over most regions, which is especially important for multimodal problems. The initial velocity of a particle can be taken as a random vector with each components in the range of $[0, v_{\max}]$. The new position can then be updated by

$$\boldsymbol{x}_i^{t+1} = \boldsymbol{x}_i^t + \boldsymbol{v}_i^{t+1}\Delta t, \tag{14.3}$$

where $\Delta t$ is the (pseudo)time increment. Since we are dealing with the discrete time increment or iteration in iterative algorithms, we can always set $\Delta t = 1$, and we have

$$\boldsymbol{x}_i^{t+1} = \boldsymbol{x}_i^t + \boldsymbol{v}_i^{t+1}. \tag{14.4}$$

It is worth pointing out that the above updating equations are based on the physical system of particles, but there is no need to consider the units of variables, from the mathematical or numerical point of view. This also applies to other algorithms that are based on SI.

Although $\boldsymbol{v}_i$ can be any values, it is usually bounded in some range $[0, \boldsymbol{v}_{\max}]$. However, the velocities cannot be too high, otherwise, it may lead to an unstable system because the high energies are associated with high velocities.

There are many variants which extend the standard PSO algorithm, and the most noticeable improvement is probably to use inertia function $\theta(t)$ so that $\boldsymbol{v}_i^t$ is replaced by $\theta(t)\boldsymbol{v}_i^t$,

$$\boldsymbol{v}_i^{t+1} = \theta \boldsymbol{v}_i^t + \alpha\epsilon_1[\boldsymbol{g}^* - \boldsymbol{x}_i^t] + \beta\epsilon_2[\boldsymbol{x}_i^{*(k)} - \boldsymbol{x}_i^t], \tag{14.5}$$

where $\theta$ takes the values between 0 and 1 in theory. In the simplest case, the inertia function can be taken as a constant, typically $\theta \approx 0.5 - 0.9$. This is equivalent to introducing a virtual mass to stabilize the motion of the particles, and thus the algorithm can usually be expected to converge more quickly.

### 14.3.1 Accelerated PSO

The standard PSO uses both the current global best $\boldsymbol{g}^*$ and the individual best $\boldsymbol{x}_i^{*(t)}$. One of the reasons of using the individual best is probably to increase the diversity in the quality solutions; however, this diversity can be simulated using some randomness. Subsequently, there is no compelling reason for using the individual best, unless the optimization problem of interest is highly nonlinear and multimodal.

A simplified version which could accelerate the convergence of the algorithm is to use the global best only. The so-called accelerated particle swarm optimization (APSO) was developed by Xin-She Yang in 2008 and then has been developed further in recent years. Thus, in APSO, the velocity vector is generated by a simpler formula

$$\boldsymbol{v}_i^{t+1} = \boldsymbol{v}_i^t + \alpha\left(\epsilon - \frac{1}{2}\right) + \beta(\boldsymbol{g}^* - \boldsymbol{x}_i^t), \tag{14.6}$$

where $\epsilon$ is a random variable with values from 0 to 1. Here, the shift 1/2 is purely out of convenience. We can also use a standard normal distribution $\alpha\epsilon_t$ where $\epsilon_t$ is drawn from $N(0, 1)$ to replace the second term. Now we have

$$\boldsymbol{v}_i^{t+1} = \boldsymbol{v}_i^t + \beta(\boldsymbol{g}^* - \boldsymbol{x}_i^t) + \alpha\epsilon_t, \tag{14.7}$$

where $\epsilon_t$ can be drawn from a Gaussian distribution or any other suitable distributions. Here, $\alpha$ is a scaling factor that controls the step size or the strength of randomness, while $\beta$ is a parameter controlling the movement of particles.

The update of the position is simply

$$\boldsymbol{x}_i^{t+1} = \boldsymbol{x}_i^t + \boldsymbol{v}_i^{t+1}. \tag{14.8}$$

In order to simplify the formulation even further, we can also write the update of the location in a single step

$$\boldsymbol{x}_i^{t+1} = (1 - \beta)\boldsymbol{x}_i^t + \beta\boldsymbol{g}^* + \alpha\epsilon_t. \tag{14.9}$$

The typical values for this accelerated PSO are $\alpha \approx 0.1 - 0.4$ and $\beta \approx 0.1 - 0.7$, though $\alpha \approx 0.2$ and $\beta \approx 0.5$ can be taken as the initial values for most unimodal objective functions. It is worth pointing out that the parameters $\alpha$ and $\beta$ should in general be related to the scales of the independent variables $\boldsymbol{x}_i$ and the search domain. Surprisingly, this simplified APSO can have global convergence under appropriate conditions.

A further improvement to the accelerated PSO is to reduce the randomness as iterations proceed. This means that we can use a monotonically decreasing function such as

$$\alpha = \alpha_0 e^{-\gamma t} \tag{14.10}$$

or

$$\alpha = \alpha_0 \gamma^t \qquad (0 < \gamma < 1), \tag{14.11}$$

where $\alpha_0 \approx 0.5 - 1$ is the initial value of the randomness parameter. Here, $t$ is the number of iterations or time steps. $0 < \gamma < 1$ is a control parameter. For example, in most implementations, we can use $\gamma = 0.9 - 0.99$. Obviously, other non-increasing function forms $\alpha(t)$ can also be used. In addition, these parameters should be fine-tuned to suit your optimization problems of interest, and such parameter tuning is essential for all evolutionary algorithms.

A demo implementation for APSO can be found at the Mathworks website.[1]

### 14.3.2 Binary PSO

In the standard PSO, the positions and velocities take continuous values. However, many problems are combinatorial and their variables only take discrete values. In some cases, the variables can only be 0 and 1, and such binary problems require modifications of the PSO algorithm. Kennedy and Eberhart in 1997 presented a stochastic approach to discretize the standard PSO, and they interpreted the velocity in a stochastic sense.

First, the continuous velocity $\boldsymbol{v}_i = (v_{i1}, v_{i2}, \dots, v_{ik}, \dots, v_{iD})$ is transformed using a sigmoid transformation

$$S(v_{ik}) = \frac{1}{1 + \exp(-v_{ik})} \qquad (k = 1, 2, \dots, D), \tag{14.12}$$

which applies to each component of the velocity vector $\boldsymbol{v}_i$ of particle $i$. Obviously, when $v_{ik} \to \infty$, we have $S(v_{ik}) \to 1$, while $S_{ik} \to 0$ when $v_{ik} \to -\infty$. However, as the variations at the two extremes are very slow, a stochastic approach is introduced.

Secondly, a uniformly-distributed random number $r \in (0, 1)$ is drawn, then the velocity is converted to a binary variable by the following stochastic rule:

$$x_{ik} = \begin{cases} 1 & \text{if } r < S(v_{ik}), \\ 0 & \text{otherwise.} \end{cases} \tag{14.13}$$

1 http://www.mathworks.co.uk/matlabcentral/fileexchange/29725-accelerated-particle-swarm-optimization.

In this case, the value of each velocity component $v_{ik}$ is interpreted as a probability for $x_{ik}$ taking the value 1. Even for a fixed value of $v_{ik}$, the actual value of $x_{ik}$ is not certain before a random number $r$ is drawn. In this sense, the binary PSO (BPSO) differs significantly from the standard continuous PSO.

In fact, since each component of each variable takes only 0 and 1, this BPSO can work for both discrete and continuous problems if the latter is coded in the binary system. As the probability of each bit/component taking one is $S(v_{ik})$, and the probability of taking zero is $1 - S(v_{ik})$, the joint probability $p$ of a bit change can be computed by

$$p = S(v_{ik})[1 - S(v_{ik})]. \tag{14.14}$$

One of the advantages of this binary coding and discritization is to enable binary representations of even continuous problems. For example, Kennedy and Eberhart provided an example of solving the second De Jong function and found that a 24-bit string 110111101110110111101001 corresponds to the optimal solution value of 3 905.929 932 from this representation. However, a disadvantage is that the Hamming distance from other local optima is large; therefore, it is unlikely that the search will jump from one local optimum to another. This means that BPSO can get stuck in a local optimum with premature convergence. New remedies are still under active research.

Various studies show that PSO algorithms can outperform genetic algorithms and other conventional algorithms for solving many optimization problems. This is partially due to that fact that the broadcasting ability of the current best estimates gives a better and quicker convergence toward the optimality. However, PSO algorithms do have some disadvantages such as premature convergence. Further developments and improvements are still under active research.

It is worth pointing out that the discretization approaches outlined here to convert PSO into BPSO can be used to convert all other algorithms into their binary counterparts. For example, we can use the above approaches to convert the CS algorithm into a binary CS algorithm to be introduced later.

## 14.4 Firefly Algorithm

FA was developed by Xin-She Yang in 2008, which was based on the flashing patterns and behavior of tropical fireflies. FA is simple, flexible, and easy to implement.

The flashing light of fireflies is an amazing sight in the summer sky in the tropical and temperate regions. There are about 2000 firefly species, and most fireflies produce short and rhythmic flashes. The pattern of flashes is often unique for a particular species. The flashing light is produced by a process of bioluminescence, and the true functions of such signaling systems are still being debated. However, two fundamental functions of such flashes are to attract mating partners (communication), and to attract potential prey. In addition, flashing may also serve as a protective warning mechanism to remind potential predators of the bitter taste of fireflies.

The rhythmic flash, the rate of flashing, and the amount of time form part of the signal system that brings both sexes together. Females respond to a male's unique pattern of flashing in the same species, while in some species such as *Photuris*, female fireflies can eavesdrop on the bioluminescent courtship signals and even mimic the mating flashing pattern of other species so as to lure and eat the male fireflies who may mistake the flashes as a potential suitable mate. Some tropic fireflies can even synchronize their flashes, thus forming emerging biological self-organized behavior.

We know that the light intensity at a particular distance $r$ from the light source obeys the inverse square law. That is, to say, the light intensity $I$ decreases as the distance $r$ increases in terms of $I \propto 1/r^2$. Furthermore, the air absorbs light which becomes weaker and weaker as the distance increases. These two combined factors make most fireflies visible to a limited distance, usually several hundred meters at night, which is good enough for fireflies to communicate. In addition, the attractiveness $\beta$ of a firefly is in the eye of the other fireflies, which varies with their distance. To avoid singularity, we use the following form of attractiveness

$$\beta = \beta_0 \exp[-\gamma r^2], \tag{14.15}$$

where $\beta_0$ is the attractiveness at distance $r = 0$ and $\gamma$ is the light absorption coefficient. Here, $r$ can be defined as the Cartesian distance between the two fireflies of interest. However, for other optimization problems such as routing, the time delay along a route can also be used as the "distance." Therefore, $r$ should be interpreted in the most wide and appropriate sense, depending on the type of problem.

The flashing light can be formulated in such a way that it is associated with the objective function to be optimized, which makes it possible to formulate new optimization algorithms.

Now we can idealize some of the flashing characteristics of fireflies so as to develop firefly-inspired algorithms. For simplicity in describing the standard FA, we now use the following three idealized rules:

- All fireflies are unisex so that one firefly will be attracted to other fireflies regardless of their sex.
- Attractiveness is proportional to their brightness, thus for any two flashing fireflies, the less brighter one will move toward the brighter one. The attractiveness is proportional to the brightness and they both decrease as their distance increases. If there is no brighter one than a particular firefly, it will move randomly.

**Algorithm 14.2** Pseudocode of the firefly algorithm (FA).

Objective function $f(\boldsymbol{x})$, $\boldsymbol{x} = (x_1, \dots, x_D)^{\mathrm{T}} \in \mathbb{R}^D$.
Generate an initial population of $n$ fireflies $\boldsymbol{x}_i$ $(i = 1, 2, \dots, n)$.
Light intensity $I_i$ at $\boldsymbol{x}_i$ is determined by $f(\boldsymbol{x}_i)$.
Define light absorption coefficient $\gamma$.
**while** ($t$ < MaxGeneration),
**for** $i = 1 : n$ (all $n$ fireflies)
    **for** $j = 1 : n$ (all $n$ fireflies) (inner loop)
    **if** $(I_i < I_j)$
        Move firefly $i$ toward $j$.
    **end if**
        Vary attractiveness with distance $r$ via $\exp[-\gamma r^2]$.
        Evaluate new solutions and update light intensity.
    **end for** $j$
**end for** $i$
Rank the fireflies and find the current global best $\boldsymbol{g}_*$.
**end while**
Postprocess results and visualization.

- The brightness of a firefly is affected or determined by the landscape of the objective function.

For a maximization problem, the brightness can simply be proportional to the value of the objective function. Other forms of brightness can be defined in a similar way to the fitness function in genetic algorithms.

Based on these three rules, the basic steps of the FA can be summarized as the pseudocode shown in Algorithm 14.2.

The movement of a firefly $i$ attracted to another more attractive (brighter) firefly $j$ is determined by

$$\boldsymbol{x}_i^{t+1} = \boldsymbol{x}_i^t + \beta_0 e^{-\gamma r_{ij}^2}(\boldsymbol{x}_j^t - \boldsymbol{x}_i^t) + \alpha\, \boldsymbol{\epsilon}_i^t, \tag{14.16}$$

where the second term is due to the attraction and $\beta_0$ is the attractiveness at zero distance $r = 0$. The third term is randomization with $\alpha$ being the randomization parameter and $\boldsymbol{\epsilon}_i^t$ is a vector of random numbers drawn from a Gaussian distribution at time $t$. Other studies also use randomization in terms of $\boldsymbol{\epsilon}_i^t$ that can easily be extended to other distributions such as a Lévy distribution.

From the above equation, we can see that mutation is used for both local and global search. When $\boldsymbol{\epsilon}_i^t$ is drawn from a Gaussian distribution and Lévy distribution, it produces mutation on a larger scale. On the other hand, if $\alpha$ is chosen to be a very small value, then mutation can be very small, and thus limited to a subspace. However, during the update in the two loops in FA, ranking as well as selection is used.

The author has provided a simple demonstration code of the FA that can be found at the Mathworks website.[2]

One novel feature of FA is that attraction is used, and this is the first of its kind in any SI-based algorithms. Since local attraction is stronger than long-distance attraction, the population in FA can automatically subdivide into multiple subgroups, and each group can potentially swarm around a local mode. Among all the local modes, there is always a global best solution which is the true optimality of the problem. FA can deal with multimodal problems naturally and efficiently.

From Eq. (14.16), we can see that FA degenerates into a variant of differential evolution when $\gamma = 0$ and $\alpha = 0$. In addition, when $\beta_0 = 0$, it degenerates into simulated annealing (SA). Furthermore, when $\boldsymbol{x}_j^t$ is replaced by $\boldsymbol{g}^*$, FA also becomes the accelerated PSO. Therefore, DE, APSO, and SA can be considered as special cases of the FA, and thus FA can have the advantages of all these three algorithms. It is no surprise that FA can be versatile and efficient, and perform better than other algorithms such as GA and PSO.

## 14.5 Cuckoo Search

CS is one of the latest nature-inspired metaheuristic algorithms, developed in 2009 by Xin-She Yang and Suash Deb. CS is based on the brood parasitism of some cuckoo species. In addition, this algorithm is enhanced by the so-called Lévy flights, rather than by simple isotropic random walks. Recent studies show that CS is potentially far more efficient than PSO and genetic algorithms.

Cuckoo are fascinating birds, not only because of the beautiful sounds they can make, but also because of their aggressive reproduction strategy. Some species such as the *ani* and *Guira* cuckoos lay their eggs in communal nests, though they may remove others' eggs to increase the hatching probability of their own eggs. Quite a number of species engage the obligate brood parasitism by laying their eggs in the nests of other host birds (often other species).

There are three basic types of brood parasitism: intraspecific brood parasitism, cooperative breeding, and nest takeover. Some host birds can engage direct conflict with the intruding cuckoos. If a host bird discovers the eggs are not their owns, they will either get rid of these alien eggs or simply abandon its nest and build a new nest elsewhere. Some cuckoo species such as the New World brood-parasitic *Tapera* have evolved in such a way that female parasitic cuckoos are often very specialized in the mimicry in color and pattern of the eggs of a few chosen host species. This reduces the probability of their eggs being abandoned and thus increases their reproductivity.

2 http://www.mathworks.co.uk/matlabcentral/fileexchange/29693-firefly-algorithm.

In addition, the timing of egg-laying of some species is also amazing. Parasitic cuckoos often choose a nest where the host bird just laid its own eggs. In general, the cuckoo eggs hatch slightly earlier than their host eggs. Once the first cuckoo chick is hatched, the first instinct action it will take is to evict the host eggs by blindly propelling the eggs out of the nest, which increases the cuckoo chick's share of food provided by its host bird. Studies also show that a cuckoo chick can also mimic the call of host chicks to gain access to more feeding opportunities.

On the other hand, various studies have shown that flight behavior of many animals and insects has demonstrated the typical characteristics of Lévy flights with power-law-like characteristics. A recent study by Reynolds and Frye (2007) shows that fruit flies or *Drosophila melanogaster* explore their landscape using a series of straight flight paths punctuated by a sudden 90° turn, leading to a Lévy-flight-style intermittent scale-free search pattern. Studies on human behavior such as the Ju/'hoansi hunter-gatherer foraging patterns also show the typical feature of Lévy flights. Even light can be related to Lévy flights. Subsequently, such behavior has been applied to optimization and optimal search, and results show its promising capability.

### 14.5.1 CS Algorithm

For simplicity in describing the standard CS, we now use the following three idealized rules:

1) Each cuckoo lays one egg at a time, and dumps it in a randomly chosen nest.
2) The best nests with high-quality eggs will be carried over to the next generations.
3) The number of available host nests is fixed, and the egg laid by a cuckoo is discovered by the host bird with a probability $p_a \in (0, 1)$. In this case, the host bird can either get rid of the egg, or simply abandon the nest and build a completely new nest.

As a further approximation, this last assumption can be approximated by replacing a fraction $p_a$ of the $n$ host nests with new nests (with new random solutions). For a maximization problem, the quality or fitness of a solution can simply be proportional to the value of the objective function. Other forms of fitness can be defined in a similar way to the fitness function in genetic algorithms and other evolutionary algorithms.

From the implementation point of view, we can use the following simple representations that each egg in a nest represents a solution, and each cuckoo can lay only one egg (thus representing one solution), the aim is to use the new and potentially better solutions (cuckoos) to replace a not-so-good solution in the nests. Obviously, this algorithm can be extended to the more complicated case where each nest has multiple eggs representing a set of solutions. Here,

**Algorithm 14.3** Pseudocode of the Cuckoo Search (CS) via Lévy flights for a minimization problem.

```
Objective function f(x),  x = (x_1, ..., x_D)^T
Generate initial population of n host nests x_i
while (t < MaxGeneration) or (stop criterion)
   Get a cuckoo randomly
   Generate a solution by Lévy flights (e.g. Eq. (14.18))
   Evaluate its solution quality or objective value f_i
   Choose a nest among n (say, j) randomly
   if (f_i < f_j),
      Replace j by the new solution i
   end
   A fraction (p_a) of worse nests are abandoned
   New nests/solutions are built/generated by Eq. (14.17)
   Keep best solutions (or nests with quality solutions)
   Rank the solutions and find the current best
   Update t ← t + 1
end while
Postprocess results and visualization
```

we will use the simplest approach where each nest has only a single egg. In this case, there is no distinction between an egg, a nest or a cuckoo, as each nest corresponds to one egg which also represents one cuckoo.

Based on these three rules, the basic steps of the CS can be summarized as the pseudocode shown in Algorithm 14.3.

CS uses a balanced combination of a local random walk and the global explorative random walk, controlled by a switching parameter $p_a$. The local random walk can be written as

$$\boldsymbol{x}_i^{t+1} = \boldsymbol{x}_i^t + \alpha s \otimes H(p_a - \epsilon) \otimes (\boldsymbol{x}_j^t - \boldsymbol{x}_k^t), \tag{14.17}$$

where $\boldsymbol{x}_j^t$ and $\boldsymbol{x}_k^t$ are two different solutions selected randomly by random permutation, $H(u)$ is a Heaviside function, $\epsilon$ is a random number drawn from a uniform distribution, and $s$ is the step size. Here, the $\otimes$ means the entry-wise product of two vectors.

On the other hand, the global random walk is carried out by using Lévy flights

$$\boldsymbol{x}_i^{t+1} = \boldsymbol{x}_i^t + \alpha L(s, \lambda), \tag{14.18}$$

where

$$L(s, \lambda) \sim \frac{\lambda \Gamma(\lambda) \sin(\pi\lambda/2)}{\pi} \frac{1}{s^{1+\lambda}} \quad (s > 0). \tag{14.19}$$

Here, $\alpha > 0$ is the step size scaling factor, which should be related to the scales of the problem of interest. It is worth pointing out that here $L(s, \lambda)$ is a random step size drawn from the Lévy distribution; it is a random number generator, not direct algebraic calculations, so we use ~ to highlight this subtle difference.

A simple demonstration code of the CS is provided by the author, and the code can be found at the Mathworks website.[3]

### 14.5.2 Lévy Flight

If a step size $S_i$ is drawn from a normal distribution with zero mean and unit variance, we have

$$S_i \sim \frac{1}{\sqrt{2\pi}} \exp\left[-\frac{x^2}{2}\right], \tag{14.20}$$

and the sum of $N$ consecutive steps forms a random walk

$$W_N = \sum_{i=1}^{N} = S_1 + S_2 + \cdots + S_N, \tag{14.21}$$

which is equivalent to

$$W_N = W_{N-1} + S_N. \tag{14.22}$$

This means that the next step $W_N$ (or state) depends only on the existing state or step $W_{N-1}$ and the transition move (step) $S_N$. These kinds of random walks can also be considered as a diffusion process or Brownian motion. In a simple two-dimensional (2D) case, a random walk with steps being drawn from a Gaussian normal distribution is shown in Figure 14.2 where 100 consecutive steps are shown, starting at the origin (0,0) (marked with •).

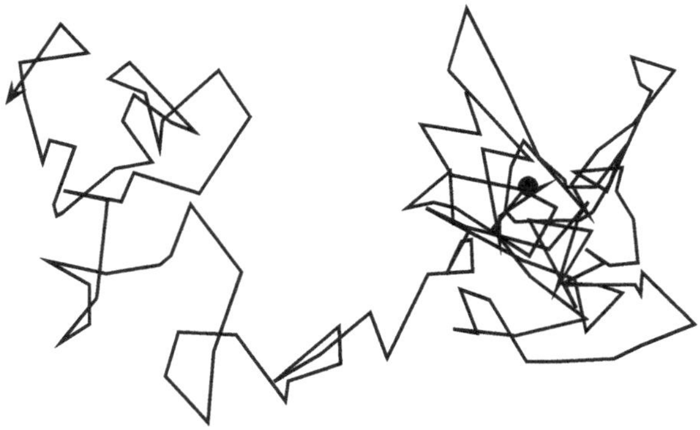

**Figure 14.2** Random walk in 2D with a Gaussian step-size distribution and the path of 100 steps starting at the origin (0, 0) (marked with •).

---

3 http://www.mathworks.co.uk/matlabcentral/fileexchange/29809-cuckoo-search–cs–algorithm.

Loosely speaking, Lévy flights are a random walk whose step sizes are drawn from a Lévy distribution, often in terms of a simple power-law formula $L(s) \sim |s|^{-1-\beta}$ where $0 < \beta \leq 2$ is an index. Mathematically speaking, a simple version of Lévy distribution can be defined as

$$L(s, \gamma, \mu) = \begin{cases} \sqrt{\frac{\gamma}{2\pi}} \exp\left[-\frac{\gamma}{2(s-\mu)}\right] \frac{1}{(s-\mu)^{3/2}}, & 0 < \mu < s < \infty \\ 0 & \text{otherwise,} \end{cases} \tag{14.23}$$

where $\mu > 0$ is a minimum step and $\gamma$ is a scale parameter. Clearly, as $s \to \infty$, we have

$$L(s, \gamma, \mu) \approx \sqrt{\frac{\gamma}{2\pi}} \frac{1}{s^{3/2}}. \tag{14.24}$$

This is a special case of the generalized Lévy distribution.

In general, Lévy distribution should be defined in terms of Fourier transform

$$F(k) = \exp[-\alpha|k|^{\beta}] \qquad (0 < \beta \leq 2), \tag{14.25}$$

where $\alpha$ is a scale parameter. The inverse of this integral is not easy, as it does not have analytical form, except for a few special cases.

For the case of $\beta = 2$, we have

$$F(k) = \exp[-\alpha k^2], \tag{14.26}$$

whose inverse Fourier transform corresponds to a Gaussian distribution. Another special case is $\beta = 1$, and we have

$$F(k) = \exp[-\alpha|k|], \tag{14.27}$$

which corresponds to a Cauchy distribution

$$p(x, \gamma, \mu) = \frac{1}{\pi} \frac{\gamma}{\gamma^2 + (x - \mu)^2}, \tag{14.28}$$

where $\mu$ is the location parameter, while $\gamma$ controls the scale of this distribution.

For the general case, the inverse integral

$$L(s) = \frac{1}{\pi} \int_0^{\infty} \cos(ks) \exp[-\alpha|k|^{\beta}] \mathrm{d}k \tag{14.29}$$

can be estimated only when $s$ is large. We have

$$L(s) \to \frac{\alpha\, \beta\, \Gamma(\beta) \sin(\pi\beta/2)}{\pi|s|^{1+\beta}}, \qquad s \to \infty. \tag{14.30}$$

Here, $\Gamma(z)$ is the Gamma function

$$\Gamma(z) = \int_0^{\infty} t^{z-1} \mathrm{e}^{-t} \mathrm{d}t. \tag{14.31}$$

In the case when $z = n$ is an integer, we have $\Gamma(n) = (n - 1)!$.

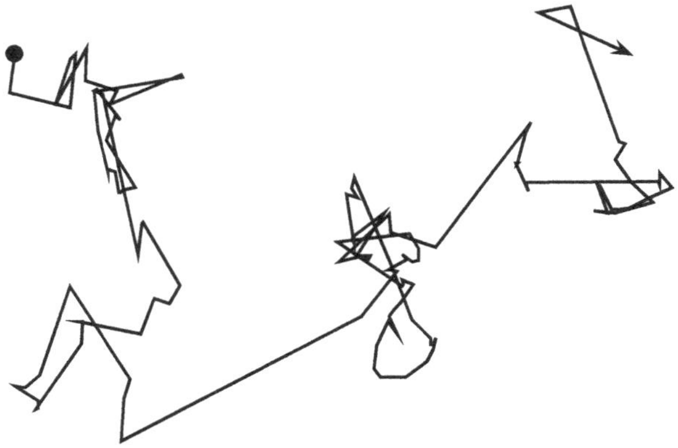

**Figure 14.3** Lévy flights in consecutive 100 steps starting at the origin (0, 0) (marked with •).

Lévy flights are more efficient than Brownian random walks in exploring unknown, large-scale search space. There are many reasons to explain this efficiency, and one of them is due to the fact that the variance of Lévy flights

$$\sigma^2(t) \sim t^{3-\beta} \qquad (1 \leq \beta \leq 2), \tag{14.32}$$

increases much faster than the linear relationship (i.e. $\sigma^2(t) \sim t$) of Brownian random walks. Figure 14.3 shows the path of Lévy flights of 100 steps starting from (0, 0) with $\beta = 1$. It is worth pointing out that a power-law distribution is often linked to some scale-free characteristics, and Lévy flights can thus show self-similarity and fractal behavior in the flight patterns.

From the implementation point of view, the generation of random numbers with Lévy flights consists of two steps: the choice of a random direction and the generation of steps which obey the chosen Lévy distribution. The generation of a direction should be drawn from a uniform distribution, while the generation of steps is quite tricky. There are a few ways of achieving this, but one of the most efficient and yet straightforward ways is to use the so-called Mantegna algorithm for a symmetric Lévy stable distribution. Here "symmetric" means that the steps can be positive and negative.

A random variable $U$ and its probability distribution can be called stable if a linear combination of its two identical copies (or $U_1$ and $U_2$) obeys the same distribution. That is, $aU_1 + bU_2$ has the same distribution as $cU + d$ where $a, b > 0$ and $c, d \in \mathbb{R}$. If $d = 0$, it is called strictly stable. Gaussian, Cauchy, and Lévy distributions are all stable distributions.

In Mantegna's algorithm, the step length $s$ can be calculated by

$$s = \frac{u}{|\nu|^{1/\beta}}, \tag{14.33}$$

where $u$ and $v$ are drawn from normal distributions. That is,

$$u \sim N(0, \sigma_u^2), \qquad v \sim N(0, \sigma_v^2), \tag{14.34}$$

where

$$\sigma_u = \left\{ \frac{\Gamma(1+\beta)\sin(\pi\beta/2)}{\Gamma[(1+\beta)/2]\ \beta\ 2^{(\beta-1)/2}} \right\}^{1/\beta}, \qquad \sigma_v = 1. \tag{14.35}$$

This distribution (for $s$) obeys the expected Lévy distribution for $|s| \geq |s_0|$ where $s_0$ is the smallest step. In principle, $|s_0| \gg 0$, but in reality $s_0$ can be taken as a sensible value such as $s_0 = 0.1-1$.

### 14.5.3 Advantages of CS

CS has two distinct advantages over other algorithms such as GA and SA, and these advantages are: efficient random walks and balanced mixing. Since Lévy flights are usually far more efficient than any other random-walk-based randomization techniques, CS can be very efficient in global search. In fact, recent studies show that CS can have guaranteed global convergence. In addition, the similarity between eggs can produce better new solutions, which is essentially fitness-proportional generation with a good mixing ability. In other words, CS has varying mutation realized by Lévy flights, and the fitness-proportional generation of new solutions based on similarity provides a subtle form of crossover. In addition, selection is carried out by using $p_a$ where the good solutions are passed onto next generations, while not so good solutions are replaced by new solutions. Furthermore, simulations also show that CS can have auto-zooming ability in the sense that new solutions can automatically zoom into the region where the promising global optimality is located.

Alternatively, we can look at the simplest case of Eq. (14.18) alone, which is essentially the generalized SA in the framework of Markov chains. In Eq. (14.17), if $p_a = 1$ and $\alpha s \in [0, 1]$, CS can degenerate into a variant of differential evolution. Furthermore, if we replace $\boldsymbol{x}_j^t$ by the current best solution $\boldsymbol{g}^*$, then Eq. (14.17) can further degenerate into APSO. This means that SA, DE, and APSO can be considered as special cases of CS in this simplified scenario, and that is one of the reasons why CS is so efficient.

In essence, CS has strong mutation at both local and global scales, while good mixing is carried out by using solution similarity, which also plays the role of equivalent crossover. Selection is done by elitism, that is, a good fraction of solutions will be passed onto the next generation. Without the explicit use of $\boldsymbol{g}_*$, CS may also overcome the premature convergence drawback as observed in PSO.

## 14.6 Bat Algorithm

The metaheuristic bat algorithm (BA) was developed by Xin-She Yang in 2010b. It was inspired by the echolocation behavior of microbats. It is the first algorithm of its kind to use frequency tuning.

Bats are fascinating animals. They are the only mammals with wings and they also have advanced capability of echolocation. It is estimated that there are about 1000 different species which account for up to 20% of all mammal species. Their size ranges from the tiny bumblebee bats (of about 1.5–2 g) to the giant bats with a wingspan of about 2 m and weight up to about 1 kg. Microbats typically have a forearm length of about 2.2–11 cm. Most bats use echolocation to a certain degree; among all the species, microbats are a famous example as microbats use echolocation extensively, while megabats do not.

Most microbats are insectivores. Microbats use a type of sonar, called echolocation, to detect prey, avoid obstacles, and locate their roosting crevices in the dark. These bats emit a very loud sound pulse and listen for the echo that bounces back from the surrounding objects. Their pulses vary in properties and can be correlated with their hunting strategies, depending on the species. Most bats use short, frequency-modulated signals to sweep through about an octave, while others more often use constant-frequency signals for echolocation. Their signal bandwidth varies with species, and often increases by using more harmonics.

Studies show that microbats use the time delay from the emission and detection of the echo, the time difference between their two ears, and the loudness variations of the echoes to build up the three-dimensional scenario of the surrounding. They can detect the distance and orientation of the target, the type of prey, and even the moving speed of the prey such as small insects. Indeed, studies suggested that bats seem to be able to discriminate targets by the variations of the Doppler effect induced by the wing-flutter rates of the target insects.

Though each pulse only lasts a few thousands of a second (up to about 8–10 ms), however, it has a constant frequency which is usually in the region of 25–150 kHz. The typical range of frequencies for most bat species are in the region between 25 and 100 kHz, though some species can emit higher frequencies up to 150 kHz. Each ultrasonic burst may last typically 5–20 ms, and microbats emit about 10–20 such sound bursts every second. When hunting for prey, the rate of pulse emission can be sped up to about 200 pulses per second when they fly near their prey. Such short sound bursts imply the fantastic ability of the signal processing power of bats. In fact, studies show the equivalent integration time of the bat ear is typically about 300–400 μs.

If we idealize some of the echolocation characteristics of microbats, we can develop various bat-inspired algorithms or the BA. For simplicity, we now use the following approximate or idealized rules:

1) All bats use echolocation to sense distance, and they also "know" the difference between food/prey and background barriers.
2) Bats fly randomly with velocity $\boldsymbol{v}_i$ at position $\boldsymbol{x}_i$. They can automatically adjust the frequency (or wavelength) of their emitted pulses and adjust the rate of pulse emission $r \in [0, 1]$, depending on the proximity of their target.
3) Although the loudness can vary in many ways, we can assume that the loudness varies from a large (positive) $A_0$ to a minimum value $A_{\min}$.

Another obvious simplification is that no ray tracing is used in estimating the time delay and three-dimensional topography. Though this might be a good feature for the application in computational geometry, however, we will not use this, as it is more computationally extensive in multidimensional cases.

Each bat is associated with a velocity $\boldsymbol{v}_i^t$ and a location $\boldsymbol{x}_i^t$, at iteration $t$, in a $D$-dimensional search or solution space. Among all the bats, there exists a current best solution $\boldsymbol{x}_*$. Therefore, the above three rules can be translated into the updating equations for $\boldsymbol{x}_i^t$ and velocities $\boldsymbol{v}_i^t$:

$$f_i = f_{\min} + (f_{\max} - f_{\min})\beta, \tag{14.36}$$

$$\boldsymbol{v}_i^t = \boldsymbol{v}_i^{t-1} + (\boldsymbol{x}_i^{t-1} - \boldsymbol{x}_*)f_i, \tag{14.37}$$

$$\boldsymbol{x}_i^t = \boldsymbol{x}_i^{t-1} + \boldsymbol{v}_i^t, \tag{14.38}$$

where $\beta \in [0, 1]$ is a random vector drawn from a uniform distribution. The frequency is tuned in a range of $f_{\min}$ to $f_{\max}$.

The loudness and pulse emission rates are regulated by the following equations:

$$A_i^{t+1} = \alpha A_i^t \tag{14.39}$$

and

$$r_i^{t+1} = r_i^0[1 - \exp(-\gamma t)], \tag{14.40}$$

where $0 < \alpha < 1$ and $\gamma > 0$ are constants. In essence, here $\alpha$ is similar to the cooling factor of a cooling schedule in SA.

Based on the above approximations and idealized rules, the basic steps of the BA can be summarized as the schematic pseudocode, shown in Algorithm 14.4.

BA has been extended to multiobjective bat algorithm (MOBA) by Yang, and preliminary results suggested that it is very efficient. The author has provided a demo code of the BA that can be found at the Mathworks website.[4]

---

4 http://www.mathworks.co.uk/matlabcentral/fileexchange/37582-bat-algorithm–demo-.

**Algorithm 14.4** Pseudocode of the bat algorithm (BA).

Initialize the bat population $\boldsymbol{x}_i$ and $\boldsymbol{v}_i$ $(i = 1, 2, \ldots, n)$
Initialize frequencies $f_i$, pulse rates $r_i$, and the loudness $A_i$
**while** ($t$ < Max number of iterations)
  Generate new solutions by adjusting frequency,
  Update velocities and locations/solutions [(14.36)–(14.38)]
  **if** (rand > $r_i$)
    Select a solution among the best solutions
    Generate a local solution around the selected best solution
  **end if**
  Generate a new solution by flying randomly
  **if** (rand < $A_i$ or $f(\boldsymbol{x}_i) < f(\boldsymbol{x}_*)$)
    Accept the new solutions
    Increase $r_i$ and reduce $A_i$
  **end if**
  Rank the bats and find the current best $\boldsymbol{x}_*$
**end while**

In the BA, frequency tuning essentially acts as mutation, while the selection pressure is relatively constant via the use of the current best solution $\boldsymbol{x}_*$ found so far. There is no explicit crossover; however, mutation varies due to the variations of loudness and pulse emission. In addition, the variations of loudness and pulse emission rates also provide an auto-zooming ability so that exploitation becomes intensive as the search moves are approaching the global optimality.

## 14.7 Flower Pollination Algorithm

The flower pollination algorithm (FPA), or simply flower algorithm, was developed by Xin-She Yang in 2012, inspired by the flower pollination process of flowering plants. It has been extended to multiobjective optimization problems and found to be very efficient.

It is estimated that there are over a quarter of a million types of flowering plants in nature and that about 80% of all plant species are flowering species. It still remains a mystery how flowering plants came to dominate the landscape from the Cretaceous period. Flowering plants have been evolving for at least more than 125 million years and flowers have become so influential in evolution, it is unimaginable what the plant world would look like without flowers. The main purpose of a flower is ultimately reproduction via pollination. Flower pollination is typically associated with the transfer of pollen, and such transfer is often linked with pollinators such as insects, birds, bats, and other animals. In fact, some flowers and insects have co-evolved into a very specialized flower–pollinator partnership. For example, some flowers can only attract and can only depend on a specific species of insects or birds for successful pollination.

Pollination can take two major forms: abiotic and biotic. About 90% of flowering plants belong to biotic pollination. That is, pollen is transferred by pollinators such as insects and animals. About 10% of pollination takes abiotic form which does not require any pollinators. Wind and diffusion help pollination of such flowering plants, and grass is a good example of abiotic pollination. Pollinators, or sometimes called pollen vectors, can be very diverse. It is estimated there are at least about 200 000 varieties of pollinators such as insects, bats, and birds. Honeybees are a good example of pollinators, and they have also developed the so-called flower constancy. That is, these pollinators tend to visit exclusively certain flower species while bypassing other flower species. Such flower constancy may have evolutionary advantages because this will maximize the transfer of flower pollen to the same or conspecific plants, and thus maximizing the reproduction of the same flower species. Such flower constancy may be advantageous for pollinators as well, because they can be sure that nectar supply is available with their limited memory and minimum cost of learning, switching, or exploring. Rather than focusing on some unpredictable but potentially more rewarding new flower species, flower constancy may require minimum investment costs and more likely guaranteed intake of nectar.

Flower pollination can be achieved by self-pollination or cross-pollination. Cross-pollination, or allogamy, means pollination can occur from pollen of a flower of a different plant, while self-pollination is the fertilization of one flower, such as peach flowers, from pollen of the same flower or different flowers of the same plant, which often occurs when there is no reliable pollinator available. Biotic, cross-pollination may occur at long distance, and the pollinators such as bees, bats, birds, and flies can fly a long distance, thus they can be considered as the global pollination. In addition, bees and birds may behave as Lévy flight behavior with jumps or flight distance steps obeying a Lévy distribution. Furthermore, flower constancy can be considered as an increment step using the similarity or difference of two flowers.

From the biological evolution point of view, the objective of the flower pollination is the survival of the fittest and the optimal reproduction of plants in terms of numbers as well as the most fittest. This can be considered as an optimization process of plant species. All the above factors and processes of flower pollination interact so as to achieve optimal reproduction of the flowering plants. Therefore, this may motivate us to design new optimization algorithms.

For simplicity in describing the FPA, we use the following four rules:

1) Biotic and cross-pollination can be considered as a process of global pollination process, and pollen-carrying pollinators move in a way which obeys Lévy flights (Rule 1).
2) For local pollination, abiotic pollination and self-pollination are used (Rule 2).
3) Pollinators such as insects can develop flower constancy, which is equivalent to a reproduction probability that is proportional to the similarity of two flowers involved (Rule 3).
4) The interaction or switching of local pollination and global pollination can be controlled by a switch probability $p \in [0, 1]$, with a slight bias toward local pollination (Rule 4).

These basic rules can be summarized as the basic steps as shown in Algorithm 14.5. In order to formulate proper updating formulae, we have to convert the above rules into mathematical equations. For example, in the global pollination step, flower pollen gametes are carried by pollinators such as insects, and pollen can travel over a long distance because insects can often fly and move in a much longer range. Therefore, Rule 1 and flower constancy can be represented mathematically as

$$\boldsymbol{x}_i^{t+1} = \boldsymbol{x}_i^t + \gamma L(\lambda)(\boldsymbol{g}_* - \boldsymbol{x}_i^t), \tag{14.41}$$

where $\boldsymbol{x}_i^t$ is the pollen $i$ or solution vector $\boldsymbol{x}_i$ at iteration $t$, and $\boldsymbol{g}_*$ is the current best solution found among all solutions at the current generation/iteration. Here, $\gamma$ is a scaling factor to control the step size.

Here, $L(\lambda)$ is a random step size that corresponds to the strength of the pollination. Since insects may move over a long distance with various distance steps, we can use a Lévy flight to mimic this characteristic efficiently. That is, we draw random numbers or steps $L$ from a Lévy distribution

$$L \sim \frac{\lambda\Gamma(\lambda)\sin(\pi\lambda/2)}{\pi} \frac{1}{s^{1+\lambda}} \quad (s \gg s_0 > 0). \tag{14.42}$$

Here, $\Gamma(\lambda)$ is the standard gamma function, and this distribution is valid for large steps $s > 0$. This step is essentially a global mutation step, which enables to explore the search space more efficiently. In addition, $s_0$ is a step size constant. In theory, $s_0$ should be sufficiently large (based on mathematical approximations), but in practice $s_0 = 0.001 - 0.1$ can be used.

For the local pollination, both Rule 2 and Rule 3 can be represented as

$$\boldsymbol{x}_i^{t+1} = \boldsymbol{x}_i^t + \epsilon(\boldsymbol{x}_j^t - \boldsymbol{x}_k^t), \tag{14.43}$$

where $\boldsymbol{x}_j^t$ and $\boldsymbol{x}_k^t$ are pollen from different flowers of the same plant species. This essentially mimics the flower constancy in a limited neighborhood. Mathematically speaking, if $\boldsymbol{x}_j^t$ and $\boldsymbol{x}_k^t$ come from the same species or are selected from

**Algorithm 14.5** Pseudocode of the proposed Flower Pollination Algorithm (FPA, or simply Flower Algorithm).

```
Objective min or max f(x), x = (x_1, x_2, ..., x_D) ∈ ℝ^D
Initialize randomly a population of n flowers, and set p ∈ [0, 1]
Find the best solution g_* in the initial population
while (t < MaxGeneration)
    for i = 1 : n (all n flowers in the population)
      if rand < p,
        Draw a step vector L from a Lévy distribution
        Global pollination via x_i^{t+1} = x_i^t + γL(g_* − x_i^t)
      else
        Draw ε from a uniform distribution in [0,1]
        Do local pollination via x_i^{t+1} = x_i^t + ε(x_j^t − x_k^t)
      end if
      Evaluate new solutions
      If new solutions are better, update them in the population
    end for
      Find the current best solution g_*
end while
Output the best solution found
```

the same population, this equivalently becomes a local random walk if we draw $\epsilon$ from a uniform distribution in [0,1]. In essence, this is a local mutation and mixing step, which can help the algorithm to converge.

In principle, flower pollination activities can occur at all scales, both local and global. But in reality, adjacent flower patches or flowers in the not-so-far-away neighborhood are more likely to be pollinated by local flower pollen than those far away. In order to mimic this feature, we can effectively use a switch probability (Rule 4) or proximity probability $p$ to switch between common global pollination to intensive local pollination. To start with, we can use a naive value of $p = 0.5$ as an initial value. A preliminary parameter showed that $p = 0.8$ may work better for most applications.

The author has provided a demo code of the basic FPA that can be downloaded from the Mathworks website.[5]

Selection is achieved by selecting the best solutions and passing onto the next generation. It also explicitly uses $\boldsymbol{g}_*$ to find the best solution as both selection and elitism. There is no explicit crossover, which is also true for many other algorithms such as PSO and harmony search.

5 http://www.mathworks.co.uk/matlabcentral/fileexchange/45112.

All the algorithms mentioned in this chapter have been applied to a diverse range of real-world applications, and there are many specialized research articles and reviews about them. Interested readers can refer to more specialized literature in academic journals.

## 14.8 Other Algorithms

Many other algorithms have appeared in the literature, including artificial immune system, bacteria foraging optimization, cultural algorithm, eagle strategy, gravitational search, harmony search, memetic algorithm, plant propagation algorithm, shuffled frog leaping algorithm, and many others. However, as this is not the main focus of this book, we will not delve into more details about these algorithms. Interested readers can refer to more advanced literature.

As nature-inspired algorithms have become hugely popular, the relevant literature is vast at all levels. There are dozens of research level books and thousands of journal papers. Interested readers can refer to more advanced edited books listed in the bibliography section of this book.

## Exercises

**14.1** In the PSO, write Eqs. (14.2) and (14.4) in a matrix form and discuss the behavior using dynamical system theory.

**14.2** In the BA, can the frequency $f_i$ in Eq. (14.36) be negative?

**14.3** In the FA, discuss the role and values of parameter $\alpha$, $\beta_0$, and $\gamma$.

**14.4** Use any programming to generate 100 random numbers that obey a Lévy distribution.

## Bibliography

Clerc, M. and Kennedy, J. (2002). The particle swarm – explosion, stability and convergence in a multidimensional complex system. *IEEE Transactions on Evolutionary Computation* 6 (1): 58–73.

Davies, N.B. and Brooke, M.L. (1991). Co-evolution of the cuckoo and its hosts. *Scientific American* 264 (1): 92–98.

Dorigo, M. (1992). *Opimization, learning and natural algorithms*. PhD thesis. Politecnico di Milano, Italy.

Engelbrecht, A.P. (2005). *Fundamentals of Computational Swarm Intelligence*. Hoboken, NJ: Wiley.

Geem, Z.W., Kim, J.H., and Loganathan, G.V. (2001). A new heuristic optimization algorithm: harmony search. *Simulation* 76 (1): 60–68.

Gutowski, M. (2001). Lévy flights as an underlying mechanism for global optimization algorithms. *ArXiv Mathematical Physics e-Prints* https://arxiv.org/abs/math-ph/0106003 (accessed 28 February 2018).

He, X.S., Yang, X.S., Karamanoglu, M. et al. (2017). Global convergence analysis of the flower pollination algorithm: a discrete-time Markov chain approach. *Procedia Computer Science* 108 (1): 1354–1363.

Karaboga, D. and Basturk, B. (2008). On the performance of artificial bee colony (ABC) algorithm. *Applied Soft Computing* 8 (1): 687–697.

Kaveh, A. and Talatahari, S. (2010). A novel heuristic optimization method: charged system search. *Acta Mechanica* 213 (3–4): 267–289.

Kennedy, J. and Eberhart, R.C. (1995). Particle swarm optimization. In: *Proceedings of IEEE International Conference on Neural Networks*, 1942–1948 Piscataway, NJ: IEEE.

Kennedy, J. and Eberhart, R.C. (1997). A discrete binary version of the particle swarm algorithm. *International Conference on Systems, Man, and Cybernetics*, Orlando, FL (12–15 October 1997), vol. 5, 4104–4109. Piscataway, NJ: IEEE Publications.

Kennedy, J., Eberhart, E.C. and Shi, Y. (2001). *Swarm Intelligence*. London: Academic Press.

Mantegna, R.N. (1994). Fast, accurate algorithm for numerical simulation of Lévy stable stochastic processes. *Physical Review E* 49: 4677–4683.

Moritz, R.F. and Southwick, E. (1992). *Bees as Superorganisms*. Berlin: Springer.

Nolan, J.P. (2009). *Stable Distributions: Models for Heavy-Tailed Data*. Washington, DC: American University.

Pavlyukevich, I. (2007). Lévy flights, non-local search and simulated annealing. *Journal of Computational Physics* 226 (2): 1830–1844.

Reynolds, A.M. and Frye, M.A. (2007). Free-flight odor tracking in drosophila is consistent with an optimal intermittent scale-free search. *PLoS One* 2 (4): e354–e363.

Richardson, P. (2008). *Bats*. London: Natural History Museum.

Waser, N.M. (1986). Flower constancy: definition, cause and measurement. *The American Naturalist* 127 (5): 596–603.

Yang, X.-S. (2008). *Nature-Inspired Metaheuristic Algorithms*. Frome, UK: Luniver Press.

Yang, X.S. (2010a). Firefly algorithm, stochastic test functions and design optimisation. *International Journal of Bio-Inspired Computation* 2 (2): 78–84.

Yang, X.S. (2010b). A new metaheuristic bat-inspired algorithm. In: *Nature-Inspired Cooperative Strategies for Optimization* (NICSO 2010), SCI 284, 65–74. Heidelberg: Springer.
Yang, X.S. (2010c). *Engineering Optimization: An Introduction with Metaheuristic Applications.* Hoboken, NJ: Wiley
Yang, X.S. (2011). Bat algorithm for multi-objective optimisation. *International Journal of Bio-Inspired Computation* 3 (5): 267–274.
Yang, X.S. (2012). Flower pollination algorithm for global optimization. In: *Unconventional Computation and Natural Computation.* Lecture Notes in Computer Science, vol. 7445, 240–249. Berlin: Springer.
Yang, X.S. (2014a). *Cuckoo Search and Firefly Algorithm: Theory and Applications,* Studies in Computational Intelligence, vol. 516. Heidelberg: Springer.
Yang, X.S. (2014b). *Nature-Inspired Optimization Algorithms.* London: Elsevier Insight.
Yang, X.S. and Deb, S. (2009). Cuckoo search via Lévy flights. In: *Proceedings of World Congress on Nature & Biologically Inspired Computing* (NaBic 2009), Coimbatore, India 210–214. New York: IEEE Publications.
Yang, X.S. and Deb, S. (2010). Engineering optimization by cuckoo search. *International Journal of Mathematical Modelling and Numerical Optimisation* 1 (4): 330–343
Yang, X.S. and Deb, S. (2014). Cuckoo search: recent advances and applications. *Neural Computing and Applications* 24 (1): 169–174.
Yang, X.S. and He, X.S. (2018). Why the firefly algorithm works?. In: *Nature-Inspired Algorithms and Applied Optimization* (ed. X.S. Yang), Studies in Computational Intelligence, vol. 744, 245–259. Cham: Springer Nature.
Yang, X.S., Karamanoglu, M., and He, X.S. (2014). Flower pollination algorithm: a novel approach for multiobjective optimization. *Engineering Optimization* 46 (9): 1222–1237.
Yang, X.S. and Papa, J.P. (2016). *Bio-Inspired Computation and Applications in Image Processing.* London: Academic Press.

# Appendix A

# Notes on Software Packages

This appendix provides a brief note about the software packages related to optimization, data mining, and machine learning.

## A.1 Software Packages

As optimization concerns many applications, most software packages and programming languages will have some sort of optimization capabilities. There is a vast list of software packages, it requires a lengthy book to cover most of it, which is not our intention here.

There are so many available software packages, it is difficult to choose an appropriate one for beginners. The choice can largely depend on the availability of a package or programming language, ease of usage, and the personal expertise. For example, wikipedia has some extensive lists of

- optimization software,[1]
- data mining and machine learning,[2]
- deep learning software.[3]

Interested readers can refer to them for more details. Here, we will only highlight a few software packages that are either open sources, free or commercial, or easy to learn and use.

It is worth pointing out that sometime there is no clear distinction between the functions of optimization, data mining, and machine learning because many software packages integrate some of these algorithms into a single framework or toolbox.

Commonly used packages and easily available/accessible packages include R, Python, Matlab, Microsoft Excel Solver, OpenSolver, Maple, Mathematica, COIN, Octave, and others.

---

1 https://en.wikipedia.org/wiki/List_of_optimization_software.
2 https://en.wikipedia.org/wiki/Category:Data_mining_and_machine_learning_software.
3 https://en.wikipedia.org/wiki/Comparison_of_deep_learning_software.

*Optimization Techniques and Applications with Examples*, First Edition. Xin-She Yang.

## A.2 Matlab Codes in This Book

There are some Matlab demo codes for most of the nature-inspired algorithms discussed in this book. You can find them from the website of this book at Wiley.com. They are also available from Matlab file exchanges,[4] including

- accelerated particle swarm optimization,[5]
- firefly algorithm,[6]
- cuckoo search,[7]
- flower pollination algorithm,[8]
- multiobjective bat algorithm.[9]

It is worth pointing out that these codes are demo and incomplete codes. The reason is that such demo codes focus on the essential steps of the algorithms without any messy implementation of handling constraints. However, the performance of such concentrated demo codes may be reduced as the proper constraint-handling is an important part of practical applications. These codes should still work reasonably well for solving function optimization problems. This gives readers an opportunity to understand the basic algorithms and potentially improve them.

## A.3 Optimization Software

As we mentioned earlier, many software packages and programming languages have implemented some limited optimization capabilities, while commercial software packages tend to have well-tested toolboxes. It is not our intention to provide a comprehensive list of toolboxes and functionalities, we thus only intend to provide some flavor and diversity of a few software packages or programming languages.

- Matlab: The optimization toolboxes of Matlab includes linear programming `linprog`, integer programming `intlinprog`, nonlinear programming such as `fminsearch` and `fmincon`, quadratic programming `quadprog`, and multiobjective optimization by genetic algorithm `gamultiobj`.

---

4 https://uk.mathworks.com/matlabcentral/profile/authors/2652824-xin-she-yang.
5 https://uk.mathworks.com/matlabcentral/fileexchange/29725-accelerated-particle-swarm-optimization.
6 https://uk.mathworks.com/matlabcentral/fileexchange/29693-firefly-algorithm.
7 https://uk.mathworks.com/matlabcentral/fileexchange/29809-cuckoo-search–cs–algorithm.
8 https://uk.mathworks.com/matlabcentral/fileexchange/45112-flower-pollination-algorithm.
9 https://uk.mathworks.com/matlabcentral/fileexchange/45118-multi-objective-bat-algorithm–moba.

- Octave has many functionalities similar to Malab, and it is an open-source package. Its optimization toolbox `optim` has implemented linear programming, quadratic programming, nonlinear programming, and linear least squares.
- R has a relatively general purpose optimization solver `optimr` with `optim()` using conjugate gradient, Nelder–Mead method, Broyden–Fletcher–Goldfarb-Shanno (BFGS) method, and simulated annealing. It also has a quadratic programming `solve.QP()` and least-squares solver `solve.qr()` as well as metaheuristic optimization such as the firefly algorithm.
- Python does have good optimization capabilities via `scipy.optimize()`, which includes the BFGS method, conjugate gradient, Newton's method, trust-region method, and least-square minimization.
- Mathematica is a commercial symbolic computation package, it has powerful functionalities for optimization, including nonlinear constrained global optimization `NMiminize` or `NMaximize`, linear programming and integer programming `LinearProgramming`, Knapsack solver `KnapsackSolve`, traveling salesman problem `FindShortesTour`, and others.
- Maple is mainly a symbolic and numerical computing tool with some functions for optimization such as `Minimize`, linear programming `LPSolve`, and nonlinear programming `NLPSolve`.
- Microsoft Excel Solver can do linear programming, integer programming, generalized reduced gradient, evolutionary algorithms (via genetic algorithm), but its standard version is usually limited to the number of variables less than 200. Its Premium Solver has a higher limit, less than about 2000 decision variables. On the other hand, the OpenSolver is free and has no limit on the number of variables. Its core algorithmic engine is COIN-OR linear and integer programming optimizers, thus OpenSolver is very powerful and efficient.

Other powerful optimization tools include the computational infrastructure for Operations Research (COIN-OR), also known as common optimization interface for OR. Many software packages use it as a core optimization engine.

## A.4 Data Mining Software

There are equally a wide spectrum of software packages for data mining and big data analytics. For example, RapidMiner is a very powerful tool.

- Matlab: Matlab has some well-tested data mining tools such as hierarchy clustering `cluster`, $k$-means clustering `kmeans`, $k$ nearest neighbor clustering `knnsearch`, fuzzy data clustering `findcluster`, and others.

- Octave has some statistical functions and regression analysis such as logistic regression `logistic-regress()`.
- R has many functions for processing and visualizing data, including linear regression `lm()`, $k$-means clustering `kmeans()`, hierarchy clustering `hclust()`, and text mining `tm()`. It also has a framework for handling big data via RHadoop, SparkR, or H2O packages.
- Python has a module Scikit learn `sklearn` for data mining and machine learning, which can do $k$-means clustering and basic clustering as well as various statistical analysis.
- Mathematica can also do data mining with interface with Excel and databases, it can also do clustering via `FindClusters` with various distance metrics.

Maple has limited data mining capabilities such as text mining and interfacing with SQL. In addition, some good open-source data mining tools include RapidMiner, DataMelt, OpenNN, and others.

## A.5 Machine Learning Software

There are many powerful machine learning, especially deep learning, software packages. Good examples include the well-known AlexNet and Google's TensorFlow as well as Microsoft's cognitive toolkit.

- Matlab: Matlab has many functions that can be used for machine learning from simple regression of polynomials `polyfit` and nonlinear regression `nlinfit` to the method of nonlinear least-squares `lsqnonlin`, and nonlinear curve fitting in general `lsqcruvefit`. It has also well-documented support vector regression `fitrsvm`, artificial neural network `nntool`, and support vector machine classifier `fitcsvm`. In recent years, it has deep learning capabilities, including the deep nets `alexnet`, transfer learning `convnets`, Google net `googlenet`, in addition to their many image-processing functionalities.
- Octave has limited image-processing functions, and some of the open-source deep learning toolboxes for Matlab can be potentially portable to Octave.
- R has various projects and packages for machine learning, including regression and Lasso `lasso2`, elastic net regularization `elasticnet`, the single hidden layer neural network `nnet`, deep learning `deepnet`, restricted Boltzmann machine `RcppDL`, and even an interface to `tensorflow`. It also has support vector machines `svm()` and many others.
- Python can do regression by ordinary least squares `ols()`. Scikitlearn in Python can do dimensionality reduction, model selection, and classification by support vector machines, and supervised learning by neural networks.

- Mathematica has powerful machine learning capabilities, including the optimal nonlinear fit to data `FindFit`, least-squares method `LeastSquares`, automatic training and classification using neural networks and support vector machine `Classify`, feature extraction `FeatureExtraction`, and many others such as Naive Bayesian networks, decision tree, and random forests.

Obviously, the above lists are just a snapshot of the current software packages that are related to optimization, data mining, and machine learning. There are a vast range of other packages that we have not mentioned. Interested readers can refer to the relevant lists on the web listed earlier in this appendix.

# Appendix B

# Problem Solutions

## Solutions for Chapter 1

**1.1** From $f(x) = \sin(x)/x$ and differentiation rule (quotient rule or chain rule), we have

$$f'(x) = \frac{\cos(x)}{x} - \frac{\sin(x)}{x^2}.$$

The second derivative is

$$f''(x) = \frac{2\sin(x)}{x^3} - \frac{x\sin(x) + 2\cos(x)}{x^2}.$$

**1.2** The gradient vector of $f$ is

$$\begin{aligned} \boldsymbol{G} &= \left(\frac{\partial f}{\partial x}, \frac{\partial f}{\partial y}, \frac{\partial f}{\partial z}\right) \\ &= (2y + 2x, 2x + 2y + 3z, 3y + 3z^2). \end{aligned}$$

The Hessian matrix is

$$\boldsymbol{H} = \begin{pmatrix} f_{xx} & f_{xy} & f_{xz} \\ f_{yx} & f_{yy} & y_{yz} \\ f_{zx} & f_{zy} & f_{zz} \end{pmatrix} = \begin{pmatrix} 2 & 2 & 0 \\ 2 & 2 & 3 \\ 0 & 3 & 6z \end{pmatrix},$$

which is indeed symmetric.

**1.3** Using integration by parts three times, we have

$$\int_0^\infty x^3 \mathrm{e}^{-x}\mathrm{d}x = \left[-(6 + 6x + 3x^2 + x^3)\mathrm{e}^{-x}\right]_0^\infty = 0 - [-6\mathrm{e}^{-0}] = 6.$$

**1.4** The eigenvalues of $\boldsymbol{A}$ can be obtained by

$$\det\begin{vmatrix} 2-\lambda & 3 \\ 3 & 4-\lambda \end{vmatrix} = 0,$$

which gives

$$(2-\lambda)(4-\lambda) - 3^2 = 0$$

*Optimization Techniques and Applications with Examples*, First Edition. Xin-She Yang.

or

$$\lambda^2 - 6\lambda - 1 = 0,$$

whose solutions are

$$\lambda_1 = 3 + \sqrt{10}, \quad \lambda_2 = 3 - \sqrt{10}.$$

Their corresponding eigenvectors are

$$\boldsymbol{u}_1 = \begin{pmatrix} 1 \\ \frac{1+\sqrt{10}}{3} \end{pmatrix}, \quad \boldsymbol{u}_2 = \begin{pmatrix} 1 \\ \frac{1-\sqrt{10}}{3} \end{pmatrix}.$$

**1.5** The eigenvalues can be obtained by

$$\det \begin{vmatrix} 2-\lambda & 2 & 0 \\ 2 & 2-\lambda & 3 \\ 0 & 3 & 6-\lambda \end{vmatrix} = 0,$$

which gives

$$\begin{aligned} (2-\lambda)\begin{vmatrix} 2-\lambda & 3 \\ 3 & 6-\lambda \end{vmatrix} - 2\begin{vmatrix} 2 & 3 \\ 0 & 6-\lambda \end{vmatrix} + 0\begin{vmatrix} 2 & 2-\lambda \\ 0 & 3 \end{vmatrix} \\ = (2-\lambda)[(2-\lambda)(6-\lambda) - 9] - 2[2(6-\lambda) - 0] + 0 \\ = (3-\lambda)(\lambda^2 - 7\lambda - 6) = 0. \end{aligned}$$

Thus, we have three eigenvalues $\lambda_1 = 3$ and $\lambda_{2,3} = (7 \pm \sqrt{73})/2$.

Since two eigenvalues are positive and one is negative, the matrix is indefinite.

**1.6** It is easy to observe that both terms $(x-1)^2$ and $x^2y^2$ are always nonnegative. Thus, the minimum must occur at $x - 1 = 0$ and $y = 0$. Alternatively, the first derivatives of $f$ are

$$\frac{\partial f}{\partial x} = 2(x-1) + 2xy^2 = 0, \quad \frac{\partial f}{\partial y} = 0 + 2x^2y = 0.$$

The second condition gives $y = 0$ or $x = 0$. Substituting $y = 0$ into the first condition, we have $x = 1$. However, the other solution $x = 0$ does not satisfy the first condition. Thus, the only valid solution is $x_* = 1$ and $y_* = 0$.

In addition, the Hessian matrix of $f(x, y)$ is

$$\boldsymbol{H} = \begin{pmatrix} f_{xx} & f_{xy} \\ f_{yx} & f_{yy} \end{pmatrix} = \begin{pmatrix} 2y^2 + 2 & 4xy \\ 4xy & 2x^2 \end{pmatrix}.$$

At $x_* = 1$ and $y_* = 0$, we have

$$\boldsymbol{H}_* = \begin{pmatrix} 2 & 0 \\ 0 & 2 \end{pmatrix},$$

which is positive definite (because both eigenvalues are 2). Therefore, the point at $(1, 0)$ corresponds to a minimum.

## Solutions for Chapter 2

**2.1** The simplified expressions are as follows:

- $f(n) = O(20n^3 + 20\log(n) + n\log(n)) = O(20n^3) = O(n^3)$.
- $g(n) = O(5n^{2.5} + n^2 + 10n) = O(5n^{2.5}) = O(n^{2.5})$.
- $f(n) + 2g(n) = O(n^3) + 2O(n^{2.5}) = O(n^3)$.

**2.2** The answers are as follows:

- $f(x) = x^4$ is convex because $f''(x) = 12x^2 \geq 0$ for $\forall x$. However, it is not strictly convex as $f''(x) = 0$ at $x = 0$. The derivative of $f(x) = |x|$ at $x = 0$ does not exist. However, we know its $f''(x) = 0$ for $x < 0$, and $f''(0) = 0$ for $x > 0$. Thus, $|x|$ is convex, not strictly convex.
- From $f(x) = e^{-x}$, we know that $f''(x) = e^{-x}$ is positive for all $x$. So $e^{-x}$ is strictly convex.
- For $f(x) = 1/x$, we have $f''(x) = 2/x^3$. Thus, $1/x$ is convex when $x > 0$ as $f''(x) > 0$. For $g(x) = 1/x^4$, we have $g''(x) = 20/x^6$, which is always positive except $x = 0$ (singularity). This means is convex on $(-\infty, 0)$ and $(0, \infty)$, not on $(-\infty, +\infty)$.
- $f(x, y, z) = x^2 + y^2 + z^2$ are the simple sum of three convex functions, so it is convex. It can also be tested that its Hessian has three positive eigenvalues.

**2.3** The mean or expectation $E(X)$ can be calculated by

$$E(X) = \sum_{x=0}^{\infty} xp(x) = \sum_{x=0}^{\infty} x\frac{\lambda^x e^{-\lambda}}{x!}$$
$$= 0 \times e^{-\lambda} + 1 \times (\lambda e^{-\lambda}) + 2 \times \left(\frac{\lambda^2 e^{-\lambda}}{2!}\right) + 3 \times \left(\frac{\lambda^3 e^{-\lambda}}{3!} + \cdots\right)$$
$$= \lambda e^{-\lambda}\left[1 + \lambda + \frac{\lambda^2}{2!} + \frac{\lambda^3}{3!} + \cdots\right] = \lambda e^{-\lambda} e^{+\lambda} = \lambda.$$

**2.4** From $p(x) = x/\sigma^2 e^{-x^2/(2\sigma^2)}$ when $x \geq 0$, we have its cumulative probability distribution

$$F(x) = \int_{-\infty}^{x} p(u)du = \frac{1}{\sigma^2}\int_0^x ue^{-u^2/(2\sigma^2)}du$$
$$= \frac{1}{\sigma^2}\int_0^x \frac{2\sigma^2}{2} e^{-u^2/(2\sigma^2)} d\left(\frac{u^2}{2\sigma^2}\right) = -\left[e^{-u^2/(2\sigma^2)}\right]_0^x$$
$$= 1 - \exp\left(-\frac{x^2}{2\sigma^2}\right).$$

Similarly, its mean can be calculated by

$$<x>= \int_0^{\infty} xp(x)dx = \frac{1}{\sigma^2}\int_0^{\infty} x^2 e^{-x^2/(2\sigma^2)}dx.$$

Using $\int_0^\infty x^2 e^{-x^2/(2\sigma^2)} = \sigma^3 \sqrt{\pi/2}$, we have

$$<x>= \frac{1}{\sigma^2}\sqrt{\frac{\pi}{2}}\sigma^3 = \sigma\sqrt{\pi/2}.$$

**2.5** A simple Monte Carlo simulation is to generate some $n$ random numbers that are uniformly distributed in a square with $a = 2r$. Then, count the number $m$ of points that are inside a unit circle $r = 1$. As the area of the unit circle is $\pi$ for $r = 1$, the area of the square is 4. Thus, the ratio $4m/n$ should be an estimate of $\pi$ when $n$ is sufficiently large. As the accuracy of Monte Carlo is $O(1/\sqrt{n})$, $n = 10^4$ should be used to get two significant digits. This can be easily realized by any programming language.

## Solutions for Chapter 3

**3.1** Since $f'(x) = 3x^2 - 15x + 12 = 3(x-1)(x-4) = 0$, we know its stationary points are $x* = 1$ and $x* = 4$. In addition, we have $f''(x) = 6x - 15$. At $x* = 1$, $f''(1) = -9$, thus $f(1) = 12.5$ is a local maximum. At $x* = 4$, we have $f''(4) = +9$, thus $f(4) = -1$ is a local minimum. However, they are not global maximum or minimum because $f(-10) = -1863$ and $f(10) = 377$. Thus, the global maximum occurs at $x = 10$, while the global minimum occurs at $x = -10$.

**3.2** The partial derivatives of $f(x, y)$ are

$$\frac{\partial f}{\partial x} = 4x^3 + 4xy = 0, \quad \frac{\partial f}{\partial y} = 2x^2 + 2y = 0,$$

which give $x = 0$ and $y = 0$. In addition, the Hessian matrix of $f(x, y)$ is

$$H = \begin{pmatrix} f_{xx} & f_{xy} \\ f_{yx} & f_{yy} \end{pmatrix} = \begin{pmatrix} 12x^2 + 4y & 4x \\ 4x & 2 \end{pmatrix},$$

which at (0, 0) becomes

$$H = \begin{pmatrix} 0 & 0 \\ 0 & 2 \end{pmatrix}.$$

Its eigenvalues are 0 and 2, thus $H$ at (0,0) is positive semidefinite. The optimal solution at (0,0) is a local minimum.

Alternatively, if we rewrite $f(x, y) = (x^2 + y)^2$, we can conclude that $x^2 + y = 0$ gives the minimum. However, this minimum occurs at not just one point, it occurs on a whole curve traced by $y = -x^2$. This highlights that we should be careful when solving stationary conditions. In fact, apart from $x = 0$ and $y = 0$, both partial derivatives have a factor $(x^2 + y)$, which corresponds to solutions at $y = -x^2$. In this case, the Hessian becomes

$$H = \begin{pmatrix} 12x^2 + 4y & 4x \\ 4x & 2 \end{pmatrix} = \begin{pmatrix} 8x^2 & 4x \\ 4x & 2 \end{pmatrix},$$

whose eigenvalues are 0 and $8x^2 + 2$, which is still positive semidefinite.

**3.3** As explained in the previous question, the solutions correspond to a spatial curve, not a single point. Though it is possible to use a software package to find a solution, it may be difficult to find this curve.

## Solutions for Chapter 4

**4.1** You can solve it using the method of Lagrange multiplier. The stationary conditions should lead to three solutions

$$(x_*, y_*) = (1, -1), \quad (-2, 2), \quad \left(\frac{-1}{2}, \frac{-7}{4}\right).$$

The solution (−1/2,−7/4) corresponds to a maximum $f_{\max} = 81/16$, while the two solutions (1,−1) and (−2,2) correspond to two minima with $f_{\min} = 0$.

Alternatively, you can rewrite $f(x, y) = (x + y)^2$ subject to $y - x^2 + 2 = 0$. It is obvious that minima should occur at $x + y = 0$. Thus, $y = -x$ and $y - x^2 + 2 = 0$ should give $x = 1$ or $x = -2$. How do you find the maximum at $(-1/2, -7/4)$ in this case?

**4.2** This function is not differentiable at $(0, 0)$, and this singularity corresponds to a kink point. Thus, it may pose some difficulty using any gradient-based method. Fortunately, the kink point at $(0, 0)$ corresponds to the minimum point $f_{\min} = 0$, as $|x| + |y| > 0$ and $e^{-x^2-y^2} > 0$ if $x \neq 0$ and $y \neq 0$.

As we intend to find the maximum value of $f$, we can simply exclude the only singularity at $(0, 0)$, and the function is differentiable in the rest of the domain. In addition, $f$ is symmetric with respect to $x$ and $y$, so we can simply only consider one quadrant $x > 0, y > 0$. The stationary conditions lead to

$$\frac{\partial f}{\partial x} = [1 - 2x(x + y)]e^{-x^2-y^2} = 0,$$

$$\frac{\partial f}{\partial y} = [1 - 2y(x + y)]e^{-x^2-y^2} = 0.$$

The symmetry implies that the optimality requires $x_* = y_*$, or

$$1 - 4x_*^2 = 0,$$

or $x_* = \pm 1/2$. Therefore, the maximum $f_{\max} = e^{-1/2}$ occurs at $(1/2, 1/2)$, $(1/2, -1/2)$, $(-1/2, 1/2)$, and $(-1/2, -1/2)$.

**4.3** This problem can be solved using a slack variable $t$ to turn the inequality $x^2 + y^2 \leq 1$ into an equality $x^2 + y^2 + t^2 = 1$. Then, we can use the method of Lagrange multipliers to solve it. Alternatively, since the inequality corresponds to a region inside a unit circle, we can solve the unconstrained problem $\min f(x, y) = x^2 + 5y^2 + xy$, which gives $f_{\min} = 0$ at (0,0). As this point is inside the unit circle (satisfying the inequality automatically), it is thus the global minimum inside the feasible domain.

## Solutions for Chapter 5

**5.1** The conversion is related to the epigraph method by introducing an extra variable $t$ such that

$$\text{minimize } t,$$
$$\text{subject } f(\boldsymbol{x}) - t \le 0, \quad g(\boldsymbol{x}) \le 0, \quad h(\boldsymbol{x}) = 0.$$

It is straightforward to verify that the new problem with a linear objective is indeed equivalent to the original problem

$$\text{minimize } f(\boldsymbol{x}),$$
$$\text{subject } g(\boldsymbol{x}) \le 0, \qquad h(\boldsymbol{x}) = 0.$$

**5.2** The constrained optimization problem can be written as an unconstrained one

$$\min L = x^2 + (y-1)^2 - \mu[\log(9 - x^2 - y^2) + \log(y - x^2 - 2)],$$

which can be solved by setting

$$\frac{\partial L}{\partial x} = 2x + 2\mu x \left[\frac{1}{9 - x^2 - y^2} + \frac{1}{y - x^2 - 2}\right] = 0,$$

$$\frac{\partial L}{\partial y} = 2(y-1) + \mu \left[\frac{2y}{9 - x^2 - y^2} - \frac{1}{y - x^2 - 2}\right] = 0,$$

whose solution will depend on $\mu$. After some lengthy algebra and discussions of different cases of $\mu$, we can conclude that the optimal solution is $f_{\min} = 1$ at $x_* = 0$ and $y_* = 2$. Even with an additional constraint is added, the optimal solution remains the same because this new constraint is satisfied automatically.

**5.3** It is easy to verify that the optimal solution is $f_* = 0$ at $\boldsymbol{x}_* = (0, 0, \dots, 0)$. A good program should produce an approximate solution with a certain accuracy, say $10^{-5}$.

## Solutions for Chapter 6

**6.1** In linear programming, all problems are linear, which is a special case of convex functions. Therefore, LP is a special case of convex optimization. However, this does not mean that all methods that work well for convex optimization can be applied to solve LP. In fact, LP problems with integer variables can be very challenging to solve.

**6.2** We can use the graph method to show that the optimal solution occur at $x_* = 1$ and $y_* = 6$ with $P_{\max} = 27$. If the constraint $2x + 3y \leq 20$ is replaced by $2x + 3y \leq 40$, the optimal solution will not change because this constraint is not active and it lies outside the feasible domain. This becomes clearer when plotting these constraints on a graph.

**6.3** Using the following notations

$$a = \begin{pmatrix} 3 \\ 4 \end{pmatrix}, \quad \boldsymbol{x} = \begin{pmatrix} x \\ y \end{pmatrix},$$

we have the primal

$$\text{maximize } f(\boldsymbol{x}) = a^{\mathrm{T}}\boldsymbol{x},$$

subject to

$$\boldsymbol{A}\boldsymbol{x} \leq \boldsymbol{b}, \quad \boldsymbol{x} \geq 0,$$

where

$$\boldsymbol{A} = \begin{pmatrix} 8 & 7 \\ 2 & 3 \\ 1 & 4 \end{pmatrix}, \quad \boldsymbol{b} = \begin{pmatrix} 50 \\ 20 \\ 25 \end{pmatrix}.$$

The optimal primal value is $f_{\max} = 27$ at $(1, 6)$.

Its dual problem can be written as

$$\text{minimize } g(\boldsymbol{u}) = \boldsymbol{b}^{\mathrm{T}}\boldsymbol{u}, \quad \boldsymbol{u} = \begin{pmatrix} u_1 & u_2 & u_3 \end{pmatrix}^{\mathrm{T}},$$

subject to

$$\boldsymbol{A}^{\mathrm{T}}\boldsymbol{u} \geq a, \quad \boldsymbol{u} \geq 0.$$

We leave it as a further exercise to show that the optimal solution to this dual problem is $g_{\min} = 27$ at $\boldsymbol{u} = (0.1704,\ 0.7478,\ 0.1409)^{\mathrm{T}}$. It clearly shows that the duality gap is zero as $f_{\max} = g_{\min}$.

## Solutions for Chapter 7

**7.1** The optimal integer solution to the integer problem can be obtained by branch and bound, and we have

$$x_* = 5, \quad y_* = 1, \quad P_* = 19.$$

If we use the LP relaxation, the relaxed LP has the optimal solution at

$$\left(\frac{275}{47}, \frac{50}{47}\right) \approx (5.8511, 1.0638), \quad P_{\max} = \frac{1025}{47}.$$

If we round them down, we have $(5, 1)$, which is the optimal solution. This is a special, lucky case. In general, this simple rounding up will not work.

**7.2** First, we can write the problem as the following standard form:

$$\text{maximize } P(\boldsymbol{x}) = \boldsymbol{c}^{\mathrm{T}}\boldsymbol{x},$$

subject to

$$\boldsymbol{A}\boldsymbol{x} \le \boldsymbol{b}, \quad \boldsymbol{A} = \begin{pmatrix} 115 & 70 & 60 & 35 \\ 14 & 20 & 15 & 5 \\ 6 & 10 & 12 & 2 \end{pmatrix}, \quad \boldsymbol{b} = \begin{pmatrix} 7000 \\ 1500 \\ 600 \end{pmatrix},$$

where $\boldsymbol{c} = (40,\ 50,\ 60,\ 20)^{\mathrm{T}}$ and $\boldsymbol{x} = (x_1,\ x_2,\ x_3,\ x_4)^{\mathrm{T}}$. In addition, we have

$$x_i \ge 10 \quad (i = 1, 2, 3, 4).$$

You can use the standard simplex method. Alternatively, you can use a software package such as Excel Solver or Matlab to solve it. The optimal solution $P_{\max} = 4280$ with $\boldsymbol{x} = (10, 10, 17, 118)^{\mathrm{T}}$.

**7.3** In the previous problem, in order to incorporate $x_i \ge 10$, we can rewrite the primal problem as

$$\text{maximize} \boldsymbol{c}^{\mathrm{T}}\boldsymbol{x},$$

subject to

$$\boldsymbol{B}\boldsymbol{x} \le \boldsymbol{g},$$

$$\boldsymbol{B} = \begin{pmatrix} 115 & 70 & 60 & 35 \\ 14 & 20 & 15 & 5 \\ 6 & 10 & 12 & 2 \\ -1 & 0 & 0 & 0 \\ 0 & -1 & 0 & 0 \\ 0 & 0 & -1 & 0 \\ 0 & 0 & 0 & -1 \end{pmatrix}, \quad \boldsymbol{g} = \begin{pmatrix} 7000 \\ 1500 \\ 600 \\ -10 \\ -10 \\ -10 \\ -10 \end{pmatrix}, \quad \boldsymbol{x} = \begin{pmatrix} x_1 \\ x_2 \\ x_3 \\ x_4 \end{pmatrix}.$$

The dual problem of the previous problem can be written as

$$\text{minimize } Q = \boldsymbol{g}^{\mathrm{T}}\boldsymbol{y}, \quad \boldsymbol{y} = [y_1,\ y_2, \dots, y_7]^{\mathrm{T}},$$

subject to

$$\boldsymbol{B}^{\mathrm{T}}\boldsymbol{y} \ge \boldsymbol{c}, \quad \boldsymbol{y} \ge 0.$$

If we solve it using a software package, we should get $Q_{\min} = 4280$, which is the same as the primal optimal solution $P_{\max} = 4280$.

## Solutions for Chapter 8

**8.1** This nonlinear function can be linearized by taking logarithm, and we have

$$\ln y = \ln \exp[ax^2 + b \sin x] = ax^2 + b \sin x.$$

If the data points $y_i > 0$, then we can use standard linear regression.

**8.2** We can use the regression in terms of linear polynomials of degree $p$. For $p = 2$, we have the best-fit curve as

$$f(x) = 1.04x^2 - 2.16x - 0.32,$$

with the residual sum of squares RSS = 0.2. We can calculate the information criteria and we have $AIC_c = -6.08$, $BIC = -12.86$.

Similarly, for $p = 1$, we have $RSS = 36.55$, $AIC_c = 13.28$, and $BIC = 11.56$. In addition, for $p = 3$, we have $RSS = 0.14$, $AIC_c = 11.99$, and $BIC = -13.18$.

As we can see that $AIC_c$ is the lowest when $p = 2$, which gives the best fit. For higher $p$ values, the RSS will in general decrease, which may lead to overfit. For this example, $AIC_c$ seems to be a better indicator than BIC.

**8.3** From the above data, we can calculate the co-variance matrix

$$\boldsymbol{C} = \begin{pmatrix} 3.517 & 3.335 \\ 3.335 & 12.3 \end{pmatrix},$$

whose eigenvalues are 13.423 and 2.394. Their corresponding eigenvectors are

$$\boldsymbol{u}_1 = \begin{pmatrix} 3.19 \\ 0.9477 \end{pmatrix}, \quad \boldsymbol{u}_2 = \begin{pmatrix} -0.9477 \\ 0.319 \end{pmatrix}.$$

Though it can be argued that $x$ is the main component, however, this is a nonlinear relationship (due to $x^2$). Thus, the PCA is not strictly valid.

**8.4** Using the model

$$y = \frac{1}{1 + \exp(a + bx)},$$

starting with $a = 1$ and $b = 1$ as the initial values, we can use Newton's method to get

$$a = 1.8730, \quad b = -0.2919.$$

You can use any software tool listed in Appendix A.

## Solutions for Chapter 9

**9.1** The right choice of the number ($k$) of clusters can largely affect the final clustering results. In case of 5 data points, one extreme is to use $k = 5$, which means each point is a cluster itself. Thus, the total intra-cluster distances can be zero. But, this hardly provides any insight into the data. Alternatively, if we use $k = 1$, all these data points belong to one cluster. It does not provide any insight either.

Even we use $k = 2$ or 3, it may not make much sense. In fact, the number of data points is so small, the data does not contain enough information about its intrinsic structure, which makes the clustering difficult.

**9.2** Both R and Python have packages to do $k$-mean clustering.

**9.3** You can use Matlab, Python, R, and TensorFlow to experiment some case studies.

## Solutions for Chapter 10

**10.1** As the process obeys a Poisson model, we can write the queue system as M/M/1 if we consider the help desk as a single server.

**10.2** In the previous question, we have $\lambda = 20\text{h}^{-1}$ and $\mu = 30\text{h}^{-1}$, and we have

$$\rho = \frac{\lambda}{\mu} = \frac{2}{3},$$

the mean waiting time is

$$W_q = \frac{\lambda}{\mu(\mu - \lambda)} = \frac{20}{30 \times (30 - 20)} = \frac{1}{15} \text{ hour},$$

which is 4 minutes. The probability of no waiting at all is

$$P_0 = 1 - \rho = \frac{1}{3}.$$

The probability of waiting longer than 10 minutes is

$$P(T_w > \frac{10}{60}) = \rho \exp[-(\mu - \lambda)t] = \frac{2}{3} \exp\left[-(30 - 20) \times \frac{10}{60}\right] \approx 0.126.$$

**10.3** If 3 counters are open in the shop, we have $s = 3$, $\lambda = 100\text{h}^{-1}$ and $\mu = 30\text{h}^{-1}$ (one customer every 2 minutes). Thus, the utility measure is

$$U = \frac{\lambda}{s\mu} = \frac{100}{3 \times 30} = \frac{10}{9} > 1,$$

thus the queue length will increase unboundedly. In order to have a stable queuing system, we have to ensure $U < 1$. For $s = 4$, we have

$$U = \frac{100}{4 \times 30} = \frac{5}{6} < 1,$$

which means that all 4 counters are busy about $5/6 = 83.3\%$ of the time. In addition, we have

$$\rho = \frac{100}{30} = \frac{10}{3}.$$

The probability of no customer is (from Eq. (10.51))

$$p_0 = \left[\frac{\rho^s}{s!} \frac{s}{(s - \rho)} + \sum_{k=0}^{s-1} \frac{\rho^k}{k!}\right]^{-1} \approx 0.021,$$

and the queue length in the system is

$$L = \rho + p_0 \frac{\rho^{s+1}}{(s - 1)!(s - \rho)^2} \approx 6.62.$$

The average waiting time is 2 minutes (or 0.033 hour). Therefore, all 5 counters should be used so as to reduce $L$ to under 5. In fact, if all 5 counters are used, the queue length is 3.987. That is, there are about 4 customers in the queueing system.

**10.4** Since the arrival process of cars on the road is Poisson, we can conclude that the inter-arrival time $t$ obeys a negative exponential distribution

$$f(t) = \lambda e^{-\lambda t} = \frac{1}{T} e^{-t/T} \quad (t \geq 0, \quad \lambda > 0),$$

where $\lambda = Q/3600$ is the hourly traffic flow rate and $Q$ is the number of passing vehicles per hour. Here, $T = 1/\lambda$ is the characteristic time. The probability of $t \geq h$ is

$$P(t \geq h) = 1 - \int_0^h \lambda e^{-\lambda t} dt = e^{-\lambda h},$$

where $h$ is the so-called time headway, which is the time between successive cars. Thus, the probability of $P(t < h)$ is

$$P(t < h) = 1 - e^{-\lambda h}.$$

As it requires 15 seconds to cross the road, we can use $h = 15$ seconds to be safe. From $Q = 250$, we have $\lambda = 250/3600 = 0.069$. Thus, the probability of crossing the road without waiting is

$$P(h \geq 15) = e^{-\lambda t} = e^{-0.069 \times 15} = 0.355.$$

The probability of $h < 15$ is

$$P(h < 15) = 1 - e^{-\lambda h} = 1 - e^{-0.069 \times 15} = 0.645,$$

and the probability of $h \geq 20$ is

$$P(h \geq 20) = e^{-\lambda h} = e^{-0.069 \times 20} = 0.252.$$

Thus, the probability of finding a gap between 15 and 20 seconds is

$$P = 1 - 0.645 - 0.252 = 0.103.$$

**10.5** The arrival rate is $\lambda = 18\,\text{h}^{-1}$ and the service time is $\mu = 6\,\text{h}^{-1}$ (10 minutes per vehicle), thus we have $\rho = \lambda/\mu = 3$. Since there are $s = 5$ spaces, we have

$$U = \frac{\rho}{s} = \frac{3}{5} = 0.6.$$

Therefore, the probability of no cars in all the parking spaces is

$$p_0 = \frac{1}{\rho^5/5!(1 - 0.6) + \sum_{k=0}^{5-1} \rho^k/k!} = 0.0329,$$

and the probability of not finding a parking space is

$$p_{n>5} = \frac{p_0 \rho^{5+1}}{5!5 \times (1 - \rho/5)} \approx 0.1.$$

## Solutions for Chapter 11

**11.1** To find the Pareto of the bi-objective problem, we can use the weighted sum method to convert it to a single objective optimization problem

$$g(x) = \alpha f_1(x) + \beta f_2(x) = \alpha x^2 + \beta(x-1)^2, \quad \alpha + \beta = 1, \quad \alpha, \beta \geq 0.$$

From $g'(x) = 0$, we have

$$2\alpha x + 2\beta(x-1) = 0,$$

which gives

$$x_* = \frac{\beta}{\alpha + \beta} = \beta.$$

Thus, $f_1(x_*) = \beta^2$ and $f_2(x) = (\beta - 1)^2$. After eliminating $\beta$, we have

$$f_2 = (\sqrt{f_1} - 1)^2,$$

which is the analytical Pareto front.

**11.2** This bi-objective optimization problem looks quite complicated, however, if we use the fact that $0 \leq x_i \leq 1$, then the minimum value of $g(x)$ is 1 when $x_2 = 0, x_3 = 0, \ldots, x_D = 0$. If $f_1 = x_1^*$ becomes optimal at $x_1^*$, we have

$$f_2 = g(x)\left\{1 - [\frac{x_1^*}{g(x)}]^2\right\} = 1 - x_1^{*2} = 1 - f_1^2,$$

which is the analytical expression of its Pareto front.

## Solutions for Chapter 12

**12.1** One way to solve this problem is to solve $y = x^2 - 2$, then rewrite $f$ as

$$f = x^2 + y^2 + 2xy = (x+y)^2 = (x + x^2 - 2)^2,$$

which becomes an unconstrained problem for $x$. Its optimal solution is $x = 1$ and $y = -1$, or $x = -2$ and $y = 2$. Both solutions give $f_{\min} = 0$.

Alternatively, you can use the method of Lagrange multiplier by setting

$$L = x^2 + y^2 + 2xy + \lambda(y - x^2 + 2),$$

which should lead to the same conclusion.

**12.2** The function $f(x, y) = x^3 - 3xy^2$ is the so-called Monkey surface, which has no minimum. However, with an additional constraint, it does have a minimum. Use the method of Lagrange multipliers, we have

$$L = f(x, y) + \lambda h(x, y) = x^3 - 3xy^2 + \lambda(x - y^2 - 1).$$

The stationary conditions are

$$\frac{\partial L}{\partial x} = 3x^2 - 3y^2 + \lambda = 0, \quad \frac{\partial L}{\partial y} = 0 - 6xy + \lambda(-2y) = 0$$

and

$$\frac{\partial L}{\partial \lambda} = x - y^2 - 1 = 0.$$

The condition $-6xy - 2\lambda y = 2y(3x + \lambda) = 0$ implies that either $y = 0$ or $\lambda = -3x$.

If $y = 0$, the other two conditions give $\lambda = -3$ and $x_* = 1$. Thus, the optimal solution $f_{\min} = 1$ occurs at $x_* = 1$ and $y_* = 0$.

The other case $\lambda = -3x$ gives no real solution. The above solution is the only minimum solution.

**12.3** It is obvious that the optimal solution occurs at the boundary $x_* = 2$. We can show this formally by using a logarithmic barrier

$$L = (x-1)^2 + 1 - \mu \log(x^2 - 4),$$

which gives

$$\frac{dL}{dx} = 2(x-1) - \frac{2x\mu}{x^2 - 4} = 0$$

or

$$x^3 - x^2 - (4 + \mu) + 4 = 0.$$

This is a cubic equation, which can have a complicated analytical solution. In the limit of $\mu \to 0$, we can show that there are three solutions:

$$x_1^* = 2, \quad x_2^* = -2, \quad x_3^* = 1.$$

It is straightforward to check that the optimal solution is $f_{\min} = 2$ at $x_* = 2$.

## Solutions for Chapter 13

**13.1** The function $f(x, y) = (1 - x)^2 + 100(y - x^2)^2$ is called Rosenbrock's Banana function because the minimum lies in a banana-shaped valley. There are many software packages on the web. For example, you can use Matlab, and Microsoft Excel Solver to solve it.

**13.2** The exact role and effects of mutation, crossover, and selection in genetic algorithms have been well studied, though not all researchers agree with the results. Loosely speaking, mutation provides a mechanism to modify solutions both locally and globally, depending on the mutation strength. The mutation rate is partly associated with the mutation strength. Higher mutation rates can drive the system to evolve quickly as solutions can be far away from existing solutions; however, it may lead to a much higher diversity, which may slow down the convergence of the system. If the mutation rate is sufficiently low, the system may have many similar solutions, thus it may get stuck at the local optima. Similarly, crossover can also play an important role. If the crossover rate is high, the mixing ability of different solutions is high, thus providing enough diversity for the population. If crossover is less frequent, then it may limit the evolution speed. Therefore, the rate of crossover is usually higher, more than 0.9 in most cases, while the mutation rate is low, typically 0.05.

In addition, selection is a driving mechanism for the system to evolve towards the fitter landscape. The selection based on fitness acts like the selection pressure in the real-world evolution. Without any selection pressure, the fitness landscape can become a plateau, so the system may not converge at all.

**13.3** The cooling schedule for temperature in simulated annealing should be monotonically decreasing so as to lower the energy in the system gradually and let the system settle down (thus converge). However, if the cooling is too quick, it may cause the system to freeze too quickly, thus leading to premature convergence. In the multimodal landscape, it may be useful if the temperature is raised slightly after cooling so as to allow the system to potentially jump out of local optimum. In fact, there is a simulated tempering algorithm that uses non-monotonic temperature variations.

**13.4** The parameter $F$ in differential evolution is important as it controls the mutation strength. In one extreme, if $F$ is too small, then new solutions generated by mutation is too near to the old solution. In case of $F = 0$, they become the same as mutation has no effect. If $F$ is too large, the mutated solutions may be too far away. Thus, a moderate value $F \in [0.5, 1]$ can be used in many cases.

There is no requirement that $F$ must be fixed. In fact, $F$ can be chosen from a random distribution adaptively. For example, in the self-adaptive differential evolution, $F$ is drawn from a uniform distribution between 0.1 and 0.9.

## Solutions for Chapter 14

**14.1** From the updating equation in PSO

$$\boldsymbol{v}_i^{t+1} = \boldsymbol{v}_i^t + \alpha\epsilon_1(\boldsymbol{g}^* - \boldsymbol{x}_i^t) + \beta\epsilon_2(\boldsymbol{x}_i^* - \boldsymbol{x}_i^t),$$

$$\boldsymbol{x}_i^{t+1} = \boldsymbol{x}_i^t + \boldsymbol{v}_i^{t+1},$$

we have

$$\begin{pmatrix} \boldsymbol{x}_i^{t+1} \\ \boldsymbol{v}_i^{t+1} \end{pmatrix} = \boldsymbol{A}\begin{pmatrix} \boldsymbol{x}_i^t \\ \boldsymbol{v}_i^t \end{pmatrix} + \begin{pmatrix} 0 \\ \alpha\epsilon_1\boldsymbol{g}^* + \beta\epsilon_2\boldsymbol{x}_i^* \end{pmatrix},$$

where

$$A = \begin{pmatrix} 1 & 1 \\ -(\alpha\epsilon + \beta\epsilon_2) & 1 \end{pmatrix},$$

whose eigenvalues control the behavior of this algorithm. However, this is a random matrix with two random variables $\epsilon_1$ and $\epsilon_2$, which makes it harder to analyze.

**14.2** Though the frequency is always positive in the real-world, the virtual frequency used in the bat algorithm can be both positive and negative. This has been observed in numerical experiments, and also supported by theory of parameter analysis.

**14.3** In the firefly algorithm, the main algorithmic equation is

$$\boldsymbol{x}_i^{t+1} = \boldsymbol{x}_i^t + \beta_0 e^{-\gamma r_{ij}^2}(\boldsymbol{x}_j^t - \boldsymbol{x}_i^t) + \alpha \boldsymbol{x}_i^t,$$

where $\alpha$, $\beta_0$, and $\gamma$ are parameters. In essence, $\alpha$ controls the randomness, thus its value should be reduced gradually from a higher value to a lower value. The parameter $\beta_0$ controls the attractiveness, thus it should not be too small. In case of $\beta_0 = 0$, it loosens its ability for attracting other fireflies or solutions. $\gamma$ is a scaling parameter, which controls the range of the visibility in the landscape. If $\gamma$ is small, the visibility is large, thus the influence of the attraction is large. If $\gamma$ is large, then the visibility is reduced, and fireflies cannot see far enough. Thus, moderate values of $\gamma$ that is comparable to the inverse of length scales should be used. Some parametric studies should be carried out for new types of problems, this is also true for other algorithms. In fact, parameter tuning is needed for all algorithms.

**14.4** The proper generation of random numbers that obey a Lévy distribution can be tricky. One of the stable algorithms that works well is the Mantegna's algorithm, which uses two normally-distributed random numbers $u$ and $v$:

$$u \sim N(0, \sigma^2), \quad v \sim N(0, 1),$$

where $\sigma^2$ is a constant variance as a scaling factor. Then, the step length

$$s = \frac{u}{|v|^{1/\lambda}}$$

obeys a Lévy distribution with an exponent $\lambda$:

$$L(s) \sim \frac{1}{|s|^{1+\lambda}}, \quad |s| \gg 0.$$

# Index